International Workshop on Beamed and Unbeamed Gamma-Rays from Galaxies 2011

Journal of Physics: Conference Series Volume 355

Muonio, Finland
11-15 April 2011

ISBN: 978-1-62276-772-4
ISSN: 1742-6588

Printed from e-media with permission by:

Curran Associates, Inc.
57 Morehouse Lane
Red Hook, NY 12571

Some format issues inherent in the e-media version may also appear in this print version.

Printed by Curran Associates, Inc. (2012)

For permission requests, please contact the Institute of Physics
at the address below.

Institute of Physics
Dirac House, Temple Back
Bristol BS1 6BE UK

Phone: 44 1 17 929 7481
Fax: 44 1 17 920 0979

techtracking@iop.org

Additional copies of this publication are available from:

Curran Associates, Inc.
57 Morehouse Lane
Red Hook, NY 12571 USA
Phone: 845-758-0400
Fax: 845-758-2634
Email: curran@proceedings.com
Web: www.proceedings.com

TABLE OF CONTENTS

GEV/TEV REVIEWS AND INSTRUMENTS

RADIO OBSERVATIONS

MODELING AND GAMMA-RAY OBSERVATIONS

BLAZARS AND JETTED AGNS

NON-BLAZAR AGNS

EXTRAGALACTIC BACKGROUND LIGHT

FLARES AND FAST VARIABILITY

RADIO GALAXIES

COSMIC GAMMA-RAY BACKGROUND AND STARBURST GALAXIES

MULTIMESSENGER ASTRONOMY

Journal of Physics: Conference Series **355** (2012) 011001 doi:10.1088/1742-6596/355/1/011001

Preface

Both the number and types of extragalactic very high-energy gamma-ray emitters have pleasingly increased substantially during the last few years, due to the extremely successful Fermi mission and the achievements of ground-based gamma-ray detectors since 2003. The international scientific workshop 'Beamed and Unbeamed Gamma-rays from Galaxies' took place from 11–15 April 2011 at the Lapland Hotel Olos, Muonio, Finland with the aim of reviewing and discussing current knowledge and understanding of gamma-ray emission, both from active and other types of galaxies (e.g., starburst galaxies), particularly in the light of results delivered recently from high-energy and very-high energy gamma-ray instruments and by multi-wavelength studies.

Following the 2008 Workshop on AGN and Related Fundamental Physics in High-Energy Gamma Astronomy, this workshop consisted of invited review talks and contributed talks. 'Round table' discussions enabled the exchange of facts and ideas. Ample time was provided for discussions not only inside, but also outside the lecture theatre. As for the 2008 workshop, Lapland was chosen for the meeting as an extraordinary location which provided more than just beautiful winter scenery. The 'isolated location' led to an intimate but also open and informal atmosphere for discussions. This, together with the broad approach of the workshop, facilitated communication between theorists, phenomenologists and observers, and started new collaborations.

The scientific scope and the topics of the workshop intentionally covered a wide range of topics, including:

- blazars and jetted AGNs
- blazar and AGN physics, observations and phenomenology
- black holes, central engines, jets and environment of the central engine
- non-blazar AGNs and non-AGN galaxies: star-forming and starburst galaxies, cosmic-ray generation
- observational tools: variability, correlations, power spectra analysis and periodicity
- multi-wavelength aspects and approaches: radio, optical and X-rays along with GeV/TeV gamma-rays
- future observational opportunities
- EBL status and progress

The workshop could not have taken place in such a constructive atmosphere and exquisite location without the support of Max-Planck-Institut für Physik München, and of Tuorla Observatory of the University of Turku. We particularly acknowledge the financial support of the Cluster of Excellence 'Origin and Structure of the Universe', Garching b. München.

Robert Wagner
(On behalf of the organizers)

Journal of Physics: Conference Series **355** (2012) 011001 doi:10.1088/1742-6596/355/1/011001

Committees

Scientific Organizing Committee

C Dermer	Naval Research Laboratory, Washington, DC
H Krawczynski	Washington University, St. Louis
A Levinson	Tel Aviv University
L Maraschi	Brera Observatory
A Marscher	Boston University
I Moskalenko	Stanford University
D Paneque	Max-Planck-Institut für Physik, Munich
A Reimer	Innsbruck University
Y Rephaeli	Tel Aviv University
A Sillanpää	Tuorla Observatory
T Totani	Kyoto University
R Wagner (chair)	Max-Planck-Institut für Physik, Munich
S Wagner	Landessternwarte, Heidelberg University

Local Organizing Committee

A Biland	ETH Zurich
E Lindfors	Tuorla Observatory
S Rodriguez	Max-Planck-Institut für Physik, Munich
A Sillanpää (chair)	Tuorla Observatory
R Wagner	Max-Planck-Institut für Physik, Munich

Journal of Physics: Conference Series 355 (2012) 011001 doi:10.1088/1742-6596/355/1/011001

Participants

Ivan Agudo	IAA-CSIC
Marco Ajello	SLAC/KIPAC
Maxim Barkov	Max-Planck-Institut für Kernphysik
Josefa Becerra González	Instituto Astrofisica de Canarias
Keith Bechtol	SLAC/KIPAC
Adrian Biland	ETH Zurich
Colm Coughlan	UC Cork
Gessica de Caneva	DESY Zeuthen
Charles Dermer	Naval Research Laboratory
Marlene Doert	TU Dortmund
Immacolata Donnarumma	INAF-IASF Rome
Björn Eichmann	TU Dortmund
Lars Fuhrmann	Max-Planck-Institut für Radioastronomie
Amy Furniss	University of California Santa Cruz
Denise Gabuzda	UCC Cork
Rudy Gilmore	SISSA
Jose L. Gomez	IAA-CSIC
Juan Pablo Gómez González	CSIC/University of Valencia
Sara Rebecca Gozzini	DESY Zeuthen
Jennifer Gupta	Jodrell Bank Centre for Astrophysics
Dorothee Hildebrand	ETH Zurich
Talvikki Hovatta	Purdue University
Yoshiyuki Inoue	Kyoto University
Oliver King	Caltech
Tanja Kneiske	University of Hamburg
Stefan Larsson	Stockholm University
Amir Levinson	Tel Aviv University
Elina Lindfors	University Turku, Tuorla Observatory
Benoit Lott	CENBG/IN2P3/CNRS
Ryu Makiya	Kyoto University

Journal of Physics: Conference Series **355** (2012) 011001 doi:10.1088/1742-6596/355/1/011001

Laura Maraschi	INAF/Brera Observatory
Francesco Massaro	Smithsonian Astrophysical Observatory
Manuel Meyer	University of Hamburg
Eoin Murphy	UC Cork
Krzysztof Nalewajko	Nicolaus Copernicus Astronomical Centre
Ioannis Nestoras	Max-Planck-Institut für Radioastronomie
Andrzej Niedzwiecki	University of Lodz
Jacek Niemiec	Institute of Nuclear Physics, Polish Academy of Sciences
Kari Nilsson	FINCA
Roopesh Ojha	NASA Goddard Space Flight Center
Massimo Persic	INAF/INFN Trieste
Juri Poutanen	University of Oulu
Martin Raue	University of Hamburg
Andrea Reichstein	UC Cork
Anita Reimer	University of Innsbruck
Riho Reinthal	Tuorla Observatory
Guido Risaliti	INAF/Arcetri Observatory
Sybille Rodriguez	Max-Planck-Institut für Physik/ München
Mahito Sasada	Hiroshima University
Aimo Sillanpää	University Turku, Tuorla Observatory
Boris Stern	Inst. For Nuclear Research, Russian Academy of Science
Hajime Takami	Max-Planck-Institut für Physik, München
Joni Tammi	Metsähovi Radio Observatory
Marc Türler	ISDC, University of Geneva
Alexandra Veledina	University of Oulu
Robert Wagner	Max-Planck-Institut für Physik, München

All papers published in this volume of *Journal of Physics: Conference Series* have been peer reviewed through processes administered by the proceedings Editors. Reviews were conducted by expert referees to the professional and scientific standards expected of a proceedings journal published by IOP Publishing.

The H.E.S.S. extragalactic sky

Martin Raue
Institut für Experimentalphysik, Universität Hamburg, Hamburg, Germany
E-mail: martin.raue@desy.de

for the H.E.S.S. Collaboration

Abstract. The H.E.S.S. Cherenkov telescope array, located on the southern hemisphere in Namibia, studies very high energy (VHE; $E > 100\,\mathrm{GeV}$) γ-ray emission from astrophysical objects. During its successful operations since 2002 more than 80 galactic and extra-galactic γ-ray sources have been discovered. H.E.S.S. devotes over 400 hours of observation time per year to the observation of extra-galactic sources resulting in the discovery of several new sources, mostly AGNs, and in exciting physics results e.g. the discovery of very rapid variability during extreme flux outbursts of PKS 2155-304, stringent limits on the density of the extragalactic background light (EBL) in the near-infrared derived from the energy spectra of distant sources, or the discovery of short-term variability in the VHE emission from the radio galaxy M 87. With the recent launch of the Fermi satellite in 2008 new insights into the physics of AGNs at GeV energies emerged, leading to the discovery of several new extragalactic VHE sources. Multi-wavelength observations prove to be a powerful tool to investigate the production mechanism for VHE emission in AGNs. Here, new results from H.E.S.S. observations of extragalactic sources will be presented and their implications for the physics of these sources will be discussed.

1. Introduction

Very high energy (VHE; $E > 100\,\mathrm{GeV}$) γ-ray astrophysics of the extragalactic sky is highly dynamic field. Since the detection of the first source in 1992 (Mkn 421 by Whipple; [1]) the number of sources has been growing continuously to more than 45 known sources at the beginning of 2011 (see Fig. 1 left). The detected source population is still dominated by active galactic nuclei (AGN), but recently starburst galaxies have also been established as extragalactic VHE emitters [2, 3].

The key science questions in the field are (still) investigating the origin and physics mechanism responsible for the VHE emission in AGNs and thereby probe the conditions of particle acceleration and cooling in relativistic plasma outflows and in the vicinity of super-massive black holes (SMBH). One of the most striking observational result of the recent years was the discovery of rapid variability with minute time-scales [4, 5], which has not yet fully been integrated in the theories. Such short variability time-scales strongly impact the canonical modeling of the VHE emission in synchtrotron-self-Compton (SSC) frameworks, requiring large bulk Lorentz factors of the emission region ($\Gamma > 50$) [6]. In addition, new source classes have been established as VHE emitters e.g. radio galaxies (M 87, Centaurus A), low- and medium-frequency peaked BL Lac objects, etc. With the launch of the Fermi satellite in 2008, a wealth of new information on the extragalactic sky in the 100 MeV to 100 GeV energy regime became available, strongly influencing the science programs in the VHE domain.

The multi-wavelength (MWL) aspect, in general, plays a crucial role in investigating VHE emission from AGN: The limited angular resolution of VHE instruments usually does not allow for a precise

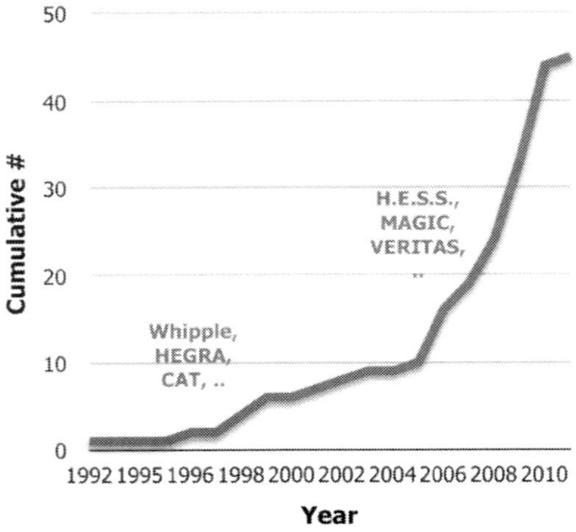

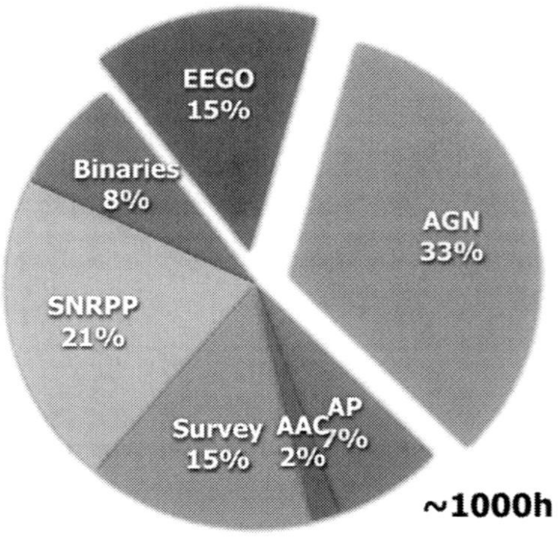

Figure 1. *Left:* Cumulative number of extragalactic VHE sources from TeVCat (`http://tevcat.uchicago.edu/`). *Right:* Typical time allocation for different observing programs in H.E.S.S. EEGO: Extended extragalactic objects; AGN: Active galactic nuclei; AP: Astroparticle; AAC: Analysis and calibration; SNRPP: Supernova remnants, pulsars, and plerions. The total observing time per year is of the order of 1000 h.

spatial determination of the origin of the emission in extragalactic objects. Investigating the MWL behavior e.g. by searching for correlations with high resolution radio observations allows one to probe spatial scales down to a few hundred Schwarzschild radii of the SMBH [7]. New insights into the structure of the magnetic fields in AGN jets also come from polarimetric observations in the optical and radio.

Extragalactic VHE sources are also successfully used to study the meta-galactic radiation and magnetic fields (e.g. [8, 9, 10]). In such studies, the extragalactic VHE sources are used as *lighthouses*, whose emission gets altered by the intervening fields. For example, in case of the extragalactic background light (EBL), an attenuation of the VHE photons from pair-production with the low energy photons from the EBL is expected. With assumptions about the intrinsic spectrum emitted at the source constraints on the intervening fields can be derived (e.g. [8, 9]).

2. H.E.S.S. status & extragalactic program

The High Energy Stereoscopic System (H.E.S.S.) is an array of four Imaging Atmospheric Cherenkov Telescopes (IACT) for the detection of very high energy (VHE; $E > 100\,\mathrm{GeV}$) γ-ray emission from astrophysical objects [11]. H.E.S.S. is located on the southern hemisphere in Namibia (23°16′18″S, 16°30′00″E) at an altitude of 1800 m. It detects the Cherenkov light emitted by extensive air showers initiated by γ-rays or charged cosmic rays entering the Earth's atmosphere using ultra-fast photomultiplier cameras with 968 pixels located in the focal plane of 13 m optical telescopes (tessellated mirror). The array is sensitive to γ-rays in the energy range from $\sim$100 GeV to $\sim$100 TeV with a typical energy resolution of $\lesssim 15\%$. H.E.S.S. has been in operation since 2002, the full four telescope array since 2004. During its 9 years of operations H.E.S.S. observations led to the discovery of over 80 new VHE sources and significant contribution to the field of VHE astrophysics and beyond.

In 2010/11 the mirrors of the H.E.S.S. telescopes are being re-coated to lower the energy threshold back to the original value of $\sim$ 100 GeV[1]. At the same time, construction of the next extension of

[1] The reflectivity of the mirrors decreases over time due to environmental effects (dust, wind, ..), which leads to an increased

Figure 2. H.E.S.S. site in January 2011. In the center of the image the dish support structure of the H.E.S.S. II telescope is visible, at the left and right edge two of the four H.E.S.S. I telescopes can be seen (Image credit: Julien Bolmont).

H.E.S.S., a very large (25 m) single telescope in the center of the array, is under way with first light expected in 2012 (see Fig. 2).

Observing the extragalactic sky is one of the key science topics in H.E.S.S. Fig. 1 right shows the typical distribution of the H.E.S.S. observation time for one season (here 2011) on the different science groups, with the extragalactic topics taking about 50% of the total time of ~1000 h. The extragalactic science in H.E.S.S. is divided into two sub-topics: (1) physics related to active galactic nuclei (AGN) and (2) extended extragalactic objects (EEGO), which includes galaxy clusters and starburst galaxies. The extragalactic program includes discovery observations, monitoring of known sources, multi wavelength campaigns (MWL), and deep studies of individual sources. In addition, extragalactic source are also used to study the intervening photon fields, mainly the extragalactic background light (EBL), and magnetic fields.

In the following, selected highlights from the past 2 years of H.E.S.S. observations of the extragalactic sky are presented.

3. HESS J1943+213: a candidate extreme BL Lac object

HESS J1943+213 is a new point-like source detected at a significance level of 7.9σ (post-trials) in the H.E.S.S. galactic plane survey. Details on the detection are reported in [12]. HESS J1943+213 is located at RA(J2000) = $19^{h}43^{m}55^{s} \pm 1^{s}_{stat} \pm 1^{s}_{sys}$, DEC(J2000) = $+21°18'8'' \pm 17''_{stat} \pm 20''_{sys}$. The source has a soft spectrum with photon index $\Gamma = 3.1 \pm 0.3_{stat} \pm 0.2_{sys}$ and a flux above 470 GeV of $(1.3 \pm 0.2_{stat} \pm 0.3_{sys}) \times 10^{-12}$ cm^{-2} s^{-1}. This source coincides with an unidentified hard X-ray source IGR J19443+2117, which was proposed to have radio and infrared counterparts. There is no *Fermi*/LAT counterpart down to a flux limit of 6×10^{-9} cm^{-2} s^{-1} in the 0.1–100 GeV energy range (95% confidence upper limit calculated for an assumed power-law model with a photon index $\Gamma = 2.0$). The data from radio to VHE gamma-rays do not show any significant variability.

The new H.E.S.S., *Fermi*/LAT, and Nançay Radio Telescope observations were combined with pre-existing non-simultaneous multi-wavelength observations of IGR J19443+2117 to investigate the likely source associations as well as the interpretation as an active galactic nucleus, a gamma-ray binary or a pulsar wind nebula (see Fig. 3). The lack of a massive stellar counterpart disfavors the binary hypothesis, while the soft VHE spectrum would be very unusual in case of a pulsar wind nebula. In addition, the distance estimates for Galactic counterparts places them outside of the Milky Way. All available observations favor an interpretation of HESS J1943+213 as an extreme, high-frequency peaked BL Lac object with a redshift $z > 0.14$, which would make it the first extragalactic source discovered

energy threshold. This is being accounted for in the analysis by evaluating the optical throughput of the system from the ring images of individual muons.

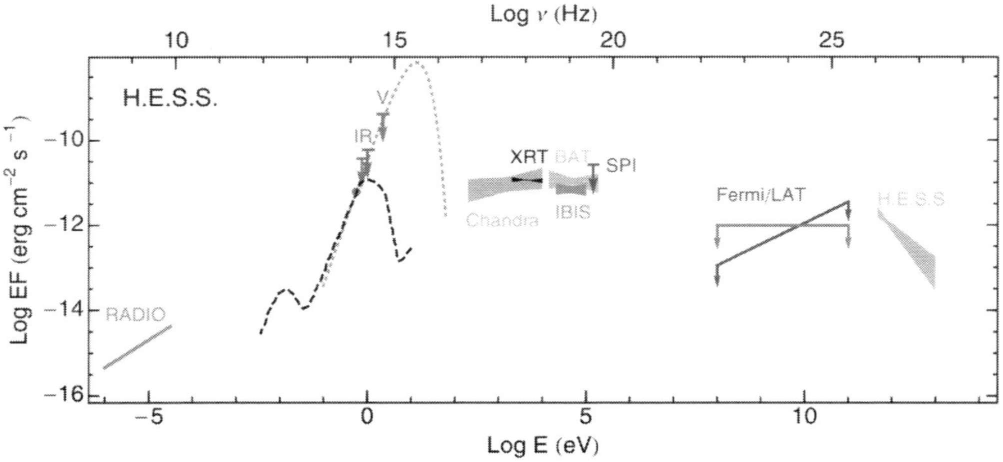

Figure 3. Spectral energy distribution of HESS J1943+213, a candidate extreme BL Lac object, and possible counterparts. The dashed black line is the spectral template of elliptical galaxies at a distance of 570 Mpc. The dotted (red) line is the blackbody spectrum of a massive star with parameters identical with the companion star in the gamma-ray binary LS 5039 at a distance of 25 kpc. The data in IR (2MASS) and visual (Swift/UVOT) bands were corrected for extinction. Further details can be found in [12].

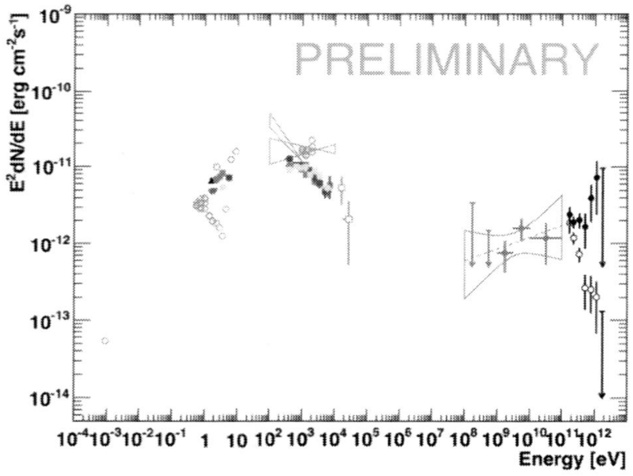

Figure 4. Average SED of 1ES 0414+009 with observations carried out between 2005 and 2009: H.E.S.S. (black fill/open circles with/without EBL correction respectively); *Fermi*/LAT (red), *Swift* BAT in 5 yrs (magenta), *Swift* XRT & UVOT (purple October 2006, blue January 2008 and cyan February 2008) and ATOM (triangles). (from [13])

serendipitously in the H.E.S.S. galactic plane survey.

4. 1ES 0414+009

The BL Lac object 1ES 0414+009 is located a redshift of $z = 0.287$ and harbors a super-massive black hole of mass 2×10^9 $M_\odot$. It has been identified as a high-frequency-peaked BL Lac (HBL) object with the synchrotron-emission peak located in the UV/soft-X-ray range.

H.E.S.S. observation of 1ES 0414+009 have been performed between 2005 and 2009 for a total of 73.7 hours, resulting in a detection with a statistical significance of 7.8 standard deviations (details are presented in [13]). The average energy spectrum measured by H.E.S.S. between 200 GeV and 2 TeV is well described by a power $dN/dE \propto E^{-\Gamma}$ with $\Gamma_{VHE} = 3.44 \pm 0.27_{stat} \pm 0.2_{sys}$ and an integral flux of

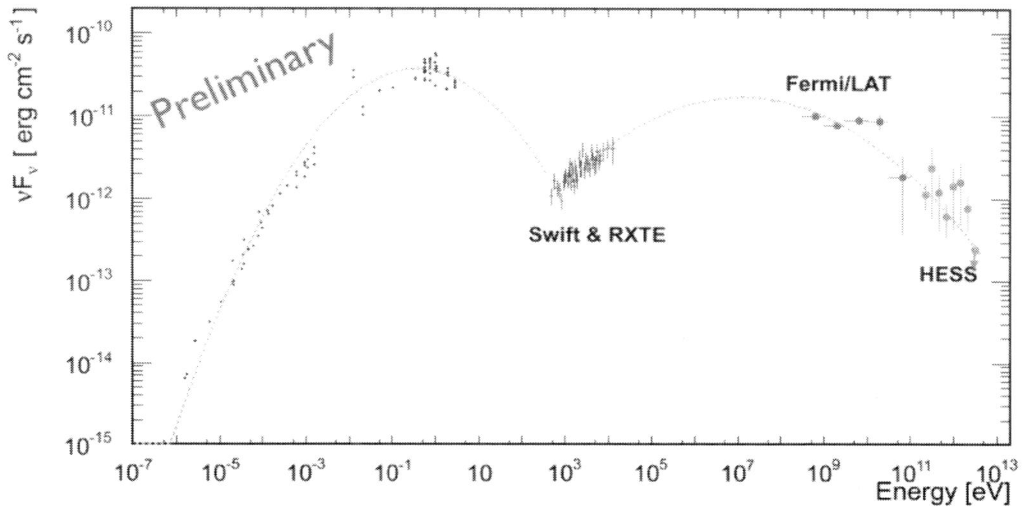

Figure 5. SED of AP Librae with contemporaneous data from *Swift*, Fermi/LAT, and H.E.S.S. Small black points in the radio to infrared are archival data from NED. Each component of the SED was fitted with with a 3rd degree polynomial in νF_ν. (from [15])

$(1.83 \pm 0.21_{stat} \pm 0.37_{sys}) \times 10^{-12}\,\mathrm{cm^{-2}s^{-1}}$ ($\sim$ 0.6% of the flux of the Crab Nebula) above an energy of 200 GeV. No indication for flux or spectral variability was found in the H.E.S.S. data-set. Observation with the Fermi/LAT (first 20 months of operations) result in a flux of $(2.3 \pm 0.2) \times 10^{-9}\,\mathrm{erg\,cm^{-2}s^{-1}}$ between 200 MeV and 10 GeV and a spectrum well described by a power-law function with $\Gamma_{HE} = 1.85 \pm 0.18$. Swift/XRT observations show an X-ray flux of $F_{2-10\,\mathrm{keV}} = (0.8-1) \times 10^{-11}\,\mathrm{erg\,cm^{-2}s^{-1}}$, and a steep spectrum $\Gamma_x = 2.2 - 2.3$. Combining X-ray with Optical-UV data a fit with log-parabolic function locates the synchrotron peak around 0.1 keV (see Fig. 4 for the SED). When corrected for EBL attenuation with a minimal EBL model [14] the VHE spectra smoothly extrapolates down to the Fermi spectrum, confirming previous constrains on the EBL density. The resulting hard intrinsic spectrum ($\Gamma \sim 1.9$), spanning from 0.2 GeV to 2 TeV, challenges simple one-zone SSC models.

5. AP Lib

The BL Lac type object AP Librae ($z = 0.049$) was observed with H.E.S.S. for 11 h of live time from May 11, 2010 to July 10, 2010 and was detected with a statistical significance of 7 standard deviations (details can be found in [15]). The energy spectrum is well described by a power law of index $\Gamma_{VHE} = 2.5 \pm 0.2$. Following the detection, additional X-ray observations where performed with *RXTE* and *Swift*. The joint X-ray spectrum from 0.3 keV to $\sim$ 12 keV is well described by a power law ($P(\chi^2) = 0.98$), a hard photon index of $\Gamma = 1.58 \pm 0.06$ and a 2 – 10 keV flux of $F_{2-10\,\mathrm{keV}} = 4.9^{+0.8}_{-0.7}\,\mathrm{erg\,cm^{-2}\,s^{-1}}$. This indicates that the second component of the spectral energy distribution, usually attributed to inverse-Compton radiation, is present at energies as low as 0.3 keV. Fermi/LAT data on the source were taken from 4 August 2008 to 4 September 2010 (25 months). The HE energy spectrum is well described by a power law with spectral index of $\Gamma = 2.1 \pm 0.1$. No evidence for spectral curvature is found. The integral flux measured with Fermi/LAT between 300 MeV and 300 GeV is $F_{0.3-300\,\mathrm{GeV}} = (1.9 \pm 0.1) \times 10^{-8}\,\mathrm{cm^{-2}s^{-1}}$. When fitting the low and the high energy components with a 3rd degree polynomial in νF_ν synchrotron and inverse-Compton peak energies are found at $E_s^{\mathrm{peak}} \simeq 0.1$ eV and $E_{IC}^{\mathrm{peak}} \simeq 10$ MeV, respectively.

The SED of AP Librae is shown in Fig. 5. One of the most striking features of the SED, compared to other VHE emitting BL Lacs, is the width of the high-energy component, spanning over 10 decades in

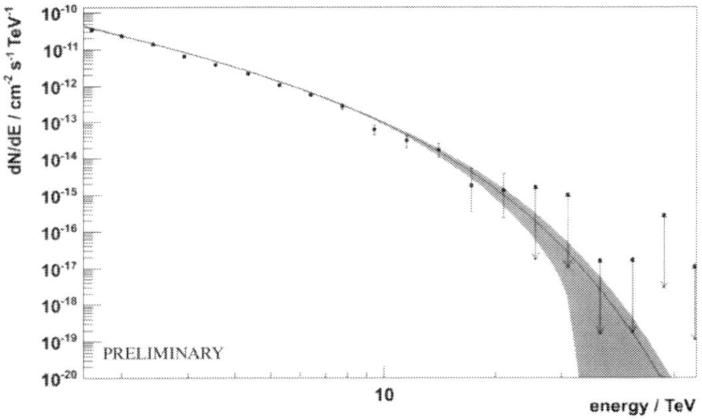

Figure 6. Differential energy spectrum of Mkn 421 as measured by H.E.S.S. in Feb. 2010 in 5.4 h of live-time. The spectral shape can be parameterized by $dN/dE = \phi_0(E/E_0)^{-\Gamma}\exp(-E/E_{cut})$, with a flux normalization $\phi_0 = (1.96 \pm 0.32)\,10^{-11}\,\mathrm{cm^{-2}s^{-1}TeV^{-1}}$, a spectral index $\Gamma = 2.05 \pm 0.22$, a cutoff energy $E_{cut} = (3.4 \pm 0.6)$ TeV, and a decorrelation energy $E_0 = 2.739$ TeV. (from [16])

energy, extending from $\simeq 0.1$ keV up to $\sim$ TeV energies. Such wide distributions pose serious challenges for synchrotron self-Compton type models.

6. Mkn 421: 2010 VHE flare

The high-frequency peaked BL Lac object Mkn 421 ($z = 0.03$) showed a strong outburst of activity in February 2010, first reported by the VERITAS collaboration (see A. Furniss, this proceedings). Following the trigger, H.E.S.S. observed Mkn 421 for a total of 5.4 h of good quality observation time from Feb. 17 to 20, 2010, at an average zenith angle of 62.4° (see [16] for details). A total of 2112 excess events (2188 on-events and 838/11 off-events) were detected yielding a significance of 86.5 standard deviations (s.d.). Significant nightly variability has been detected in the data-set. The time averaged energy spectrum is well described by a power law function with exponential cut-off $dN/dE = \phi_0(E/E_0)^{-\Gamma}\exp(-E/E_{cut})$, with flux normalization $\phi_0 = (1.96 \pm 0.32)\,10^{-11}\,\mathrm{cm^{-2}s^{-1}TeV^{-1}}$, a spectral index $\Gamma = 2.05 \pm 0.22$, cutoff energy $E_{cut} = (3.4 \pm 0.6)$ TeV, and de-correlation energy $E_0 = 2.739$ TeV (see Fig. 6). The last significant energy bin (> 3 s.d.) in the spectrum extends beyond 20 TeV (highest energy observed from this source), making these observations relevant for studies of the density of the mid-infrared of the extragalactic background light (EBL). The measured spectral shape is compatible with previous results, implying a stability of the central engine over the epochs.

7. M 87: 2010 VHE campaign

The giant radio galaxy M 87 with its proximity (16 Mpc), famous jet, and its very massive black hole ($3 - 6 \times 10^9 M_\odot$) provides a unique laboratory to investigate the origin of VHE γ-ray emission from relativistic outflows and the vicinity of super-massive black holes (SMBH). M 87 has been established as a VHE emitter since 2005 [17, 18]. The VHE emission displays strong variability on time-scales as short as a day [18]. The detection of a VHE outburst contemporaneous with a flare of the radio core, directly imaged by 43 GHz VLBA observations with sub-parsec resolution, lead to the interpretation that the VHE emission is likely generated in the close vicinity of SMBH [7]. To further investigate this exciting possibility, VHE monitoring campaigns on the source have been performed, with the results of the latest one, lead by MAGIC and VERITAS in 2010, being reported in this proceedings. A more complex MWL picture is found, which eludes an easy interpretation. Details can be found in M. Raue et al. in this proceedings.

8. PKS 2155-304: Variability studies and monitoring

PKS 2155-304 is a well-known high-frequency-peaked BL Lac object with VHE emission. It displayed an extreme flux outburst in 2006, with variability time-scales down to a few minutes [4]. To further investigate the spectral and temporal variability of the VHE emission in the source a large H.E.S.S. data-set ($\sim$ 90 h), collected between 2005 to 2007, has been analyzed for its timing and spectral properties [19]. The source was found in low flux state from 2005 to 2007 with the exception of a set of exceptional flares occurring in July 2006. The quiescent state can be characterized by a mean flux level of $(4.32 \pm 0.09_{\rm stat} \pm 0.86_{\rm syst}) \times 10^{-11}\,{\rm cm}^{-2}\,{\rm s}^{-1}$ above 200 GeV, or approximately 15% of the Crab Nebula, and a power law energy spectrum with photon index of $\Gamma = 3.53 \pm 0.06_{\rm stat} \pm 0.10_{\rm syst}$. During the flares doubling timescales of $\sim$ 2 min are found [4]. The variation of the photon index with flux was investigated over two orders of magnitude in flux, yielding different behavior at low and high fluxes, which is a new phenomenon in the VHE domain for γ-ray emitting blazars. The variability amplitude, characterized by the fractional r.m.s. $F_{\rm var}$, is strongly energy-dependent ($\propto E^{0.19\pm0.01}$). The light curve r.m.s. correlates with the flux. This is the signature of a multiplicative process which can be accounted for as a red noise with a Fourier index of $\sim$ 2.

To further investigate the long-term variability of the source at VHE (H.E.S.S.), HE (*Fermi*/LAT), and optical (ATOM, ROTSE), H.E.S.S. is performing a long-term monitoring campaign on PKS 2155-304 in 2010 and 2011 with observations every 2nd night all throughout the H.E.S.S. observability window (June/July to December; gaps due to moonlight).

9. Conclusion & Outlook

The H.E.S.S. extragalactic science program is still going strong. Exciting discoveries have been made in the past two years (HESS J1943+213 [12], AP Librae [15], 1ES 0414+009 [13], SHBL J001355.9-185406, and 1RXS J101015.9-311909 [20]). Known sources keep to deliver compelling science [19, 7, 16].

In 2012 the H.E.S.S. II telescope will start operating extending the H.E.S.S. energy range down to ~20-30 GeV and increasing the overall sensitivity by a factor ~2. H.E.S.S. II will enable to extend AGN studies to higher redshifts and to probe the transition region between the Fermi/LAT and the VHE energy range with high sensitivity.

The Cherenkov Telescope Array (CTA) [21], which is expected to start construction in 2014, will deliver a factor of 10 improvement in sensitivity over current generation instruments and an extended energy range from 20 GeV to ~100 TeV. Two sites are planned, with the northern side being dedicated to extra-galactic science and low energy thresholds.

Acknowledgments

The author acknowledges support by the LEXI program of the state of Hamburg, Germany. — The H.E.S.S. Collaboration acknowledges support of the Namibian authorities and of the University of Namibia in facilitating the construction and operation of H.E.S.S., as is the support by the German Ministry for Education and Research (BMBF), the Max Planck Society, the French Ministry for Research, the CNRS-IN2P3 and the Astroparticle Interdisciplinary Programme of the CNRS, the U.K. Science and Technology Facilities Council (STFC), the IPNP of the Charles University, the Polish Ministry of Science and Higher Education, the South African Department of Science and Technology and National Research Foundation, and by the University of Namibia. We appreciate the excellent work of the technical support staff in Berlin, Durham, Hamburg, Heidelberg, Palaiseau, Paris, Saclay, and in Namibia in the construction and operation of the equipment. — This research has made use of NASA's Astrophysics Data System.

References

[1] Punch M, Akerlof C W, Cawley M F, Chantell M, Fegan D J, Fennell S, Gaidos J A, Hagan J, Hillas A M, Jiang Y, Kerrick A D, Lamb R C, Lawrence M A, Lewis D A, Meyer D I, Mohanty G, O'Flaherty K S, Reynolds P T, Rovero A C, Schubnell M S, Sembroski G, Weekes T C and Wilson C 1992 *Nature* **358** 477–+
[2] VERITAS Collaboration, Acciari V A, Aliu E, Arlen T, Aune T *et al.* 2009 *Nature* **462** 770–772 (*Preprint* 0911.0873)
[3] Acero F, Aharonian F, Akhperjanian A G, Anton G *et al.* 2009 *Science* **326** 1080– (*Preprint* 0909.4651)

[4] Aharonian F, Akhperjanian A G, Bazer-Bachi A R *et al.* 2007 *ApJ* **664** L71–L74 (*Preprint* arXiv:0706.0797)
[5] Albert J, Aliu E, Anderhub H *et al.* 2007 *ApJ* **669** 862–883 (*Preprint* arXiv:astro-ph/0702008)
[6] Begelman M C, Fabian A C and Rees M J 2008 *MNRAS* **384** L19–L23 (*Preprint* 0709.0540)
[7] Acciari V A, Aliu E, Arlen T, Bautista M, Beilicke M *et al.* 2009 *Science* **325** 444–448 (*Preprint* http://www.sciencemag.org/cgi/reprint/325/5939/444.pdf) URL http://www.sciencemag.org/cgi/content/abstract/325/5939/444
[8] Aharonian F, Akhperjanian A G, Bazer-Bachi A R *et al.* 2006 *Nature* **440** 1018–1021
[9] Mazin D and Raue M 2007 *A&A* **471** 439–452 (*Preprint* arXiv:astro-ph/0701694)
[10] Neronov A and Vovk I 2010 *Science* **328** 73– (*Preprint* 1006.3504)
[11] Hinton J A 2004 *New Astronomy Review* **48** 331–337 (*Preprint* astro-ph/0403052)
[12] HESS Collaboration, Abramowski A, Acero F, Aharonian F, Akhperjanian A G, Anton G *et al.* 2011 *A&A* **529** A49+ (*Preprint* 1103.0763)
[13] Volpe F, Ohm S, Hauser M, Kaufmann S, Gérard L, The HESS Collaboration, Costamante L, Fegan S, Ajello M and The Fermi-LAT Collaboration 2010 *25th Texas Symposium on Relativistic Astrophysics, December 6-10, 2010, Heidelberg, Germany*
[14] Franceschini A, Rodighiero G and Vaccari M 2008 *A&A* **487** 837–852 (*Preprint* 0805.1841)
[15] Fortin P, Fegan S, Horan D, Sanchez D, The Fermi-LAT Collaboration, Giebels B, Becherini Y, Dubus G, de Naurois M, Punch M, Raue M and The HESS Collaboration 2010 *25th Texas Symposium on Relativistic Astrophysics, December 6-10, 2010, Heidelberg, Germany*
[16] Tluczykont M and The HESS Collaboration 2010 *25th Texas Symposium on Relativistic Astrophysics, December 6-10, 2010, Heidelberg, Germany*
[17] Aharonian F, Akhperjanian A, Beilicke M *et al.* 2003 *A&A* **403** L1–L5 (*Preprint* arXiv:astro-ph/0302155)
[18] Aharonian F, Akhperjanian A G, Bazer-Bachi A R *et al.* 2006 *Science* **314** 1424 (*Preprint* astro-ph/0612016)
[19] HESS Collaboration, Abramowski A, Acero F, Aharonian F, Akhperjanian A G, Anton G *et al.* 2010 *ArXiv e-prints* Accepted for publication in A&A (*Preprint* 1005.3702)
[20] Becherini Y, Behera B, Biteau J, Cerruti M, Giebels B, Lenain J P, de Naurois M, Punch M, Raue M, Sanchez D, Volpe F, Zech A and The HESS Collaboration 2010 *25th Texas Symposium on Relativistic Astrophysics, December 6-10, 2010, Heidelberg, Germany*
[21] CTA Consortium T 2010 *ArXiv e-prints* (*Preprint* 1008.3703)

Journal of Physics: Conference Series **355** (2012) 012002 doi:10.1088/1742-6596/355/1/012002

VERITAS Very High Energy Gamma-ray Observations of Galaxies

Amy Furniss for the VERITAS Collaboration

University of California Santa Cruz Institute of Particle Physics and Physics Department, 1156 High Street, Santa Cruz, CA 96064-1077

E-mail: afurniss@ucsc.edu

Abstract. Remarkable progress has been made in very high energy (VHE; E $\geq$ 100 GeV) gamma-ray astrophysics in the last decade. The VHE source catalog has increased tenfold, with a wide variety of source classes, and includes more than 45 extragalactic objects. Currently, the Very Energetic Radiation Imaging Telescope Array System (VERITAS) has made observations of numerous source candidates, both galactic and extragalactic, including more than 100 galaxies. The last 18 months of observations are especially noteworthy as they have been completed in a new array configuration, following the relocation of one of the four 12–meter telescopes. This new configuration has yielded a significant increase in the VERITAS sensitivity to gamma-ray sources above 100 GeV, resulting in several new source detections in the first few months of operations alone. To date, the VERITAS extragalactic source catalog includes various blazars, the radio galaxy M87 and the starburst galaxy M82. Highlights from the VERITAS extragalactic observation program are summarized here.

1. The VERITAS Instrument

VERITAS comprises four imaging atmospheric Cherenkov telescopes. The VERITAS array is located in southern Arizona at 1268 meters above sea level and is sensitive to gamma rays between ~100 GeV and ~30 TeV with an energy resolution of 15%. These telescopes are 12–meter dishes which reflect the light produced through cosmic-ray and gamma-ray interaction with the atmosphere onto a pixelated camera composed of 499 photomultiplier tubes. These very brief air shower events are promptly digitized, allowing moment and timing analysis for background rejection. The array has been in full operation since September of 2007 and has been used to observe very-high-energy (VHE) events from astrophysical targets with an angular resolution of ~0.1 degrees. In the current configuration, the array is capable of detecting a 1% Crab source in 28 hours. See [1] and [2] for a more general overview of the instrument.

2. The VERITAS Extragalactic Catalog

VERITAS has detected three types of extragalactic objects. BL–Lac–type blazars, specifically high-frequency-peaked (HBL), are the most commonly VHE detected galaxy type, with 14 having been detected by VERITAS at the time of this publication. Additionally, 5 intermediate-frequency-peaked (IBL) BL Lacs have been detected by VERITAS. The historically typical VHE class of BL Lac blazars is joined by the VERITAS detection of the radio galaxy M87 and the starburst galaxy M82, pertinent discoveries with regards to long posed questions about TeV gamma-ray and Galactic cosmic-ray production.

Journal of Physics: Conference Series **355** (2012) 012002 doi:10.1088/1742-6596/355/1/012002

2.1. Blazars

The VERITAS catalog of detected blazars is continually expanding with 20 detected thus far. These blazars are summarized in Table 1[1]. The VERITAS blazar observation program currently splits the available time each year between discovery and long term monitoring observations. This observation program hopes to increase both the number and type of blazars available for study with VHE instruments as well as catch previously detected VHE emitting blazars in both quiescent and flaring states. Organizing VHE observations to be coincident with other wavebands is a high priority for VERITAS and allows for a more complete view of blazar emission states, while at the same time accumulating deep exposures. Only a small selection of these multiwavelength observations will be summarized here.

Table 1. VERITAS blazar detections shown along with BL Lac type. The table is organized in order of increasing redshift. VERITAS discoveries are shown in bold.

AGN	Type	Redshift
Mrk 421	HBL	0.030
Mrk 501	HBL	0.034
1ES 2344+514	HBL	0.044
1ES 1959+650	HBL	0.102
W Comae	IBL	0.102
RGB J0710+591	HBL	0.125
H 1426+428	HBL	0.129
1ES 0806+524	HBL	0.138
1ES 0229+200	HBL	0.139
1ES 1440+122	IBL	0.16
RX J0648.7+1516	HBL	0.179
1ES 1218+304	HBL	0.182
RBS 0413	HBL	0.190
1ES0414+009	HBL	0.287
PG 1553+113	HBL	$0.43 \leq z \leq 0.47$
1ES 0502+675	HBL	0.341?
3C 66A	IBL	0.444?
PKS 1424+240	IBL	?
VER J0521+211	?	?
B2 1215+30	IBL	?

2.1.1. Detected Blazars RGB J0710+591 is an "extreme" HBL, with its synchrotron peak near 10^{19} Hz. This historically promising VHE candidate, as indicated by its hard X-ray spectrum, is a relatively low redshift (z=0.14), hard-spectrum VHE blazar. This target was the first VHE-led *Fermi*-LAT discovery. Upon detection of VHE emission by VERITAS [3], observations by instruments in other wavebands were encouraged. The collections of these data allow for the synchrotron self-Compton (SSC) modeling of the quasi-simultaneous broadband data, as can be seen in Figure 1. As is often the case for TeV emitting HBLs, a SSC model adequately describes the observed emission. More information on the multiwavelength observations and modeling can be found in [4].

[1] Full TeV publication history available through the TeV Online Catalog TeVCat: http://tevcat.uchicago.edu/

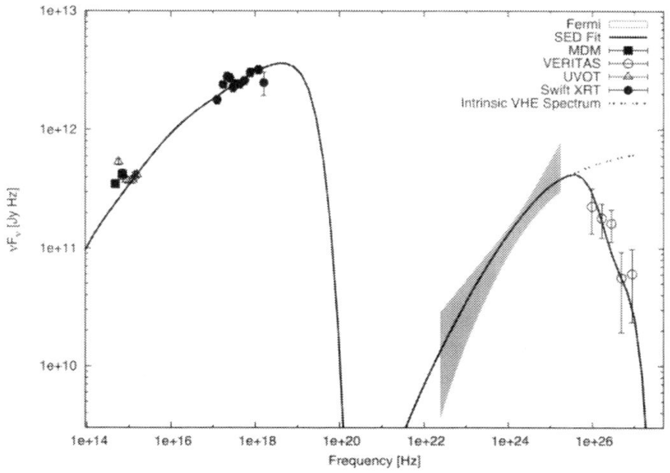

Figure 1. Synchrotron self-Compton model of the "extreme" HBL RGB J0710+591 applied to the quasi-simultaneous multiwavelength data collected in response to the VHE detection. The solid line represents the total extragalactic background light (EBL) corrected emission, while the dotted line shows the blazar intrinsic emission. It can be seen that nearly all of the observed turnover from the high energy band (300 MeV - 100 GeV) to the VHE band can be attributed to absorption by the EBL (from [4]).

PKS 1424+240 was the first *Fermi*-LAT motivated VHE discovery, which prompted simultaneous observations of the target at other wavelengths. This blazar has no determined redshift and is bordering on the IBL/HBL split with the synchrotron peak occurring at 10^{16} Hz during these observations, as can be seen by its SED in Figure 2. A SSC model is applied to the SED data for redshifts ranging from 0.05 to 0.7. This broadband SED modeling favors a redshift of less than 0.1. An upper limit on the redshift can be produced with the simple assumption that all softening of the spectrum observed between the high energy and the VHE regimes is due solely to the absorption of VHE photons on the EBL. Using the *Fermi* high-energy spectrum extended up to the VHE band and absorbing the fit with the EBL models from [5,6,7] until the spectrum matches that measured by VERITAS for a range of redshifts suggests a redshift of less than 0.66, although a lower redshift is likely, as can be reasoned from the soft X-ray spectrum observed with the *Swift*-XRT. More information on the VHE discovery, redshift constraint and SSC modeling of PKS 1424+240 can be found in [8].

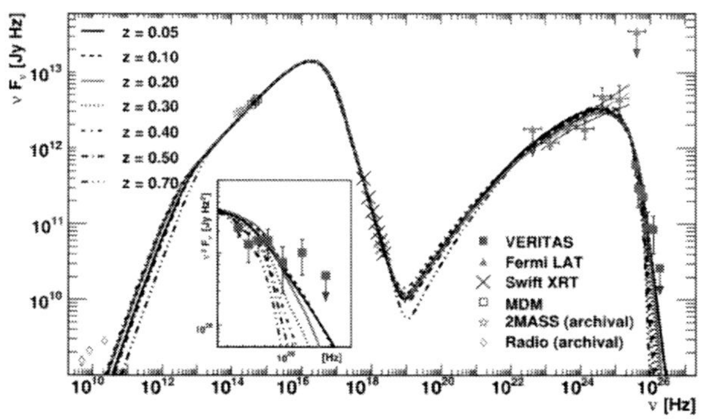

Figure 2. The contemporaneous multiwavelength data and SED modeling of PKS 1424+240 for a set of redshifts ranging from z=0.05 to 0.7. The single zone SSC model provides a good representation of the data, favoring a redshift of less than 0.1. A zoom of the VHE spectrum can be seen in the $\nu^2 F_\nu$ representation for each redshift (from [8]).

RX J0648.7+1516 (1FGL J0648.8+1516) was another *Fermi*-LAT motivated VHE discovery. This target was selected as a promising VHE candidate through the inspection of localized groups of photons collected by *Fermi*-LAT with energy greater than 50 GeV. At the time of VHE detection by VERITAS, this object was an unidentified source 6 degrees off the Galactic plane with no redshift measurement. The angular resolution of VERITAS enabled the previously unidentified Fermi source to be associated with the X-ray source RX J0648.7+1516. Follow-up spectroscopic measurements made at the Lick Observatory identified the object as a BL–Lac–type object with a redshift of 0.179. The optical spectrum can be seen in Figure 3. Additionally, the concurrent multiwavelength data gathered in response to the TeV detection allows the HBL sub-classification of the blazar. The photon spectrum above 200 GeV is well fit by a power law $dN/dE = F_0(E/E_0)^{-\Gamma}$ with a photon index Γ of $4.4 \pm 0.8_{stat} \pm 0.3_{syst}$ and a flux normalization F_0 of $(2.3 \pm 0.5_{stat} \pm 1.2_{sys}) \times 10^{-11}$ $\mathrm{TeV^{-1}cm^{-2}s^{-1}}$ at $E_0 = 300$ GeV (see Figure 4). This soft spectrum offers no constraint on the EBL density and no VHE variability is detected during the VERITAS observations between 4 March and 15 April 2010.

In addition to the newly detected *Fermi*-LAT blazars PKS 1424+240 and RX J0648.7+1516 summarized above, RBS 0413, 1ES 0502+675, VER J0521+21 and 1ES 1440+122 are also new additions to the VERITAS extragalactic catalog [9,10,11,12]. Additionally, VERITAS has recently detected VHE emission from a few previously detected TeV blazars, including 1ES 0414+009, B2 1215+30 and H1426+428. The details of the types of blazars and redshifts are given in Table 1. Continued VERITAS observations of these historical TeV blazars allows valuable insight into the variable nature of the gamma-ray emission, as in the case of 1ES 1218+304.

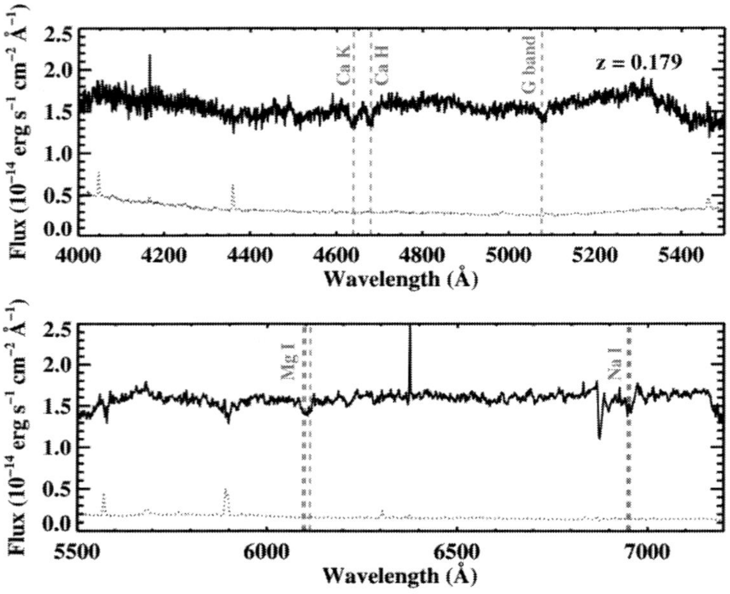

Figure 3. Spectrum of RX J0648.7+1516 showing the Ca H+K, G-band, Na I and Mg I spectral features indicating a redshift of $z = 0.179$. Since the G-band arises in stellar atmospheres, we interpret this as the redshift for the host galaxy and not an intervening absorber. The blazar was observed at Lick Observatory using the 3–meter Shane Telescope on UT night of 6 November 2010.

In early 2009, VERITAS monitoring observations of the TeV-emitting HBL 1ES 1218+304 [13] detected a short flare with a variability timescale of a single day. The light curve of VERITAS observations can be seen in Figure 5. The VHE flux was seen to increase from ~7 to ~20 percent of the Crab Nebula with no spectral change. The power law fit to the low and high states of the blazar are shown in relation to the Crab Nebula in Figure 6. The fast variability timescale detected challenges the kiloparsec jet model of hard spectrum emission, as proposed in [14]. More details on the variable nature of 1ES 1218+304 can be found in [15].

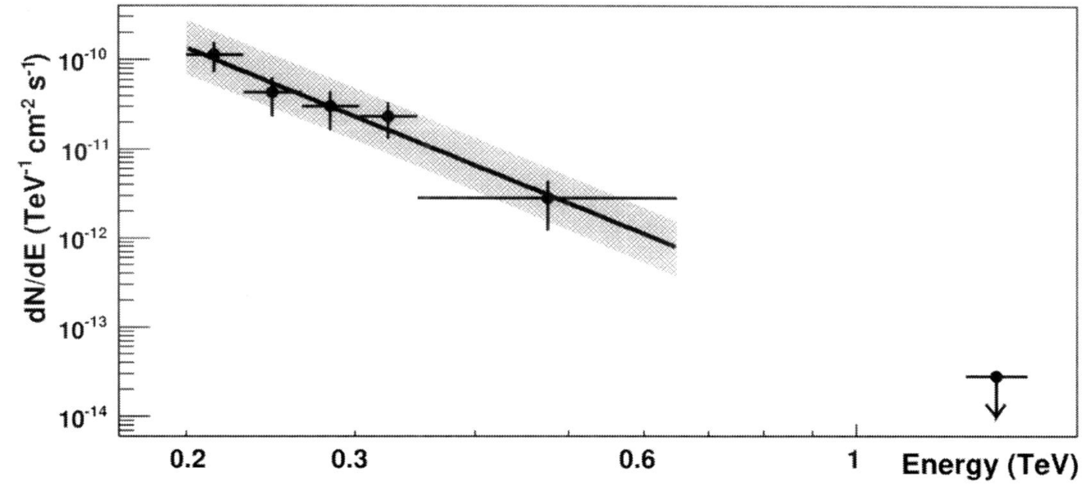

Figure 4. The differential photon spectrum of RX J0648.7+1516 between 200 and 650 GeV measured by VERITAS between 4 March and 15 April 2010 (MJD 55259–55301). The solid line shows a power-law fit to the measured flux derived with four equally log-spaced bins and a final bin boundary at 650 GeV, above which there are few on-source photons. A 99% confidence upper limit evaluated between 650 GeV and 5 TeV assuming a photon index of 4.4 is also shown. The shaded region shows the systematic uncertainty of the fit, which is dominated by 20% uncertainty on the energy scale.

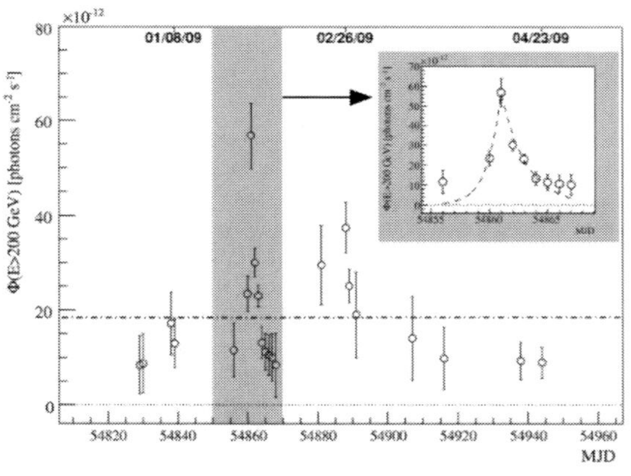

Figure 5. Light curve of the VERITAS observations of the hard-spectrum HBL 1ES 1218+304 above 200 GeV. The flaring region is highlighted, with a corresponding zoomed plot showing the day-scale flux variations (from [15]).

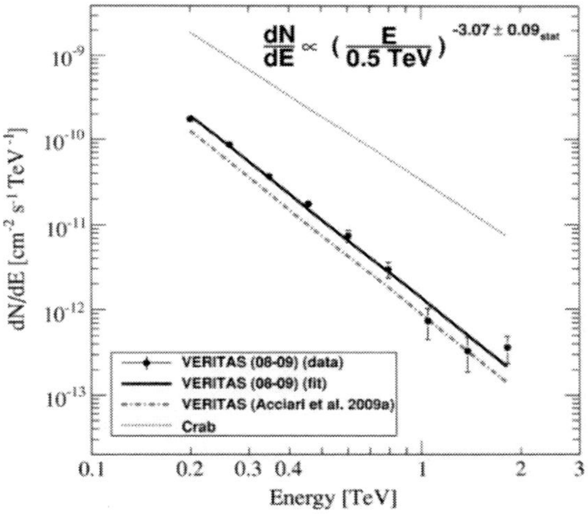

Figure 6. Spectra for the HBL 1ES 1218+304 in both the flaring and quiescent state as compared to the Crab Nebula. The hard spectrum of the blazar is seen to remain constant during both the high and low emission states (from [15]).

2.1.2. Nondetected Blazars The small field of view of the imaging atmospheric Cherenkov telescopes (IACTs) makes new source discovery from a large-scale sky survey difficult, and the hunt for VHE emitting objects has historically been guided by X-ray and high energy gamma-ray experiments such as EGRET onboard the Compton Gamma Ray Observatory [16], Swift [17], ROSAT [18] and now the *Fermi*-LAT [19].

The distribution summarizing the observation of 47 blazars selected as promising VHE candidates prior to the launch of *Fermi* is shown in Figure 7. These candidates, each observed between one and twenty hours, are mainly hard X-ray, low redshift blazars that were observed with VERITAS between 2007 and 2011. Stacking the result from the observation of these blazars results in a 4.1 standard deviation excess. These exposures often provide the best VHE upper limits to date on these targets. A similar stacking analysis for MAGIC observations of BL Lacs [20] yielded a similar result.

A significance distribution can also be seen for 21 high flux, hard-spectrum Fermi-motivated TeV candidate blazars, which were not confined solely to those with known redshifts. Each candidate had between 5 and 10 hours of VERITAS exposure, which again provide some of the best VHE upper limits to date. There is no hint of a stacked excess seen from these observations.

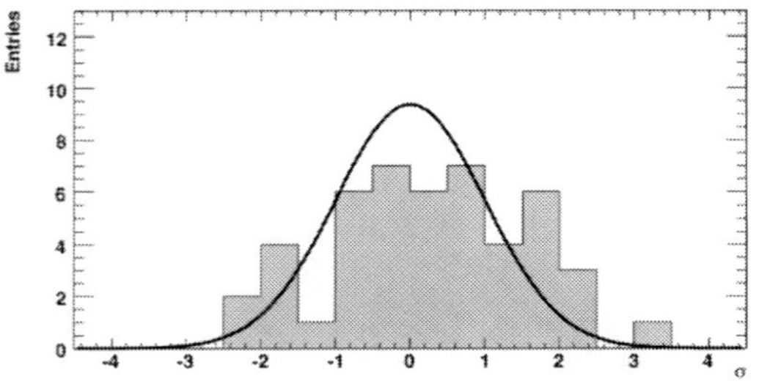

Figure 7. Significance distribution of VHE blazar candidates selected before the *Fermi* launch, restricted to blazars with known redshift.

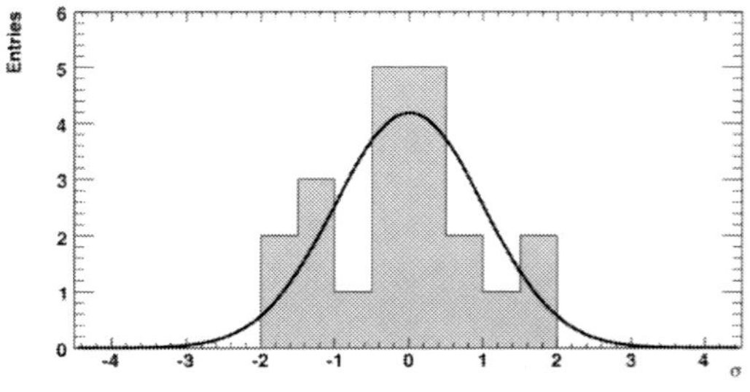

Figure 8. Significance distribution of VHE blazar candidates selected from the *Fermi* data, including blazars with no known redshift.

2.2. Non-Blazar Galaxies

2.2.1. M82 M82 is a prototypical starburst galaxy at a distance of nearly 3.9 megaparsecs in the direction of Ursa Major [21]. This close proximity makes the absorption of gamma rays by the EBL negligible. The galaxy diameter on the sky is one arcmin, which is small enough to appear as a point source for VERITAS. The central supermassive black hole is less than 3×10^7 solar masses, showing no evidence for an active nucleus [22].

The tidal forces resulting from the gravitational interaction with nearby M81 result in an active starburst region with a diameter of approximately 1000 light years [23]. Imaging from the Hubble Space Telescope shows this region to contain more than 200 massive star clusters, suggesting a star formation rate nearly 10 times that of the Milky Way Galaxy. The galaxy's high star formation rate is paired with a high supernova rate of between one and three supernovae every 10 years [24, 25].

Observations by VERITAS during the 2007 through 2009 observing seasons (totaling 137 hours of live time) resulted in the detection of 91 excess gamma-ray events above the threshold of 700 GeV [26]. This high threshold results from the average angle of VERITAS observations being 39 degrees from zenith. The sky map of the VERITAS observations is shown in Figure 9. This detection represents a 4.8 standard deviation post-trials result with a chance probability of 7.7×10^{-7}. The subsequent *Fermi*-LAT detection of high-energy gamma-ray events shows good agreement with the VERITAS result [27]. The VERITAS spectrum and power law fit can be seen in Figure 10.

The VHE emission can be explained through the production of high-energy cosmic-ray hadrons by supernovae in the active starburst region. These cosmic-ray hadrons interact with gas to produce pions which in turn produce VHE gamma rays. Cosmic ray electrons can interact with ambient photons within the galaxy to directly produce gamma-rays. A model representing these processes is also shown to be in good agreement with the VERITAS data in Figure 10. These results are also in agreement with those found through TeV detection from starburst galaxy NGC 253 by HESS [28].

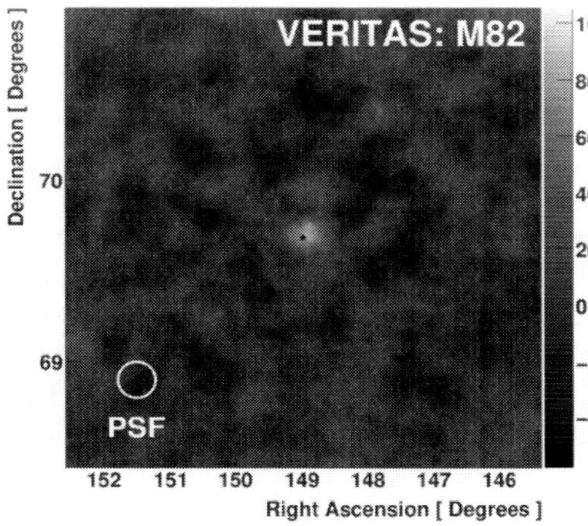

Figure 9. Sky map showing a VHE point source excess coincident with the M82 position, denoted by the black star (from [26]).

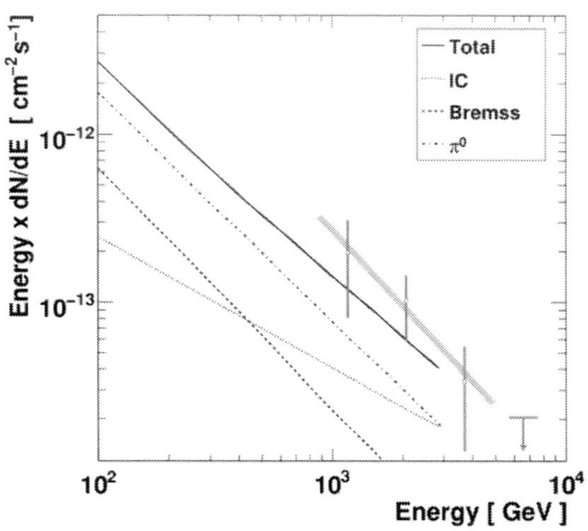

Figure 10. M82 spectrum above 700 GeV shown together with emission model lines for non-thermal gamma-ray production methods thought to be responsible for VHE emission in the starburst galaxy (from [26]).

2.2.2. M87 M87 is an active galactic nucleus with a jet oriented slightly off the Earth line of sight. This radio galaxy offers a unique laboratory for the study of jet and jet substructure physics. One unknown in the field of extragalactic gamma-ray astronomy is where the origin of the gamma-ray emission is located within the active galactic nucleus. This question is complicated by the fact that extragalactic objects appear as no more than point sources for IACTs. Imaging from radio, optical and X-ray instruments shows the M87 jet to contain a bright location at the core (near the base of the jet) and a bright knot farther out from the core, referred to as HST-1.

Simultaneous VHE and X-ray observations of this flaring TeV source in 2005 suggest that both the VHE and X-ray emission increased in the HST-1 region of the jet at the same time. This behavior is shown through the Chandra X-ray observation of increased 2-10 keV flux from the HST-1 region contemporaneous with a TeV flare, detected by HESS. The Chandra observations of the core during the same timeline show no variability. More details on this multiwavelength campaign can be found in [29]. This same variability trend was not observed with more recent multiwavelength observations of M87 flaring events, suggesting that the core is capable of emitting VHE gamma rays during flaring episodes.

Multiwavelength efforts in early 2008 including radio observations from VLBA show correlated variability across the radio, X-ray and TeV bands for the core emission. Results from

Journal of Physics: Conference Series **355** (2012) 012002 doi:10.1088/1742-6596/355/1/012002

these observations can be seen in Figure 11. These observations resulted from a well organized multiwavelength campaign involving monitoring by the very high energy gamma-ray instruments VERITAS, HESS and MAGIC, X-ray observations from Chandra and radio observations from VLBA. As can be seen in the figure, an increase in flux from the core region is apparent in both radio and X-ray observations coincident with a TeV flaring episode. More information on this flare can be found in [30].

Additionally, as is detailed in a proceedings by M. Raue, over 50 hours of monitoring shared between VERITAS, HESS and MAGIC resulted in the most detailed observation to date of a fast flare from M87 in April of 2010. In addition to the X-ray and radio observations triggered by these observations, this flare marks the first TeV flare from M87 with *Fermi*-LAT monitoring.

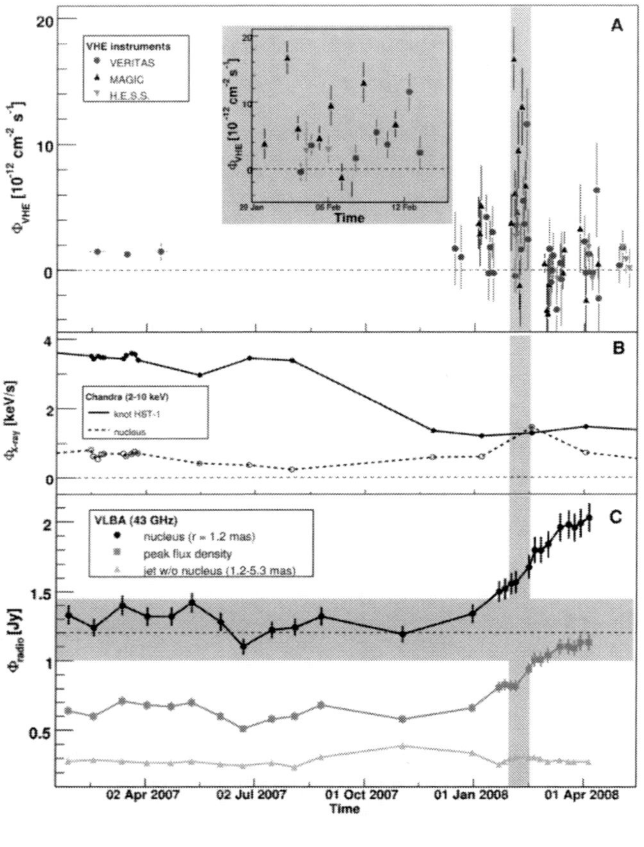

Figure 11. A light curve of the multiwavelength observations of a flare from M87. The top panel contains the VHE integral flux observed by the IACTs VERITAS, MAGIC and HESS, with the window of the flare highlighted. A zoomed plot of the flaring TeV activity is shown. The middle panel shows Chandra 2-10 keV X-ray observations of both the M87 HST-1 and core regions. An increase in flux coincident with the TeV flare can be seen from the nucleus, while the HST-1 emission remains constant. The bottom panel shows the VLBA 43 GHz radio observations of the nucleus, which exhibits an increase in flux coincident with the flaring activity observed in the other wavebands. The radio observations of the jet without the nucleus are also shown, which do not follow the increasing flux observed for the nucleus alone (from [30].

3. Conclusions

Discoveries of TeV emission from the starburst galaxies M82 by VERITAS and NGC 253 by HESS, the radio galaxy M87 as well as blazars have led to exciting new insights regarding the origin of Galactic cosmic-rays and the particle populations responsible for the gamma-ray emission, but even with these new discoveries questions still remain. What is the nature of the blazar sequence? Do blazars go through an evolutionary process? What are the constituent particles that are responsible for the non-thermal emission observed? Is this a blazar-specific question? With continued monitoring and discovery observations of galaxies by IACTs such as VERITAS, HESS and MAGIC, even more targets and a larger set of variable states will be collected. As can be seen from the observations reported above, these observations are most enlightening when simultaneous observations are made across the entire broadband spectrum, allowing a complete view of the galaxy emission. It is evident that TeV extragalactic astronomy is changing from a single-instrument's agenda to a well organized and collaborative effort. The observations that have and will be made from these efforts will help address some of the remaining questions about gamma-ray emission from galaxies.

References

[1] Weekes, T. C. et al. 2002 *Astropart. Phys.* **17** 221
[2] Holder, J. et al. 2006 *Astropart. Phys.* **25** 391
[3] Ong, R. for the VERITAS Collaboration 2009 *Astronomer Telegram* **#1941**
[4] Acciari, V. et al. 2010 *ApJL* **715** 49
[5] Gilmore, R. et al. 2009 *MNRAS* **399** 1694
[6] Franceschini, A., Rodighiero, G. & Vaccari, M. 2008 *A&A* **487** 837
[7] Finke, J. D., Razzaque, S., & Dermer, C. C. 2010 *ApJ* **712** 238
[8] Acciari, V. A. et al. 2009 *ApJL* **708** 100
[9] Ong, R. for the VERITAS Collaboration 2009 *Astronomer Telegram* **#2260**
[10] Ong, R. for the VERITAS Collaboration 2009 *Astronomer Telegram* **#2272**
[11] Ong, R. for the VERITAS Collaboration 2009 *Astronomer Telegram* **#2301**
[12] Ong, R. for the VERITAS Collaboration 2009 *Astronomer Telegram* **#2786**
[13] Albert, J. et al. 2006 *ApJL* **642** 119
[14] Bottcher, M., Dermer, C., & Finke, J. 2008 *ApJL* **679** 9
[15] Acciari, V. et al. 2010 *ApJL* **709** 163
[16] Thompson, D. et al. 1993 *ApJS* **86** 629
[17] Gehrels, N. et al. 2004 *ApJ* **611** 1005
[18] Turriziani, S. et al 2007 *A&A* **472** 699
[19] Atwood, W. et al. 2009 *ApJ* **697** 1071
[20] Aleksic, J. et al. 2011 *ApJ* **729** 115
[21] Sakai, S., & Madore, B. F. 1999 *ApJ* **526** 599
[22] Wills, K. A., Pedlar, A., Muxlow, T. W. B. & Stevens, I.R. *New Astronomy Reviews* **43** 633
[23] Yun, M. S., Ho, P. T. P., Lo, K. Y. 1994 *Nature* **372** 530
[24] Kronberg, P. P., Biermann, P., & Schwab, F. R. 1985 *ApJ* **291** 693
[25] Fenech, D. M., Muxlow, T. W. B., Beswick, R. J., Pedlar, A., Argo, M. K. 2008 *MNRAS* **391** 1384
[26] Acciari, A. et al. 2010 *Nature* **462** 770
[27] Abdo, A. et al. 2010 *ApJL* **709** 152
[28] Acero, F. et al. 2009 *Science* **326** 1080
[29] Aharonian, F. et al. 2006 *Science* **314** 1424
[30] Acciari, V. et al. 2010 *ApJ* **716** 819

Acknowledgments

VERITAS is supported by the US Department of Energy, the US National Science Foundation, and the Smithsonian Institution, by NSERC in Canada, by Science Foundation Ireland, and by STFC in the UK. We acknowledge the excellent work of the technical support staff at the FLWO and the collaborating institutions. This work was also supported by NASA grants from the Swift (NNX10AF89G) and Fermi (NNX09AU18G) Guest Investigator programs.

Journal of Physics: Conference Series **355** (2012) 012003 doi:10.1088/1742-6596/355/1/012003

Recent results from MAGIC observations of AGN

Elina Lindfors on behalf of the MAGIC collaboration
Tuorla Observatory, Department of Physics and Astronomy, University of Turku, Finland
E-mail: elilin@utu.fi

Abstract. MAGIC is a system of two 17m Cherenkov telescopes, sensitive to very high energy (VHE) gamma-rays between 50 GeV up to several tens of TeV. MAGIC observations of Active Galactic Nuclei (AGN) have resulted in the discovery of many previously unknown VHE gamma-ray emitting sources adding also new source types to the list of extragalactic gamma-ray emitters as well as new constraints for the emission models. Many of the discoveries have been triggered by the optical high state of the source as reported by the Tuorla blazar monitoring program, in several cases the targets were also listed as promising candidates by the Fermi collaboration based on their flux and spectrum. In this contribution I discuss the new discoveries as well as the new constraints for emission models of AGN derived from recent MAGIC observations.

1. Introduction

MAGIC is a system of two 17 meters Cherenkov Telescopes located in Observatorio del Roque de los Muchachos on La Palma, 2200 m above sea level. MAGIC-I has been in operation since 2004 and the stereoscopic system has been operation since 2009. MAGIC has an enhanced duty cycle $\sim$ 20% as it is able to operate in presence of moderate moonlight and twilight.

The performance of the MAGIC stereoscopic system is reported in [1]. The low energy threshold of 50 GeV (with special trigger 25 GeV) allows observations of the distant universe and overlaps with the Fermi satellite. Together with the improved sensitivity of stereoscopic system (E>250GeV 0.8% Crab in 50h), it has played a key role in most recent new VHE γ-ray emitting AGN discoveries.

The known extragalactic VHE γ-ray emitters are mostly active galactic nuclei (with the exception of two starburst galaxies, which are not discussed here) with relativistic jet. The most numerous subclass within the VHE γ-ray AGN are blazars, in which the relativistic jet points close to our line of sight. However, also a few non-blazar AGN have been detected in VHE γ-rays namely radiogalaxies M 87 [2] and Cen A [3] and the Perseus Cluster galaxies NGC 1275 and IC 310 [4, 5, 6].

The blazar class consists of two separate classes of AGN namely BL Lac objects and flat spectrum radio quasars (FSRQs). BL Lacs are more numerous in the class of VHE γ-ray emitting blazars (> 30 known) and only three FSRQs have been detected. The VHE γ-ray emitting BL Lacs typically have their synchrotron peak in UV to X-ray energies and are therefore classified as high energy peaking BL Lacs. The second peak of the SED is then located at >GeV energies, which makes them good candidates for sub-TeV and TeV emission. On the other hand some BL Lacs and all FSRQs have their first peak in optical regime, which makes the detection of VHE γ-rays from them less likely.

Figure 1. The MAGIC telescopes at dusk. Photo credit Robert Wagner.

The MAGIC extragalactic science program covers all the classes of VHE γ-ray emitting AGN. The observational strategies include systematic searches, target of opportunity observations, multiwavelength campaigns as well as monitoring of known sources. Target of opportunity observations are performed when a source is seen in a high state in optical (triggers from the Tuorla blazar monitoring program) or γ-ray (triggers from Fermi); this has proved to be more successful strategy than systematic searches (nine of the MAGIC discoveries have resulted from ToO, while three discoveries have been made in systematic searches). Target of Opportunity observations of known sources focus on the fast variability of the sources in VHE regime, as well as correlations between different wavebands, both of which puts severe constrains on the emission mechanism and emission region size. However, these questions are even better accessed with long-term multiwavelength monitoring of the strong sources, in which MAGIC is also participating.

In this contribution we give an overview of recent MAGIC AGN observations covering VHE detection of two FSRQs, which challenge the current canonical models for their emission, discoveries of new sources and finally results from multiwavelength campaigns and monitoring of known strong VHE γ-ray emitters.

2. Flat Spectrum Radio Quasars 3C 279 and PKS 1222+216

There are several scenarios for the γ-ray and VHE γ-ray emission in FSRQs. The emission can, in principle, be explained by both leptonic and hadronic models: the leptonic models traditionally rely on external Compton (EC, e.g. [7, 8]), invoking the inverse Compton scattering of external photons from accretion disk or broad line region (BLR) clouds [9, 10] , while in the hadronic models [11, 12, 13] the VHE γ-ray photons are produced by proton initiated cascades or directly through proton synchrotron radiation.

The leptonic models are very sensitive to the site of the emission: the external Compton models relying on photons originating from broad line emission clouds are not efficient if the emitting blob is outside the BLR and in the SSC models the γ-ray emission must originate from a different emission region than the main component of the synchrotron radiation in order to reproduce the observed γ-ray flux [13]. It should also be noted, that independently of the emission mechanism, the internal absorption cannot be neglected if the emission region is located inside the BLR [14, 15]. In the case of external Compton the internal absorption combined with reduced efficiency of the IC scattering occurring in the KleinNishina (KN) regime [16] should produce a strong softening of the spectrum above few tens of GeV if the external photons derive from the BLR.

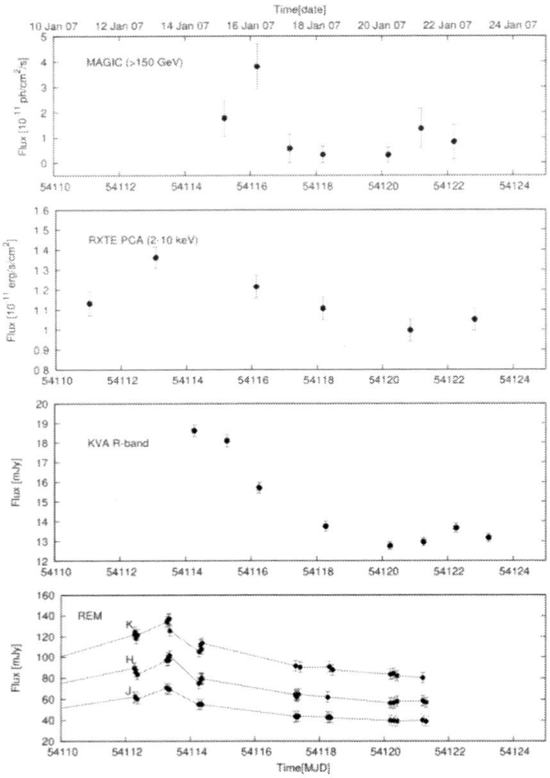

Figure 2. The light curves of 3C 279 in January 2007 from VHE γ-rays (MAGIC), X-rays (RXTE), optical (KVA) and infrared (REM). MAGIC detected $> 5\sigma$ γ-ray excess from the source only on 16th of January (MJD 54116). For the MAGIC light curve the night by night flux is calculated assuming that 3C 279 always emits γ-rays above 150 GeV.

Alternatively, the emission can be produced in regions located far beyond the broad line region, at distances at which the dominant radiation field for EC is that of the parsec-scale dusty torus [17]. In this case, internal absorption can be neglected up to $\sim$ 1 TeV but, due to the large size of the emission region, a minimum variability timescale of the order of $\sim$1 day is expected.

MAGIC observations of 3C 279 and PKS 1222+216 have shed new light on the emission mechanism and site in these two well-known FSRQs.

3C 279 was the first flat-spectrum radio quasar (FSRQ) discovered to emit very high energy (VHE, defined here $>$ 100 GeV) γ-rays [8]. With a redshift of 0.536 , 3C 279 is also the most distant of the VHE γ-ray emitting sources discovered so far. The source was re-observed by MAGIC-I (for performance see [18]) in January 2007 during a major optical flare and from December 2008 to April 2009 following an alert from the Fermi space telescope on an exceptionally high γ-ray state. In January 2007 it was observed for a total of 23.6 h, 18.6 h (seven nights) of which passed the quality selection. In December 2008 to April 2009 a total observation time of 28.1 h was accumulated over 20 days. Only one night, MJD 54116 (January 16, 2007), revealed a significant excess in the MAGIC data after off-subtraction, the significance of the excess is 5.6σ (pre-trial). The VHE γ-ray spectrum of the flare can be described by a simple power law (with the differential flux given in units of $\mathrm{TeV^{-1}\ m^{-2}\ s^{-1}}$):

$$\frac{\mathrm{d}F}{\mathrm{d}E} = (5.7 \pm 1.3) \times 10^{-7} \left(\frac{\mathrm{E}}{300\,\mathrm{GeV}}\right)^{-3.1 \pm 1.1} \tag{1}$$

The highest spectral energy point has a mean energy of 350 GeV. Flux and spectral index are comparable with the 2006 detection.

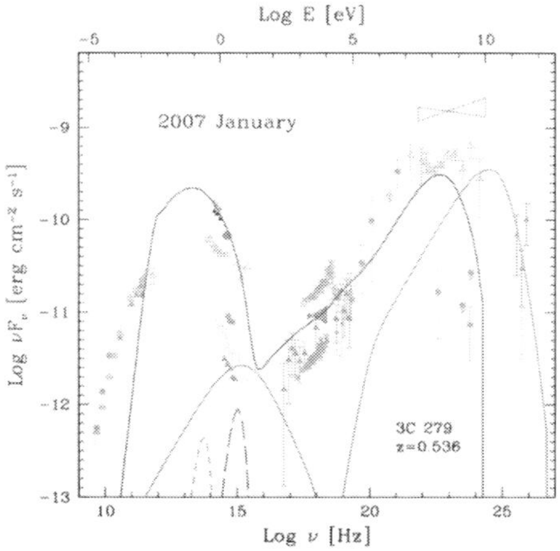

Figure 3. The spectral energy distribution of 3C 279 on January 2007 16th (MJD 54116). The red symbols show the REM (MJD 54114.4, triangles), KVA (filled circle), RXTE (bow-tie) and MAGIC (deabsorbed, triangles) data used to fit the two-zone model (see [8]). Blue line shows the emission from the zone inside BLR and red line the emission from outside the BLR. Additionally REM data from MJD 54113.3 (Maximum of the REM lightcurve) cyan and MJD 54117.3 (black) are shown. The dashed lines correspond to blackbody radiation from the IR torus (red) and BLR (blue). For comparison also historical data are shown: 1991 high state (gray: [19]), 1993 low state [20] (green: Maraschi et al. 1994) 1996 high state (orange, blue bow-tie: [21, 22])and 2003 low state (magenta: [23]).

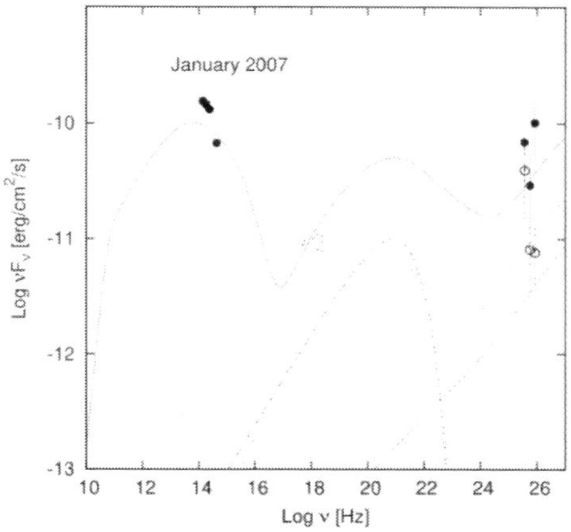

Figure 4. The observed spectral energy distribution of 16th of January 2007 modeled with the lepto-hadronic model. The data used for fit are the same as in Figure 3. The overall fit is shown with a red line with the following components at high energies: the synchrotron radiation of pair creation electrons and positrons cascaded down from the optically thick regime (magenta dotted line), synchrotron radiation from positrons from the pion decay (green dashed line), synchrotron radiation from electrons from the pion decay (blue dashed line), inverse Compton scattering (cyan dot dashed line) and proton synchrotron emission (black double dashed line). The VHE γ-ray emission is mostly the sum of the three first components while in X-rays the main contribution comes from the inverse Compton scattering (like in the purely leptonic models). The low energy bump is produced by the electron synchrotron radiation.

There is multiwavelength data from radio to X-rays available for both observing periods. The MAGIC detection took place in the decay phase of the optical-IR and X-ray flare (see Fig.2). Larionov et al. [24] associate the optical flare with a VLBA component emerging from the core at MJD 54063+-40 based on simultaneous ~300 degree rotation of the optical and VLBA 43 GHz core polarization angle. If the VHE γ-ray emission is connected to the optical flare -which in turn appears connected to the radio flare- this would also place the emission region of the VHE γ-ray flare far out in the jet.

The spectral energy distribution (SED) is difficult to reproduce with traditional one-zone SSC-EC model, because it largely overproduces the MeV emission (although there is no

Journal of Physics: Conference Series **355** (2012) 012003 doi:10.1088/1742-6596/355/1/012003

simultaneous measurements from this energy regime the required MeV flux would be ~10 higher than the historical maximum observed from the source). The SED is better reproduced by a two zone SSC-EC model (Fig.3), where the optical to X-ray emission zone is located inside the BLR, and the VHE γ-ray emission outside the BLR where soft photons from the infra-red torus serve as seed photons for Inverse Compton scattering. The SED can also be reasonably well reproduced by a hadronic model (Figure 4). The observations, results and modeling are described in detail in [25].

PKS 1222+21 (z=0.432) is a γ-ray blazar [26] with a relatively hard spectrum in the GeV range and has been included in the list of > 100 GeV emitters in the analysis of [27].It is characterized by highly superluminal jet knots with apparent velocity up to 21c [28]. On June 17, 2010, PKS 1222+21 was observed with the MAGIC telescopes for ~0.5 hr (MJD 55364.908 to MJD 55364.931), in the so-called wobble mode. These observations yield an excess corresponding to a statistical significance of 10.2σ.

The observed VHE and GeV (from simultaneous Fermi observations) γ-ray spectra are consistent with a single powerlaw with index $\sim 2.7 \pm 0.3$. This suggests that the 100 MeV-400 GeV emission originates from one single component, which must be located outside the BLR as no strong softening is seen. This result is discussed in detail in [29]and Becerra-González et al. 2011 (this volume).

3. New Discoveries

The number of known extragalactic γ-ray emitters has been increasing by ~5-10 sources per year since the current generation of instruments came online. Currently the number of known sources is 45[1]. To search for new VHE γ-ray emitting AGN MAGIC has been observing a sample of X-ray bright AGN [30, 31] , but also performing target of opportunity observations of AGN that are in high state in optical as reported by the Tuorla blazar monitoring program [2]. In addition Fermi collaboration has been providing lists of candidate TeV blazars, based on the observation of >50 GeV photons from the sources, to the IACT community; observations of these sources have resulted in many new discoveries: already 9 sources from the lists have been discovered to emit VHE γ-rays.

3.1. Optically Triggered Discoveries

The Tuorla blazar monitoring program was started in 2002 with a goal to study the optical variability of TeV candidate blazars from [32]. Since the beginning of the science observations of MAGIC, the optical lightcurves have been used to trigger Target of Opportunity observations. This has resulted in several discoveries: Mrk 180 [33],1ES 1011+496 [34], S5 0716+714 [35]and most recently B3 2247+381 [36]and 1ES 1215+303 [37].

B3 2247+381 (z= 0.119) was previously observed by MAGIC telescope as a part of the systematic search of VHE γ-rays from X-ray bright BL Lac objects [31]. The source was selected from the Nieppola et al. [38] sample with reported X-ray flux > 2 mJy. They classify the source as an intermediate peaking BL Lac object, while in [39] it is classified as high peaking and the reported X-ray flux is F(>1 keV) 0.6 mJy. MAGIC observed the source in August-September 2006 and these observations resulted in a upperlimit $F(> 140\ \mathrm{GeV}) < 1.6 * 10^{-11} cm^{-2} s^{-1}$. The source has been monitored in optical by Tuorla blazar monitoring program ever since. In autumn of 2010 the source was observed in a high optical state (brightening of the core > 50%, see Reinthal et al. this volume), which triggered MAGIC observations. MAGIC observed the source during 13 nights between September 30th, and October 30th 2010, collecting a total of

[1] http: //www.mppmu.mpg.de/ ~rwagner/sources
[2] http: //users.utu.fi/kani/1m

21.2 hours of data, of which only 2.9 hours were discarded. The source was detected at the 5.6σ level. This is the first detection of the source in VHE γ-rays.

1ES 1215+303 (also known as ON 325) is a high energy peaking BL Lac object with uncertain redshift (two values can be found in the literature: z=0.130 and z=0.237). The source was classified as a promising candidate TeV blazar by [32] and has been observed several times in VHE γ-rays prior to the observations presented here (e.g. [31]). The source was also in the Fermi bright AGN catalog [40] showing variable flux and hard spectra (Γ = 1.89 $\pm$ 0.06). In January 2011 the source was in a high state in optical, which triggered MAGIC observations in January-February 2011 for a total ~20 hours. The observations were made in the so called wobble mode in dark night and moderate moon conditions. The data were taken at zenith angles from 1 to 40 degrees. The source was clearly detected at 10σ level, the first significant detection of VHE γ-rays from 1ES 1215+303.

Multiwavelength data simultaneous and quasi-simultaneous to MAGIC observations of these two blazars were collected from radio to γ-ray frequencies including e.g. Metsähovi 37 GHz data (1ES 1215+303 only), optical, Swift and Fermi data. The analysis and interpretation of these data are ongoing and dedicated publications in preparation.

3.2. Fermi Candidate Discoveries

The Fermi Space Telescope has detected 698 γ-ray emitting AGN above energies 100 MeV [26].Some of these AGN are also observed to emit photons above > 30 GeV and are therefore considered to be good candidates for ground based γ-ray observations. In October 2009 the Fermi collaboration distributed the list of such VHE candidate sources to all the IACTs (Fermi Collaboration priv. comm.) with a goal of further increasing the number of extragalactic VHE γ-ray emitters. The second edition of this list was released in October 2010. MAGIC has observed a few of the most promising candidates from these lists (and it should be noted that both latest optically triggered discoveries were also in Fermi lists).

In July 2010 MAGIC announced the discovery of Very High Energy (VHE; >100 GeV) γ-ray emission from the source MAGIC J2001+435 [41], which is positionally consistent with the Fermi-LAT γ-ray source 1FGL J2001.1+4351, and the radio source MG4 J200112+4352, recently identified with a BL Lac object [42]. The preliminary analysis of the MAGIC data using the standard cuts optimized for soft energy spectra yielded a detection of ~120 γ-rays, corresponding to a pre-trail statistical significance of ~7 standard deviations. The observed flux is estimated to be ~20% of the Crab nebula flux above 100 GeV. The dedicated publication is in preparation.

In October 2010 MAGIC also reported the discovery of VHE γ-rays from the central galaxy of the Perseus cluster, NGC 1275 [4]. The MAGIC observations were carried out in stereoscopic mode starting in August 2010 and the preliminary analysis shows an excess at the 5σ level. Dedicated publication is in preparation. In the same field of view MAGIC also observed VHE γ-ray emission from another Fermi detected [5] Persues cluster galaxy IC 310 [6].

4. Multiwavelength campaigns and multiwavelength monitoring

Longterm multiwavelength monitoring is the best tool for studying the emission mechanisms and emission sites in AGN. However, it is difficult to organize and in many cases fixed window multiwavelength campaigns are the only viable option. MAGIC participates in long-term MWL monitoring (Mrk 421, Mrk 501 and M 87) and is also organizing fixed time multiwavelength campaigns together with other instruments.

In 2008 MAGIC organized several fixed time multiwavelength campaigns: 1ES 1011+496, Mrk 180 and 1ES 2344+514 were observed in radio, optical, X-ray, γ-rays and VHE γ-rays. For 1ES 1011+496 and Mrk 180 these observations were the first follow-up observations at VHE γ-rays after the discovery and the first MWL observations of the sources. 1ES 1011+496 was observed in a similar VHE γ-ray flux level (Reinthal et al. this volume) as during the discovery. Mrk 180 was observed in a very variable X-ray state, while 1ES 2344+514 was in a very low state in all wavebands during the campaign resulting only in marginal detection of the source by MAGIC. Detailed publications on these campaigns are in preparation.

MAGIC has been monitoring the strong VHE γ-ray blazars Mrk 421, Mrk 501 and 1ES 1959+650 since the beginning of its science operations (e.g. [43]). After the launch of the Fermi satellite, which operates in survey mode, the sources have been constantly observed in γ-ray energies >100 MeV. This allows for the first time simultaneous observations covering all of the high energy peak. Since 2009 there has been intense MWL monitoring campaigns (led by the Fermi collaboration) on Mrk 421 and Mrk 501, where the sources are observed every second day from radio to VHE γ-ray energies. The campaigns have duration of 4.5 months covering basically all of the time window when these sources are observable from earth based observatories. MAGIC has been participating in these campaigns along with Swift, RXTE, GASP-WEBT, F-GAMMA and VLBA. The results of the 2009 campaigns were recently published [44, 45]. Mrk 421 was in a low emission state, showing low flux and low multifrequency variability. The spectral energy distribution (Fig.5) is well reproduced with a leptonic (one-zone Synchrotron Self-Compton) or a hadronic model (Synchrotron Proton Blazar). Instead Mrk 501 was showing large variability during the 2009 campaign, but MAGIC observations were conducted mainly in the low state. The average spectral energy distribution of Mrk 501 is well described by the standard one-zone synchrotron self-Compton (SSC) model (Fig.6). MAGIC also participated in the 2010 and 2011 campaigns.

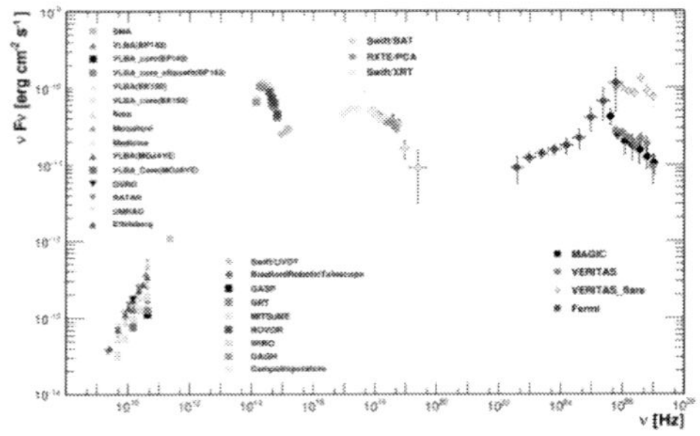

Figure 5. Spectral energy distribution of Mrk 501 averaged over all observations taken during the multifrequency campaign performed between March 15, 2009 and August 1, 2009.

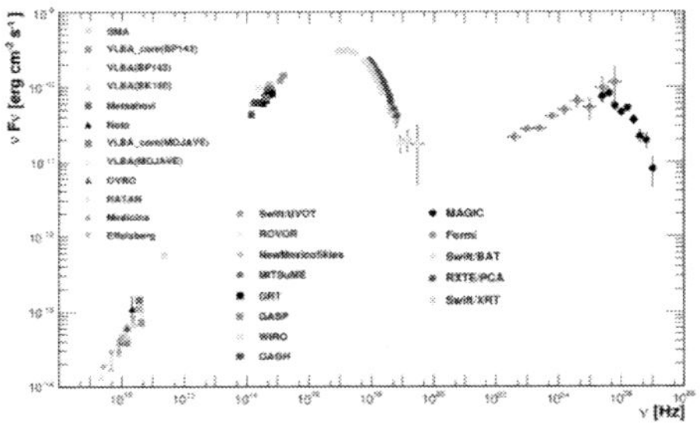

Figure 6. Spectral energy distribution of Mrk 421 averaged over all the observations taken during the multifrequency campaign from January 19, 2009 to June 1, 2009.

M 87 is a giant elliptical radio galaxy situated in the Virgo cluster at a distance of 16.7 Mpc. M 87 has been observed by MAGIC since 2005 and since 2008 MAGIC, HESS and VERITAS have been monitoring the source jointly allowing better sampling of the VHE γ-ray lightcurve. VHE γ-ray emission shows day-scale variability, which excludes many of suggested emission scenarios and puts strong limits on the size of the emission region [46, 47]. M 87 is also monitored by Chandra (X-rays), HST (optical) and VLBA. The vicinity of M 87 allows to distinguish different parts of the jet (especially the core and the component called HST-I) in these energy intervals and therefore study the location of the emission regions in the jet. This combined with the VHE γ-ray monitoring gives a direct view of the emission site of VHE γ-rays in the jet, which was demonstrated by the 2007-2008 joint campaign [48]. MAGIC also participated in 2009, 2010 and 2011 campaigns (see Raue et al. this volume).

5. Summary and Conclusions

The MAGIC AGN observation program currently consists of long-term monitoring of known strong VHE γ-ray emitters, fixed term multiwavelength observations, targeted observations of good TeV candidate sources and target of opportunity observations. In past years the long-term monitoring has also been combined with the data from other wavelengths, which allows us to study correlations between different energy intervals and put constraints on emission mechanisms in work. However, the number of sources we can detect with short observations in low state is very limited and for many sources it is more efficient to trigger MAGIC observations when the source is in a high state in optical and γ-ray. Target of opportunity observations of known VHE γ-ray emitters (e.g. 3C 279 in 2007 optical trigger, 1ES 0806+524 in 2011 optical trigger) have resulted in the detection of short term variability in these sources. Target of opportunity observations have also resulted in the discovery of new VHE γ-ray emitting blazars (e.g. PKS1222+21 in 2010 γ-ray trigger, B3 2247+381 in 2010 optical trigger, 1ES 1215+303 in 2011 optical trigger). Also targeted observations have resulted in new detections (MAGIC J2001+435 and NGC 1275 in 2010). The discoveries are still important for extending our knowledge on the population of the VHE γ-ray emitting AGN. The MAGIC AGN observing program has been very successful in discovering new sources and putting constrains on emission mechanism and emission site of the VHE γ-rays in known sources.

Acknowledgments

The work of EL is supported by grant 127740 of the Academy of Finland. The MAGIC Collaboration would like to thank the Instituto de Astrofísica de Canarias for the excellent working conditions at the Observatorio del Roque de los Muchachos in La Palma. The support of the German BMBF and MPG, the Italian INFN, the Swiss National Fund SNF, and the Spanish MICINN is gratefully acknowledged. This work was also supported by the Marie Curie program, by the CPAN CSD2007-00042 and MultiDark CSD2009-00064 projects of the Spanish Consolider-Ingenio 2010 programme, by grant DO02-353 of the Bulgarian NSF, by the YIP of the Helmholtz Gemeinschaft, by the DFG Cluster of Excellence "Origin and Structure of the Universe", by the DFG Collaborative Research Centers SFB823/C4 and SFB876/C3, and by the Polish MNiSzW grant 745/N-HESS-MAGIC/2010/0.

References

[1] Klepser S and for the MAGIC Collaboration 2011 *ArXiv e-prints* (*Preprint* 1104.0863)
[2] The HEGRA Collaboration Aharonian F and et al 2003 *Astronomy & Astrophysics* **403** L1–L5 (*Preprint* arXiv:astro-ph/0302155)
[3] The HESS Collaboration Aharonian F and et al 2009 *Astrophysical Journal Letters* **695** L40–L44 (*Preprint* 0903.1582)
[4] Mariotti M and MAGIC Collaboration 2010 *The Astronomer's Telegram* **2916** 1–+
[5] Neronov A, Semikoz D and Vovk I 2010 *Astronomy & Astrophysics* **519** L6+ (*Preprint* 1003.4615)
[6] The MAGIC Collaboration Aleksić J and et al 2010 *Astrophysical Journal Letters* **723** L207–L212 (*Preprint* 1009.2155)
[7] Hartman R C, Böttcher M, Aldering G and et al 2001 *Astrophysical Journal* **553** 683–694 (*Preprint* arXiv:astro-ph/0102127)
[8] The MAGIC Collaboration Albert J e a 2008 *Science* **320** 1752– (*Preprint* 0807.2822)
[9] Dermer C D and Schlickeiser R 1993 *Astrophysical Journal* **416** 458–+
[10] Sikora M, Begelman M C and Rees M J 1994 *Astrophysical Journal* **421** 153–162
[11] Mannheim K and Biermann P L 1992 *Astronomy & Astrophysics* **253** L21–L24
[12] Mücke A, Protheroe R J, Engel R and et al 2003 *Astroparticle Physics* **18** 593–613 (*Preprint* arXiv:astro-ph/0206164)
[13] Böttcher M, Reimer A and Marscher A P 2009 *Astrophysical Journal* **703** 1168–1175
[14] Sitarek J and Bednarek W 2008 *Monthly Notices of the Royal Astronomical Society* **391** 624–638 (*Preprint* 0807.4228)

[15] Tavecchio F and Mazin D 2009 *Monthly Notices of the Royal Astronomical Society* **392** L40–L44 (*Preprint* 0809.2467)
[16] Ghisellini G and Tavecchio F 2009 *Monthly Notices of the Royal Astronomical Society* **397** 985–1002 (*Preprint* 0902.0793)
[17] Sikora M, Moderski R and Madejski G M 2008 *Astrophysical Journal* **675** 71–78 (*Preprint* 0711.3524)
[18] The MAGIC Collaboration Albert J and et al 2008 *Astrophysical Journal* **674** 1037–1055 (*Preprint* 0705.3244)
[19] Hartman R C, Webb J R, Marscher A P and et al 1996 *Astrophysical Journal* **461** 698–+
[20] Maraschi L, Grandi P, Urry C M and et al 1994 *Astrophysical Journal Letters* **435** L91–L95
[21] Wehrle A E, Pian E, Urry C M and et al 1998 *Astrophysical Journal* **497** 178–+ (*Preprint* arXiv:astro-ph/9711243)
[22] Ballo L, Maraschi L, Tavecchio F and et al 2002 *Astrophysical Journal* **567** 50–57 (*Preprint* arXiv:astro-ph/0111499)
[23] Collmar W, Böttcher M, Burwitz V and et al 2004 *5th INTEGRAL Workshop on the INTEGRAL Universe* (*ESA Special Publication* vol 552) ed V Schoenfelder, G Lichti, & C Winkler pp 555–+ (*Preprint* arXiv:astro-ph/0410302)
[24] Larionov V M, Jorstad S G and Marscher A P e a 2008 *Astronomy & Astrophysics* **492** 389–400 (*Preprint* 0810.4261)
[25] The MAGIC Collaboration Aleksić J and et al 2011 *Astronomy & Astrophysics* **530** A4+ (*Preprint* 1101.2522)
[26] The Fermi-LAT Collaboration Abdo A A and et al 2010 *Astrophysical Journal* **715** 429–457 (*Preprint* 1002.0150)
[27] Neronov A, Semikoz D and Vovk I 2011 *Astronomy & Astrophysics* **529** A59+ (*Preprint* 1004.3767)
[28] Lister M L, Cohen M H, Homan D C and et al 2009 *Astronomy Journal* **138** 1874–1892 (*Preprint* 0909.5100)
[29] The MAGIC Collaboration Aleksić J and et al 2011 *Astrophysical Journal Letters* **730** L8+ (*Preprint* 1101.4645)
[30] The MAGIC Collaboration Albert J e a 2008 *Astrophysical Journal* **681** 944–953 (*Preprint* 0706.4453)
[31] The MAGIC Collaboration Aleksić J and et al 2011 *Astrophysical Journal* **729** 115–+ (*Preprint* 1002.2951)
[32] Costamante L and Ghisellini G 2002 *Astronomy & Astrophysics* **384** 56–71 (*Preprint* arXiv:astro-ph/0112201)
[33] The MAGIC Collaboration Albert J and et al 2006 *Astrophysical Journal Letters* **648** L105–L108 (*Preprint* arXiv:astro-ph/0606630)
[34] MAGIC Collaboration Albert J and et al 2007 *Astrophysical Journal Letters* **667** L21–L24 (*Preprint* 0706.4435)
[35] The MAGIC Collaboration Anderhub H and et al 2009 *Astrophysical Journal Letters* **704** L129–L133 (*Preprint* 0907.2386)
[36] Mariotti M and MAGIC Collaboration 2010 *The Astronomer's Telegram* **2910** 1–+
[37] Mariotti M 2011 *The Astronomer's Telegram* **3100** 1–+
[38] Nieppola E, Tornikoski M and Valtaoja E 2006 *Astronomy & Astrophysics* **445** 441–450 (*Preprint* arXiv:astro-ph/0509045)
[39] Donato D, Ghisellini G, Tagliaferri G and Fossati G 2001 *Astronomy & Astrophysics* **375** 739–751 (*Preprint* arXiv:astro-ph/0105203)
[40] The Fermi-LAT Collaboration Abdo A A and et al 2009 *Astrophysical Journal Supplement Series* **183** 46–66 (*Preprint* 0902.1340)
[41] Mariotti M 2010 *The Astronomer's Telegram* **2753** 1–+
[42] Bassani L, Landi R, Masetti N and et al 2009 *Monthly Notices of the Royal Astronomical Society* **397** L55–L59 (*Preprint* 0905.0843)
[43] Hsu C C, Satalecka K and Thom M e a 2009 *ArXiv e-prints* (*Preprint* 0907.0893)
[44] The Fermi-LAT collaboration: A A Abdo, the MAGIC collaboration: J Aleksić, GASP-WEBT consortium t and partners m w 2011 *ArXiv e-prints* (*Preprint* 1106.1348)
[45] The Fermi-LAT Collaboration Abdo A A e a, MAGIC Collaboration Aleksić J e a, et al V C A V A and et al 2011 *Astrophysical Journal* **727** 129–+
[46] The HESS Collaboration Aharonian F and et al 2006 *Science* **314** 1424–1427 (*Preprint* arXiv:astro-ph/0612016)
[47] The MAGIC Collaboration Albert J and et al 2008 *Astrophysical Journal Letters* **685** L23–L26 (*Preprint* 0806.0988)
[48] The VERITAS Collaboration, The VLBA 43 GHz M87 Monitoring Team, The HESS Collaboration and The MAGIC Collaboration 2009 *Science* **325** 444– (*Preprint* 0908.0511)

A review of the multiwavelength studies on the blazars detected by AGILE

Immacolata Donnarumma on behalf of the AGILE Team
INAF-IASF Rome, Via Fosso del Cavaliere 100, 00133, Rome, Italy
E-mail: immacolata.donnarumma@iasf-roma.inaf.it

Abstract. We report on the main results on gamma-ray blazars as obtained by AGILE during 4 years in orbit. AGILE detected several flaring blazars, mostly FSRQs, which were studied from radio to TeV energy bands thanks to the rapid dissemination of our alerts. In particular, we carried out several multifrequency campaigns resulted from the synergy with other observatories such as GASP-WEBT (GLAST-AGILE Support Programme of the Whole Earth Blazar Telescope network), REM, *Spitzer*, *Swift*, *RXTE*, *XMM*-Newton, *Suzaku*, *INTEGRAL*, MAGIC, VERITAS. Temporal and SED variabilities were studied in details thanks to the large set of simultaneous data. The most relevant properties of our sample of blazars will be presented with a particular emphasis on the spectral thermal components, time lags, spectral trends (in X-rays and gamma-rays), jet geometry and acceleration mechanism at the inner portion of the jet itself.

1. Introduction

Since its launch in 2007, AGILE (Tavani et al. 2009) has been involved in several multi-frequency campaigns to study the emission mechanisms of the brightest blazars in the gamma-ray sky. This was possible thanks to coordinated (quasi-)simultaneous observations basically obtained by the acceptance of several proposals of monitoring and/or Target of Opportunity (ToO) with X-ray observatories (*Swift*, *Suzaku*, *INTEGRAL*) and an agreement with the radio-optical monitoring provided by the GASP-WEBT (Villata et al. 2008). Moreover, a good cooperation and data sharing has been obtained with the Cherenkov observatories MAGIC and VERITAS, allowing for the exploration of GeV-TeV connection. A large set of multi-frequency data were acquired on ~ 20 objects detected in flaring states and analyzed in order to investigate possible multi-λ flux correlations and spectral energy distribution (SED) variability. Despite the limited number of the objects in our sample, the wide coverage of the SED of these blazars obtained in different states posed important constraints on the efficiency of their acceleration mechanisms, on the electron population, on the thermal emission variability, on possible time lag between optical and gamma-ray emission and finally on the dissipation region of the gamma-ray emission, which is actually still debated (Marscher et al. 2010, Tavecchio et al. 2010).

Since October 2009, AGILE is prevented to perform pointed observations due to a failure of the reaction wheel, then it now surveys a large fraction of the sky ($\sim 80\%$) rotating at 0.8 degrees/s. The sky scanning mode improved the AGILE capability to detect transient episodes from the brightest blazars. In this paper, we will summarize the main results obtained so far during these multifrequency campaigns.

2. Optical-UV observations in different γ-ray states

Simultaneous gamma-ray and Optical-UV observations are key ingredients to constrain a) the acceleration mechanism efficiency given the evidence/lack of the thermal component in FSRQs; b) the external seed photon component responsible of the IC peak by estimating the accretion disk luminosity, with a better definition of the γ-ray dissipation region; c) possible time lags between synchrotron and IC emissions in low synchrotron peak sources.

2.1. Thermal components

In order to constrain the contribution of the thermal component to the blazar SED, the optical-UV emission has to be monitored during the quiescent states of the source when, more likely, it is not hidden in the non-thermal emission. In particular, simultaneous optical-UV and gamma-ray data are crucial to disentangle among the candidate seed photons (dusty torus, accretion disk, Broad Line Region (BLR), hot corona) in the external Compton (EC) model, posing constraints on the γ-ray dissipation region. The estimate of this component was possible for 2 objects characterized by a strong γ-ray activity, 3C 454.3 and PKS 1510-089.

In particular, during the 18-month long multi-frequency campaign on 3C 454.3 the disk component was observed only during the low gamma-ray states (Fig. 19 in Vercellone et al. 2010). We remind that previous evidences of the blue bump in this blazar were found during multi-frequency campaigns performed between 2005 and 2007 as reported in Raiteri et al. 2008. The derived disk luminosity estimate of 5×10^{46} erg s^{-1} in the AGILE campaigns was injected in the model fitting of the SEDs in the higher gamma-ray states, deriving that the accretion disk emission as direct as reprocessed by the BLR was the dominant seed photon source in the IC emission.

The case of PKS 1510-089 is even more intriguing. In this object the thermal component was detected also during the high gamma-ray activity registered by AGILE in March 2008 (D'Ammando et al. 2009), showing synchrotron emission peaked in the infrared energy band. The disk component was still observed during a new high gamma-ray activity detected by AGILE in March 2009, showing a significant variation in UV (a mild in optical) as signature of disk emission enhancement with respect to the past year. However, during the maximum γ-ray activity reached in this campaign, a strong variation at optical-UV energies was also detected. Given the short time scale of this variation, this was interpreted as due to an acceleration process responsible for the shift of the synchrotron energy peak towards higher energies (Fig. 6 in D'Ammando et al. 2011 for details). This behavior is not usually observed in FSRQs being commonly observed in HBL objects.

2.2. Time Lags

Time lags between optical and gamma-ray emissions have been investigated on a few blazars ($\sim$ 4) for which a continuous monitoring in both energy bands - lasting at least one month - was available. Optical-γ-ray correlation was studied by means of a Discrete Correlation Function (DCF; Edelson & Krolik 1988; Hufnagel & Bregman 1992) in order to constrain time lag also below the time binning of the gamma-ray light curves (1-day binning) thanks to the centroid calculation (Peterson et al. 1998; Raiteri et al. 2003). The DCF method applied to our sample revealed hints of delay between optical and gamma-ray variabilities only in the case of the Intermediate BL Lac Object S5 0716+714 (Chen et al. 2008) and of 3C 454.3 (Donnarumma et al. 2009a, Vercellone et al. 2010). In detail, S5 0716+714 was observed simultaneously in optical (R) and in γ-ray energy bands between September and October 2007. A DCF analysis performed on this dataset provided a possible delay between optical and γ-ray variations of the order of 1 day (left panel in Fig.1). This 1-day time-lag was consistent with the SSC interpretation provided by the SED modeling, since it is likely associated with the light travel time of the synchrotron seed photons that scatter the energetic electrons.

3C 454.3 was the source best studied by AGILE because of its strong γ-ray activity in the last years. The 18-month long monitoring from radio to gamma-ray energy bands of this source provided a large dataset used in order to corroborate or falsify the possible 12-hour γ-optical delay found in the December 2007 campaign (Donnarumma et al. 2009a). Although the DCF analysis performed on this longer period showed a peak in correspondence with 0-day lag ($\tau = 0$, right panel in Fig.1), the asymmetric shape of DCF as probed also by the centroid distribution (insert plot in Fig.1 right panel), suggested a possible lag occurred at $\tau = -0.4^{+0.6}_{-0.8}$ (Vercellone et al. 2010 for details). This result, consistent also with the findings of Bonning et al. 2009, showed that gamma-ray variations could be delayed by few hours with respect to the optical ones, in agreement with the dominance of the EC model provided by the SED modeling: the delay is compatible with the corresponding crossing time of the external seed photons of a typical blob dimensions (Sokolov et al. 2004).

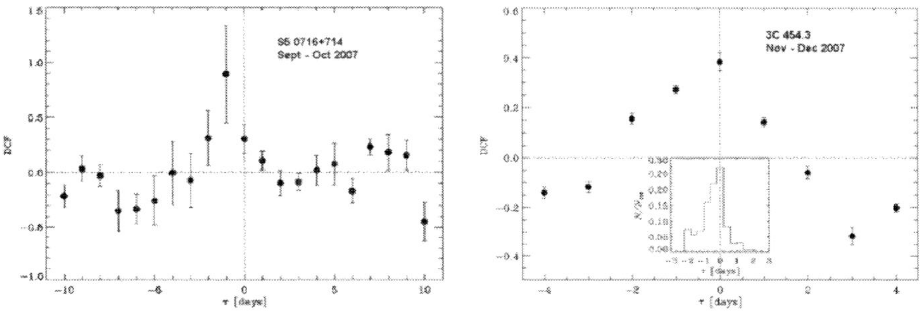

Figure 1. DCFs between optical and γ-ray light curves as performed for S5 0716+714 (left panel) and 3C 454.3 (right panel); the insert in the right panel shows the DCF centroid distribution. We remind that negative τ means that the optical variability leads the γ-ray one. These figures have been adapted from Chen et al. (2008) and Vercellone et al. (2010).

3. Soft X and γ-rays

Thanks to several ToO observations performed by *Swift*/XRT during the long term multi-λ campaign on 3C 454.3, it was possible to detect temporal and spectral variabilities in the soft X-rays during the enhanced γ-ray activity. As for the spectral variability, the data acquired so far showed a "harder-when-brighter" trend in soft X-rays, although a small deviation from this trend occurs at F(2 − 10 keV)$< 2 \times 10^{-11}$ erg cm^{-2} s^{-1} (Vercellone et al. 2010). We note that a similar trend is common in FSRQs as reported in Abdo et al. 2010a. Moreover, AGILE data suggested a possible "harder-when-brighter" trend in the γ-ray data, although we cannot draw any firm conclusion due to the larger errors in the spectral slopes. Moreover, many other ToOs observations were preformed by *Swift*, during the γ-ray "giant flares" of 3C 454.3 detected by AGILE in December 2009 (Pacciani et al. 2010) and in November 2010 (Vercellone et al. 2011). These allowed for the investigation of the spectral trend at higher fluxes, showing a saturation of the photon index with increasing flux (Fig. 2, left panel). We interpreted the "harder-when-brighter" trend as due to a higher accretion rate or a larger value of the bulk Lorentz factor coupled with a slightly lower value of the magnetic field, while the saturation at higher flux values, may be ascribed to a balance of the SSC contribution to EC from the disk (a significant component in the raising tail of the IC emission). A detailed interpretation of this trend is reported in Vercellone et al. 2011.

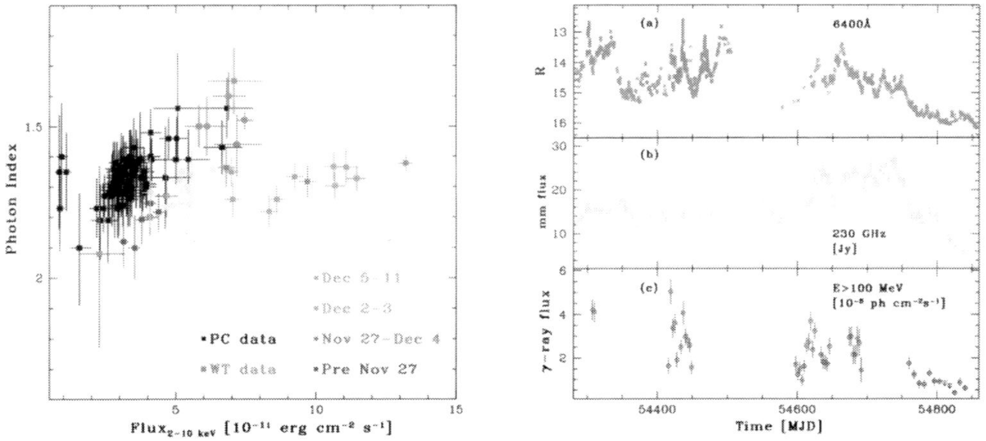

Figure 2. *Left panel*: the photon index vs $F_{2-10\mathrm{keV}}$ relation as inferred by several *Swift*/XRT observations of 3C 454.3. *Right panel*: from top to bottom Optical (R-band), millimeter (220 GHz) and γ-ray light curves of 3C 454.3, adapted from Vercellone et al. 2010.

4. Radio-Optical-γ-rays

Simultaneous radio, optical and gamma-ray observations are sheding light on the study of the jet geometry. In particular strong correlation between radio (43 GHz) and gamma-ray emissions have been observed during bright γ-ray flares of some blazars, as in the case of PKS 1510-089 (Marscher et al. 2010). In addition, long term optical polarimetric monitoring revealed that significant variations of the polarization angle occurred slightly before or in coincidence with a bright gamma-ray flare (see the case of 3C 279 reported in Abdo et al 2010b). These correlated multi-λ variabilities gave rise to several hypotheses on the jet structure and dynamics. Among them, there are the requirement of a non axis-symmetric magnetic field or a jet wobbling across the line of sight or a bending jet.

In this respect, we found a possible hint of a bending jet in 3C 454.3 on the basis of the radio (220 GHz), optical (R) and gamma-ray light curve behavior observed during the 18-month long multiwavelength campaign (Fig. 2, right panel). In detail, in 2007 (MJD$=$ 54200 $-$ 54460) the more pronounced fluxes and variability of the optical and γ -ray emissions seem to favor the inner portion of the jet as the more beamed one. On the other hand, the higher mm flux emission and its enhanced variability during 2008 (MJD$=$ 54600 $-$ 54800), together with the dimming trend in the optical and in the γ -ray bands, could indicate that the more extended region of the jet became more aligned with respect to the observer line of sight. This behavior resembles that of BL Lacertae (Villata et al. 2009), which could be interpreted in the framework of a change in orientation of a curved jet, yielding different alignment configurations within the jet itself. This scenario has been recently supported by Raiteri et al. 2011. By analyzing radio, Optical-UV, X-ray and gamma-ray simultaneous observations between April 2008 and March 2010 of 3C 454.3, they interpreted the long term multi-frequency variability in terms of an inhomogeneous bending jet, where the synchrotron radiation is produced in outer and wider jet regions - which in turn may change also their orientation- while the frequency increases. However, it is worth noting that during the latest flare (Vercellone et al. 2011), which is also the brightest ever observed in 3C 454.3, the mm (230 GHz), optical-UV, X-ray and γ-ray emissions show quasi-simultaneous variations. In this case, it is difficult to explain the correlated variabilities in terms of the bending jet model, but we have to remind that the November 2010 γ-ray flare seems quite peculiar if compared with the events previously occurred in this blazar (see Vercellone et al. 2011 for details).

5. The Gev-TeV connection

AGILE was involved in two main campaigns with Very High Energy (VHE) observatories, both performed during the June 2008 observations of the W Comae field aimed to follow a VHE γ-ray flare from this blazar announced by VERITAS. The observations started on 2008 June 9 and lasted about 1 week. During these observations the two blazars present in this field, Mrk 421 and W Comae were detected by AGILE. Mrk 421 was first detected by AGILE with its hard X-ray monitor, SuperAGILE (Feroci et al. 2007), showing an increasing hard X-ray activity, culminating in a flux level of $\sim$ 55 mCrab in the energy range 20-60 keV. No significant γ-ray emission above 100 MeV was observed by AGILE-GRID integrating on daily-time scale, however by collecting all the 1-week data, Mrk 421 was detected by the GRID ($E > 100$ MeV) with a flux of $(42 \pm 13) \times 10^{-8}$ photons cm^{-2} s^{-1}, consistent with the highest flux observed by EGRET. The soft X-ray monitoring provided by the *RXTE*/ASM together with a ToO observation performed by *Swift*/XRT showed that the source reached its historical maximum, i.e. $\sim 2.6 \times 10^{-9}$ erg cm^{-2} s^{-1}. TeV data were not acquired during the brightest X-ray flare and then also in correspondence with the γ-ray detection. However the data acquired during the pre-flare activity suggested a correlation between soft, hard X-ray and TeV emissions. On the basis of these evidences and on the relative variations observed in soft X-rays and in VHE gamma-ray light curves, the variability of the Spectral Energy Distribution of Mrk 421 in June 2008 was ascribed to an acceleration process in a standard synchrotron self Compton (SSC) leptonic model rather than to a new injection of relativistic electrons (see Donnarumma et al. 2009b for further details).
As described above, AGILE performed a ToO on W Comae field to follow a flare of W Comae detected by VERITAS between 2008 June 7 and 9, which corresponding flux was 2.5-3 times higher than the previous flare observed in March 2008 (Acciari et al. 2008). AGILE-GRID detected the source only between June 12 and 13, at a flux level of $(90 \pm 32) \times 10^{-8}$ ph s^{-1} cm^{-2}, roughly a factor of 1.5 higher than the highest flux detected by EGRET (Hartmann et al. 1999). In the rest of the observing period, AGILE-GRID provided only 3-σ upper limits ($< 60 \times 10^{-8}$ ph s^{-1} cm^{-2} for a 2-3 day time scale, see Acciari et al. 2009). Soft X-rays observations from both *Swift*/XRT and *XMM*-Newton satellites were also performed, which showed an interesting flux variability (a factor of 2) on hour time scale. On the same time scale, a slightly lower variability ($\sim$ 1.5) was also observed in UV data. Moreover, radio, Optical and near-IR observations were provided by Noto, UMRAO Metsähovi radio telescopes and the GASP-WEBT consortium: no significant variability was observed in optical $R-$band and radio (14.5, 37 GHz) data. The simultaneous γ-ray and VHE gamma-rays led to disentangle between the simple SSC model and the need of an additional source of photon field likely originating from a near nuclear dusty torus which emission peaked at 1.5×10^{14} Hz. Further details on the multifrequency campaign are provided in Acciari et al. 2009.

6. Future plans

AGILE accumulated four years of data, covering the whole sky. The forthcoming AGILE 2nd source catalog (Bulgarelli et al. in preparation) will contain a high number of extragalactic sources both associated with low energy counterparts and still unidentified. Moreover, a deep investigation on $\sim$ 60 extra-galactic ($|b| > 10$) source candidates derived by the Quick look analysis is now ongoing. Among our objectives on these data and the future ones, there is the plan to extend our spectral analysis below 100 MeV as it was already done in the case of Crab (Vittorini et al. 2011) and 3C 454.3 (Vercellone et al. 2011). In particular, the inspection of the energy range (50-200 MeV) would be particularly relevant in the study of the so-called MeV blazar, with IC emission peaking at MeV energies. Future multi-frequency campaigns on these objects will be planned to understand if they belong to a peculiar class of objects or they are a transitory class defined by the location of the γ-ray dissipation regions (Sikora et al. 2002). In this respect, during a recent AGILE multi-λ campaign the high z (2.507) blazar PKS 1830-211 (Donnarumma et al. 2011), previously classified as a MeV blazar (De Rosa et al. 2005), was highly variable in gamma-rays without any significant counterpart at lower frequencies (near-IR, Optical, X-ray energy bands). This peculiar behavior was interpreted in a framework of a 2-component leptonic model in which an important role was also attributed to a blob-cloud interaction. It will be interesting to investigate if this gamma-ray only flaring activity is a common behavior among the MeV blazars.

Acknowledgements

This work was supported by ASI grants I/040/10/0, the agreement ASI-INAF I/009/10/0, and by the Italian Ministry of University and Research (PRIN 2005025417).

References

[1] Abdo, A. A., et al. 2010a, ApJ, 716, 835
[2] Abdo, A. A., et al. 2010b, Natur, 463, 919
[3] Acciari, V. A., et al. 2008, ApJ, 684, 73L
[4] Acciari, V. A., et al. 2009, ApJ 707, 612
[5] Bonning, E. W., et al. 2009, ApJ, 697, 81L
[6] Bulgarelli, A. et al. in preparation
[7] Chen, A. W., et al. 2008, A&A, 489, 37L
[8] D'Ammando, F., et al. 2009,A&A, 508, 18
[9] D'Ammando, F., et al. 2011, A&A, 529, 145
[10] De Rosa A., et al, 2005, A&A 438, 121
[11] Donnarumma, I., et al. 2009a, ApJ, 707, 1115
[12] Donnarumma, I., et al. 2009b, ApJ, 691, 13L
[13] Donnarumma, I., et al. 2011, ApJ, 736, 30L
[14] Edelson, R. A. & Krolik, J. H. 1988, ApJ, 333, 646
[15] Feroci, M., et al. 2007, Nucl. Instrum. Meth. A, 581, 728
[16] Hartmann, R.C., et al. 1999, ApJS, 123, 79
[17] Hufnagel, B. R. & Bregman, J. N. 1992, ApJ, 386, 473
[18] Marscher, A.P., et al. 2010, ApJ, 710, 126L
[19] Pacciani, L., et al. 2010, ApJ, 716, L170
[20] Peterson, B. M., et al. 1998, PASP, 110, 660
[21] Raiteri, C. M. et al. 2003, A&A, 402, 151
[22] Raiteri, C. M. et al. 2008, A&A, 491, 755
[23] Raiteri, C. M. et al. 2011, arXiV:1107.1093v1
[24] Sikora M. & Moderski R. 2002, ApJ, 577, 78
[25] Sokolov, A., et al., 2004, ApJ, 613, 725
[26] Tavani, M. et al. 2009, A&A, 502 995
[27] Tavecchio, F., et al. 2010, MNRAS, 405, 94L
[28] Vercellone, S. et al. 2010, ApJ, 712, 405
[29] Vercellone, S. et al. 2011, arXiV: 1106.5162v1
[30] Villata, M.,et al. 2008, A&A , 481, 79L
[31] Villata, M., et al. 2009, A&A, 501, 455
[32] Vittorini, V., et al. 2011, ApJ, 732, 22L

Journal of Physics: Conference Series **355** (2012) 012005 doi:10.1088/1742-6596/355/1/012005

Gamma-ray Laue lenses under development for deep AGN observations

F. Frontera, E. Virgilli, V. Liccardo, V. Valsan

University of Ferrara, Physics Department, Via Saragat, 1, 44100 Ferrara, Italy

E-mail: frontera@fe.infn.it, virgilli@fe.infn.it, liccardo@fe.infn.it, valsan@fe.infn.it

V. Carassiti, F. Evangelisti, S. Squerzanti

INFN, Section of Ferrara, Via Saragat, 1, 44100 Ferrara, Italy

E-mail: carassiti@fe.infn.it, evangelisti@fe.infn.it, squerzanti@fe.infn.it

G. Risaliti

INAF, Astronomical Observatory of Arcetri, 50125 Firenze, Italy

E-mail: risaliti@arcetri.inaf.it

Abstract. We will review the status of the Laue lens development of space astrophysics and their importance for facing many open issues on AGNs.

1. Introduction

Broad band X-/low energy gamma–ray missions, like *BeppoSAX*, *Rossi X–ray Time Explorer*, *INTEGRAL*, have clearly shown that, in order to understand better the physics underlying several classes of Galactic and extragalactic sources, two main requirements on the instrumentation are crucial: a) to cover a broad energy band (from a fraction of keV to several hundred keV); b) to achieve a high spectrum sensitivity on time scales at least shorter than the source flux variability time scales.

To meet both requirements, it is crucial to use focusing telescopes. The current experimental scenario is the following: a) low energy X–ray (0.1-10 keV) telescopes are available and are well tested in space; b) medium energy X–ray (up to 70/80 keV) telescopes, based on multilayer mirrors, are mature and will be tested in space in the next future (American 'NUSTAR' mission, Japanese 'ASTRO–H' mission); c) high energy X–ray (>70/100 keV) telescopes are under development.

But, why extend the band to energies higher then 70/80 keV? For a review of the science goals at high energies, see, e.g., Ref. [1]. Here we shortly discuss some open issues on Active Galactic Nuclei (AGNs), that could be settled with broad band observations that cover the 511 keV annihilation line.

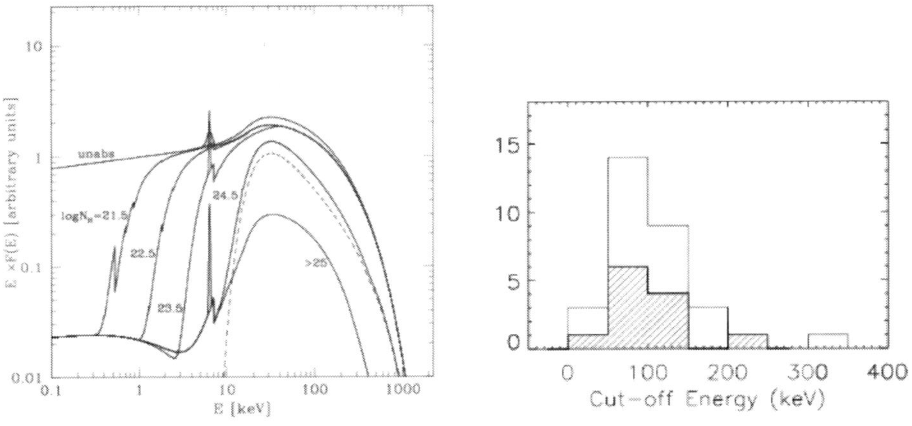

Figure 1. *Left panel*: CXB synthesis model proposed by Gilli et al. [2], in which RQ AGNs with different absorption and with power–law spectra with a high energy cutoff are assumed. A power–law photon index distribution with mean value of 1.9 and a fixed value of high energy cutoff E_F are assumed. Reprinted from Ref. [2].
Right panel: Cutoff energy distribution of a sample of RQ AGNs observed with the INTEGRAL satellite. Reprinted from Ref. [3].

1.1. Radio Quiet (RQ) AGNs for cosmology

Many open issues still exist about this class of sources (mainly type 1 and 2 Seyfert). Extremely important is to establish the contribution of this class of source to the cosmic X–ray background. Issues that are still not clear include the relative size of unabsorbed ($\log N_H < 21.5$), Compton–thin ($21.5 < \log N_H < 24.5$) and Compton-thick ($\log N_H > 25$) RQ AGNs, the luminosity function and evolution of each of the above populations with energy, the distribution of high energy cutoff (E_F) of each population.

Currently, in the synthesis models of the Cosmic X–ray Background (CXB) [2], it is assumed a combination of unobscured, Compton thin and Compton thick RQ-AGN populations with assumptions about the size of each sub–population, a scatter in the photon index power–law spectrum with a high energy cutoff E_F kept fixed (see Fig. 1). Are the assumption parameters right? Is it right to assume a fixed E_F? On the basis of the INTEGRAL results recently reported by Ref. [3], in the local Universe, this does not seem to be the case (see Fig. 1).

Soft gamma–ray observations (> 100 keV) of unprecedented sensitivity such as those possible with focusing optics, can test and improve the synthesis models.

1.2. Radio Loud (RL) AGN physics and their contribution to hard CXB

In the case of RL–AGNs (Blazars), it is well known that they show two humps in their Spectral Energy Distribution (SED) (e.g., Ref. [4]), one interpreted as synchrotron emission, the other as Inverse Compton. As can be also seen from Fig. 2, it is found that the photon peak energy of the second hump decreases with the luminosity increase, with low luminosity sources identified with BL Lacs and high luminosity sources with Flat Spectrum Radio Quasars (FSRQ). For the brightest sources the second hump is just around 400 keV, while, around this photon energy, the less luminous have their spectral dip between first and second hump. This means that the most powerful and distant blazars can be more easily picked up through sensitive soft gamma–ray observations, that become crucial to test the emission models of RL–AGNs. The soft gamma–ray

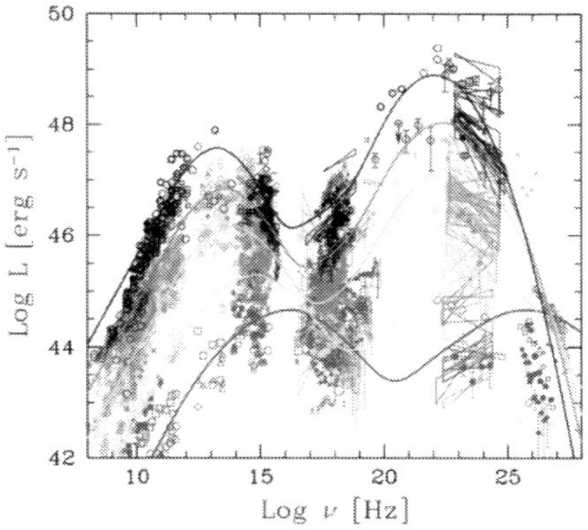

Figure 2. Luminosity distribution of blazars, also called blazar sequency. Solid lines are the phenomenological SED. The data are the SED of real sources detected in the first 3 months of survey by the Fermi satellite (LBAS catalog: Abdo et al. 2009), divided (different colors) into bins of different-ray luminosities in the 0.110 GeV energy band: $\log L < 45.5$ (blue), $45.5 < \log L < 46.5$ (green) $46.5 < \log L < 47.5$ (red), $\log L > 48.5$ (black). Reprinted from Ref. [4].

band between 100 and 600 keV is also important to evaluate the contribution of RL–AGN to the hard X–ray CXB, an issue recently faced by Ref. [5] using the 25–55 keV spectral data obtained with the *Swift* BAT instrument and thus by extrapolating to higher energies the observed and derived (e.g., X–ray luminosity function) properties at low energies.

To perform studies of these sources beyond 100 keV energies, focusing telescopes are needed.

1.3. Polarization of the gamma–ray emission from AGNs

In spite of its importance to establish the emission physics, the polarization level of the X–/gamma–ray emission from celestial X–sources is a still open issue. The only source, for which polarization measurement positive results are reported, is the Crab Nebula [6, 7]. Recently with INTEGRAL a highly polarized signal (polarization fraction of $67 \pm 30\%$) has been discovered in the soft gamma–ray emission (> 400 keV) from Cygnus X–1 [8]. This discovery opens a new window in the gamma–ray astronomy, showing that with more sensitive instruments the observation of polarized emission could be detected even from extragalactic sources, like RL–AGNs, from which polarized emission is predicted but never observed.

1.4. Dark matter in Galaxies

Positron production can occur in a variety of cosmic explosions and acceleration sites. It is expected to be the result of the presence of antimatter but it is also expected in the case of dark matter (see, e.g., Ref [9]). Thus the observation of the characteristic 511 keV annihilation line provides a powerful tool to probe the nature of dark matter. The found evidence of a diffuse annihilation line emission from the Galactic Center region with INTEGRAL [10] has also been interpreted as due to collisionally excited dark matter, that de-excites by e^+e^- pair emission [11]. In order to confirm the INTEGRAL results, to establish the diffuse or discrete distribution of the annihilation radiation, a nd possibly to extend the search of annnihilation line to external galaxies and AGNs, much more sensitivity is needed, only achievable with focusing gamma–ray

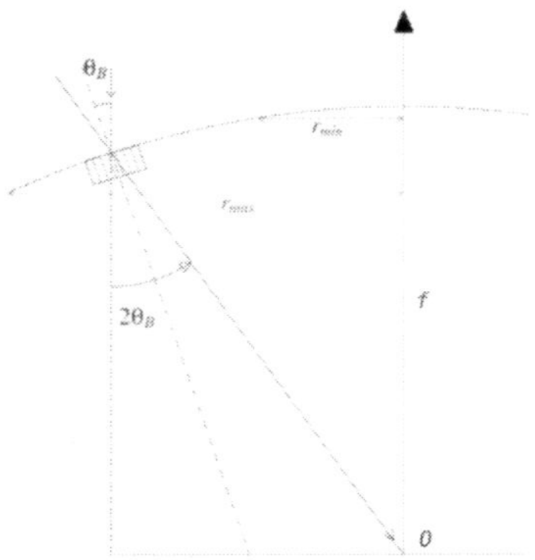

Figure 3. Geometry of a Laue lens.

telescopes.

2. Requirements for soft gamma–ray focusing telescopes

The requirements for hard X–/soft gamma–ray focusing telescopes can be summarized as follows: a) Broad bandpass (at least 80/100–600 keV); b) Continuum sensitivity 2 orders of magnitude better than the current instrumentation, like the ISGRI instrument aboard the INTEGRAL satellite, with a design goal of 10^{-8} ph/(cm^2 s keV) in a bandwidth $\Delta E = E/2$ for an observation time $T = 10^5$ s; c) Imaging capability, with angular resolution better than 1 arcmin (design goal 20 arcsec). We propose as focusing instrument in the soft gamma–ray band a Laue lens (for an exhaustive review, see Ref. [1]).

3. Laue lens principle

Laue lenses use the interference between the periodic nature of the electromagnetic radiation and a periodic structure such as the matter in a crystal. In a Laue lens, the photons pass through the full crystal, using its entire volume for interacting coherently. In order to be diffracted, an incoming gamma–ray must satisfy the Bragg condition:

$$2d_{hkl}\sin\theta_B = n\frac{hc}{E} \qquad (1)$$

where d_{hkl} (in Å) is the spacing of the lattice planes (hkl), n is the diffraction order, $hc = 12.4$ keV·Å and E is the energy (in keV) of the gamma-ray photon.

A Laue lens for astrophysical applications is made of a large number of crystals, that are disposed in such a way they will concentrate the incident radiation onto a common focal spot. A convenient way to visualize the geometry of a crystal lens is to consider it as a spherical cup covered with crystal tiles having their diffracting planes perpendicular to the sphere (see Fig. 3). The focal spot is on the symmetry axis at a distance $f = R/2$ from the cup, with R being the radius of the sphere of which the spherical cup is a part; f is the lens *focal length*.

The main requirement for the disposition of the crystals in the lens is to get a smooth dependence of the lens effective on photon energy. This result is achieved using ring-like configurations for long focal lengths ($>$ 20 m) and spiral-like configurations for smaller ones.

4. Crystals for Laue lenses

Different crystal materials are suitable to be used for Laue lenses. Among them, we mention Silicon, Copper, Germanium, GaAs, Gold, etc. Two crystal structures are suitable, mosaic

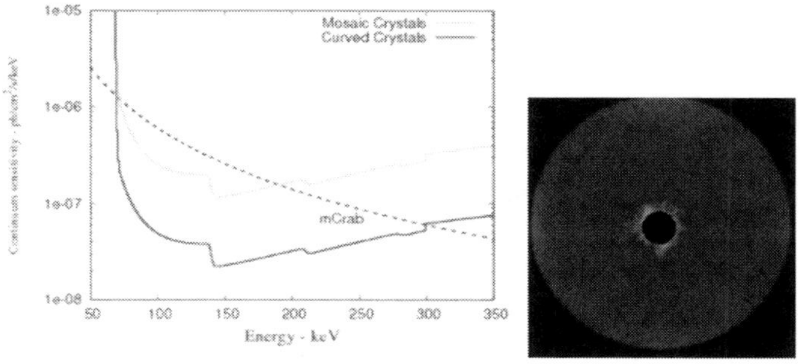

Figure 4. *Left panel*: 3σ Sensitivity of a 20 m focal length Laue lens made of flat mosaic crystal tiles of Cu(111) with 30 arcsec mosaicity and 70–300 keV nominal energy passband, compared with the corresponding lens in the case that the crystals are properly curved. The time measurement assumed is 10^5 s, while the energy bins have an amplitude $\Delta E = E/2$.
Right panel: Difference between the PSF measured and that obtained with a Monte Carlo code by assuming a perfect positioning of the crystal tiles in the lens. Reprinted from Ref. [12].

crystals and perfect curved crystals.

- *Mosaic crystals*
  Mosaic crystals are made of misaligned perfect microcrystals, with misalignment distribution well described by a Gaussian function. The FWHM of the Gaussian is the *crystal mosaicity*.
- *Curved crystals*
  Curved cystals have important advantages with respect to mosaic cystals: a) No limit to their diffraction efficiency, while for mosaic crystals the maximum theoretical value of the diffraction efficiency is 0.5; b) Better lens focusing for a proper crystal curvature.
  It is found that for the same focal length, the angular resolution of a lens made of curved crystals significantly improves. As an example, for 20 m focal length, the angular resolution increases from 3.5 arcmin to 20 arcsec, with the source image area that decreases by a factor about 100. As a consequence, also the lens sensitivity increases with respect to flat mosaic crystals. As an example, in Fig. 4, left panel, we compare the sensitivity of a 20 m focal length Laue lens made of curved crystals of Cu(111) with 30 arcsec mosaicity with that made of flat mosaic crystals tiles (size 15×15 mm^2) of the same material and mosaicity. To achieve the design goal of 10^{-8} ph/(cm^2 s keV) in 10^5 s of observation time, in addition to the use of bent crystals, that allow the narrowest PSF and the background level minimization, in the same lens more crystal materials have to be used, each optimized in the energy band where it shows its highest reflectivity.
  Techniques to curve cystals have already been developed. There are different production methods to get curved crystals, among which the indentation of one of the two wafer faces. The last technique has been developed at the University of Ferrara with very satisfactory results for Silicon and Germanium.

5. Lens assembly technology development

At the University of Ferrara we have already developed a crystal assembling technology suitable for building lenses with moderately short ($\leq$10–15 m) focal lengths. Using this technology, a first lens prototype with 6 m focal length has already been developed and tested. It makes use of flat mosaic crystals of Cu(111). The first light of the first prototype is shown in Fig. 4, right panel, where the black field region is the expected image size. The corona still visible in this image is the result of the cumulative errors done during the lens assembly process. A new prototype is being developed that takes into account the experience gained from the first one.

For longer focal length lenses, new crystal assembly technologies are needeed. To this end, a new project, "LAUE", supported by Italian space agency ASI, has been started with the contribution of the Italian industry. A lens petal made of curved crystals with 20 m focal length is being developed and an accommodation study of a lens made of petals for a satellite mission is being performed.

6. Conclusions

The hard X-/gamma-ray band covered by Laue lenses is crucial for high energy astrophysics. Given their much higher expected sensitivity (two orders of magnitude better than the current instrumentation), Laue lenses can face key importance astrophysical issues still open on AGNs.

A big effort is now in progress for the development of focusing Laue lenses. The development of curved crystals improves sensitivity and angular resolution by about an order of magnitude with respect to Laue lenses with mosaic crystals. A project "LAUE", supported by the Italian space agency ASI, is on going with production also of the curved crystals. We do not expect to need high focal lengths (>20 m) for extending the band up to 600 keV, while an energy passband up to 300 keV can be easily obtained with 10 m focal length. Thus the possibility of a broad band (1-600 keV) satellite mission with both multilayer mirrors and Laue lenses is becoming a firm perspective.

Acknowledgments

The lens development is the result of a big effort of several people. We wish to thank all of them. We wish to thank the significant efforts by DTM in Modena, Thales-Alenia Space, Italy in Milan and Turin, CNR/IMEM Institute in Parma, INAF/IASF in Bologna, Sensor and Semoconductor Laboratory (LSS) of the University of Ferrara. In addition we wish to thank all the other members of the Ferrara team interested to this development. We acknowledge the support of ASI, that has permitted the past and present activity in this field.

References

[1] Frontera F and Von Ballmoos P 2010 *X-ray Optics and Instrumentation* **2010** 109740– (*Preprint* 1007.4308)
[2] Gilli R, Comastri A and Hasinger G 2007 *Astronomy and Astrophysics* **463** 79–96
[3] Molina M, Bassani L, Malizia A, Stephen J B, Bird A J, Dean A J, Panessa F, de Rosa A and Landi R 2009 *Monthly Notices of the Royal Astronomical Society* **399** 1293–1306 (*Preprint* 0906.2909)
[4] Ghisellini G 2011 *ArXiv e-prints* (*Preprint* 1104.0006)
[5] Ajello M, Costamante L, Sambruna R M, Gehrels N, Chiang J, Rau A, Escala A, Greiner J, Tueller J, Wall J V and Mushotzky R F 2009 *The Astrophysical Journal* **699** 603–625
[6] Weisskopf M C, Silver E H, Kestenbaum H L, Long K S and Novick R 1978 *The Astrophysical Journal Letters* **220** L117–L121
[7] Dean A J, Clark D J, Stephen J B, McBride V A, Bassani L, Bazzano A, Bird A J, Hill A B, Shaw S E and Ubertini P 2008 *Science* **321** 1183–
[8] Laurent P, Rodriguez J, Wilms J, Cadolle Bel M, Pottschmidt K and Grinberg V 2011 *Science* **332** 438– (*Preprint* 1104.4282)
[9] Arkani-Hamed N, Finkbeiner D P, Slatyer T R and Weiner N 2009 *Physical Review D* **79** 015014–+ (*Preprint* 0810.0713)
[10] Weidenspointner G, Skinner G, Jean P, Knödlseder J, von Ballmoos P, Bignami G, Diehl R, Strong A W, Cordier B, Schanne S and Winkler C 2008 *Nature* **451** 159–162
[11] Finkbeiner D P and Weiner N 2007 *Physical Review D* **76** 083519–+ (*Preprint* arXiv:astro-ph/0702587)
[12] Frontera F, Loffredo G, Pisa A, Nobili F, Carassiti V, Evangelisti F, Landi L, Squerzanti S, Caroli E, Stephen J B, Andersen K H, Courtois P, Auricchio N, Milani L and Negri B 2008 *Society of Photo-Optical Instrumentation Engineers (SPIE) Conference Series* (*Society of Photo-Optical Instrumentation Engineers (SPIE) Conference Series* vol 7011)

International Workshop on Beamed and Unbeamed Gamma-Rays from Galaxies IOP Publishing
Journal of Physics: Conference Series **355** (2012) 012006 doi:10.1088/1742-6596/355/1/012006

From Radio with Love: an overview of the role of radio observations in understanding high-energy emission from active galaxies

Roopesh Ojha

Astrophysics Science Division, NASA/Goddard Space Flight Center, Code 661, Astroparticle Physics Laboratory, Greenbelt, MD 20771, USA

E-mail: `Roopesh.Ojha@gmail.com`

Abstract. The gamma-ray satellite *Fermi* and the ground based TeV facilities MAGIC, VERITAS and HESS have ushered in a new era in the observation of high-energy emission from active galaxies. The energy budgets of these objects have a major contribution from gamma-rays and it is simply not possible to understand their physics without high-energy observations. Though the exact mechanisms for high-energy production in galaxies remains an open question, gamma-rays typically result from interactions between high-energy particles. Via different interactions these *same* particles can produce radio emission. Thus the non-thermal nature of gamma-ray emission practically guarantees that high-energy emitters are also radio loud. Aside from their obvious role as a component of multiwavelength analysis, radio observations provide two crucial elements essential to understanding the source structure and physical processes of high-energy emitters: very high timing resolution and very high spatial resolution. A brief overview of the unique role played by radio observations in unraveling the mysteries of the high energy Universe is presented here.

1. Introduction

A curious dichotomy is associated with the state of our understanding of the physics of active galaxies. In the decades since their discovery, increasingly detailed (in many senses) observational data, ever more sophisticated modeling and profound theoretical ideas have combined to produce a remarkably coherent basic picture of what might go on in these objects. There is the exceptionally luminous core powered by accretion onto a supermassive blackhole. The trapped gas settles into an accretion disk which heats up due to dissipative processes that move matter inwards and angular momentum outwards. Hovering above the accretion disk is a hot corona which can inverse-Compton scatter photons to X-ray energies. There is the cold atomic material that is excited by radiation from the accretion disk and produces a range of emission lines particularly at optical/UV frequencies and the absorption features seen in the UV and X-ray. And in a small but important fraction of cases, highly collimated 'jets' emerge and travel at relativistic speeds possibly providing a link between the central black hole and the galactic environment (e.g.[1], [2], [3]). It is almost possible to believe we essentially understand AGN with only small details to be cleared up.

However, one simply has to take a step back to realize that just about every piece of this neat picture is rather less than clear: details of the central engine (such as the role of spin), accretion and disk formation are unknown. The production, acceleration and collimation of jets and how they remain collimated are open questions, indeed even their composition (positron-electron or proton-electron) is not known. The origin of the high energy emission remains a mystery. Half a century after their discovery, the fundamental questions about AGN remain unanswered.

2. Fermi meets Jansky

Since the seminal detection of γ-rays from AGN (particularly blazars) by the EGRET instrument on the Compton Gamma Ray Observatory (e.g. [4]), it has become clear that high energy radiation forms a major part of the energy budget of AGN and observations at these energies are essential to understanding them. The Large Area Telescope instrument aboard the *Fermi* Gamma Ray Space Observatory (hereafter *Fermi/LAT*; [5]) and three major ground based TeV facilities MAGIC, HESS and VERITAS, have dramatically opened up this part of the electromagnetic spectrum to observation with sensitivity, field of view, range of cadence, and resolution that is orders of magnitude better than before.

Another crucial observational breakthrough relates to a basic characteristic of AGN: they radiate throughout the electromagnetic spectrum (and possibly emit non-electromagnetic messengers, see below) and the radiation at all wavebands varies in ways that appear to be linked in very complex ways to variations at other wavebands. Thus *simultaneous* and *multiwavelength* observations are the only way AGN behavior can be constrained. The availability of the above high energy instruments, the presence of X-ray and UV instruments on satellites (e.g. *Swift*; [6]), backed up by optical, submillimeter and radio facilities on the ground have made it feasible to monitor AGN simultaneously across the spectrum for the first time since their discovery!

The role of radio observations in understanding the high energy emission from AGN is much greater than the obvious need for radio data to construct the spectral energy density (SED) of AGN. Gamma rays typically result from interactions between high energy particles (see below for a brief discussion) and, with different interactions, these *same* particles produce radio emission. Thus, the non-thermal nature of γ-ray emission almost guarantees that they will be radio emitters. This is incredibly useful because radio observations provide two handles that let us get a grip on the source structure and physical processes of AGN. The first is timing resolution. Radio telescopes are able to monitor both rapid flares and slow changes in flux. Thus, they are used to search for periodicities from minutes to years constraining physical processes responsible for high energy radiation from AGN. The second useful and indeed unique handle provided by radio observations is very high spatial resolution. The technique of radio interferometry, particularly Very Long Baseline Interferometry (VLBI), provides the highest angular resolution possible of any observational technique, often better than a milliarcsecond. This means that VLBI monitoring of AGN jets provides the *only direct measure* of relativistic motion in AGN jets, probing intrinsic jet parameters such as speed, Doppler factor, opening and inclination angles, all of which are essential constraints on jet physics and tests of the relativistic-beam model (e.g. [7]). Further, the unmatched resolution of VLBI makes identification of the location and extent of emission regions possible at a level of detail not available with any other observational technique (e.g. [8]).

Given that resolution depends directly on frequency, it is ironic that the magic of interferometry, as well as the likely origin of high energy and radio emission from the same population of particles, have made radio observations uniquely useful in studying the origins of emission from the other end of the electromagnetic spectrum. Counterintuitive as it might seem, radio and high energy studies of AGN are closely linked and essential to maximizing the scientific output from each other.

3. Origin and nature of high energy emission from AGN

AGN remain, by far, the largest class of identified *Fermi/LAT* detections [9] confirming the finding originally made by EGRET that the SEDs of blazars are often dominated by gamma-ray emission. While this link is clear, the underlying mechanism(s) is less so with a variety of models attempting to explain the origin of high frequency emission (see [10] for a review). Briefly, the spectral energy (νF_ν) distributions (SED) of blazars have two maxima. It is fairly well established that synchrotron emission from relativistic electrons in the jet produce the low energy peak in the radio-to-infrared band [11], [12]. The origin of the high energy hump remains a major open question. It could arise from inverse-Compton scattering of the same electron population off synchrotron photons, the so-called Synchrotron Self Compton model [13], [14]. Or it could arise from inverse-Compton scattering of photons external to the jet by the relativistic electrons within the jet. Depending on the possible sources of the external photons (e.g. cosmic microwave background radiation, BLR, etc) there are several flavors of this External Compton model [15]. An entirely different class of models suggest this high energy component of the SED arises from hadronic processes involving high energy protons which produce neutral and charged pions that decay into electrons and positrons which in turn produce γ-ray photons and neutrinos e.g. [16] [17]. Dynamic, quasi-simultaneous observations across the electromagnetic spectrum are needed to construct SEDs that can be used to distinguish between these models and understand emission from AGN. If the hadronic models are correct, there is the intriguing possibility of going beyond traditional photon based astronomy to studying AGN physics using neutrinos as messengers thanks to neutrino detectors such as ANTARES, KM3Net and IceCube.

4. High cadence, broadband, multifrequency monitoring of total and polarized flux density and spectral evolution

Rapid variability of the total and polarized flux density in all wavebands is a defining property of AGN and one that has great potential to constrain physical parameters and discriminate between different models attempting to understand AGN. Explanations for variability include (but are not limited to) accretion disk instabilities [18], the shock-in-jet models [19], [20], the collision of relativistic plasma shells traveling at different speeds [21], lighthouse effects arising from rapidly rotating plasma in the vicinity of a rotating black hole [22] and magneto-centrifugal acceleration of beams of particles [23]. Monitoring the variability of their SEDs at all wavebands can discriminate between these different scenarios/models. It can also yield information about linear scales that cannot be reached even by interferometry. For a few recent high impact results see [24], [25], [26], [27], [28],

As has been clear from early in the mission lifetime of EGRET, monitoring at radio wavelengths is particularly important for understanding gamma-ray emission from and the physics of AGN. The relative timing of gamma-ray and radio flares can constrain both the mechanisms responsible for them as well as the location where they originate e.g. [29]. Flux density monitoring to measure the evolution of the radio emission with time indirectly reveals when relativistic electrons are being injected into the jet. AGN are highly compact and have a large optical depth to synchrotron self-absorption. Hence, radio observations are best made at the higher radio frequencies in the optically thin part of the radio spectrum if we are to probe close to the center of these sources where the gamma-ray emission is likely generated [30]. Early work with EGRET data [31] suggested that high radio frequency flares precede gamma-ray flares but the sparseness of the EGRET data made this result tentative. If true, this would imply that the gamma-ray emission results from shock enhanced electron energies and magnetic field strengths and support the idea that the IC seed photons originate from within the jet rather than externally. Further, this putative timelag between the radio and gamma-ray flares (and the relativistic speeds of the jets) implies that the gamma-ray emission originates parsecs downstream of the base of the radio jet. This leads to the exciting possibility that these shocked regions, these birthplaces of gamma-ray emission, could be directly imaged using high radio frequency VLBI. In combination with kinematic parameters gleaned from multi-epoch VLBI (see below), such constraints on the size and flux density of shocked regions would allow quantitative modeling of inverse Compton models for gamma-ray emission.

However, before the launch of *Fermi*, a preponderance of theorists seemed inclined towards an external source for the seed photons that produce gamma-ray emission by IC scattering off relativistic jet electrons, implying production within about a tenth of a parsec of the central black hole. Two good reasons for this view are the short time-scales of flux variability which are easier to explain close to the black hole and the availability of potential external sources of seed photons such as the BLR or the dust torus [32]. The post-Fermi radio data continues to challenge this notion though not in a completely consistent way. For example, [8] have used VLBI and single-dish radio observations with optical and gamma-ray data on OJ287 to show that the gamma-ray flare originates > 14 parsecs from the radio core. In general gamma-ray emission appears to correlate well with ejection and apparent speed of VLBI components [33] and with the passage of these components through the millimeter core [34]. Meanwhile optical polarization observations suggest that gamma-rays could be produced all along the inner jet, from the core to some 20 parsecs downstream [35].

Another important use of such monitoring data is the study of the radio-gamma connection in AGN. For example, [36] have used both archival and concurrent radio data to study the statistical significance of the correlation between gamma-ray emission and radio emission from AGN. They find a positive correlation of high significance for all types of AGN (and for FSRQs and BL Lacs separately) with the significance greater for concurrent data. They also find a possible dependence of the correlation on the gamma-ray energy band.

This importance of high cadence, broadband multifrequency monitoring in general and radio monitoring in particular has been appreciated for some time and dedicated radio monitoring programs such as the centimeter band flux and polarization monitoring program at the University of Michigan [37], the 22/37 GHz monitoring program of the Metsähovi Radio Observatory [38] and the simultaneous 1-22 GHz monitoring with RATAN-600 [39] have been making important contributions for decades. The scientific dividend from such studies got a boost from the launch of *Fermi* and the availability of the ground based TeV experiments leading to the launch of dedicated new programs such as FGAMMA, the OVRO program and TANAMI.

The F-GAMMA (Fermi-GST AGN Multifrequency Monitoring Alliance) program monitors the flux density and polarization of a sample of ~ 60 *Fermi/LAT* sources at monthly intervals [40]. It observes with radio, millimeter, submillimeter, infrared and optical instruments including the 100 m Effelsberg telescope in Germany and the 30 m IRAM telescope at Pico Veleta, Spain. A much larger sample of over 1500 blazars is monitored about twice a week by the 40 m Owens Valley Radio Telescope (OVRO; [41]).

All of the above monitoring programs use instruments located in the northern hemisphere and thus their coverage does not extend very far into the southern hemisphere. The Tracking AgN with Austral Milliarcsecond Interferometry (TANAMI; [42], [43]) program fills this gap with two separate and complementary monitoring programs that concentrate on the sky south of -30 degrees declination. The first has used a 30 m dish at Ceduna in South Australia to regularly monitor sources at 6.7 GHz since August 2007 generating well-sampled lightcurves. The other program has used the Australia Telescope Compact Array (ATCA) since October 2007 to monitor at a lower cadence (~ 8 epochs a year) but in 4 frequency bands in the 4.8 to 40 GHz range to follow the evolution of the radio spectrum.

5. High resolution (VLBI), multi-frequency monitoring of parsec scale jet structure

Clearly observations at all wavebands are needed to understand AGN. However, multi-epoch VLBI observations provide the only direct measure of relativistic motion in AGN jets. So they are the only means of measuring intrinsic jet parameters such as speed, Doppler factor, opening angle, inclination angle, magnetic field strength and magnetic gradients that are necessary inputs to models of jet behavior. They are also the only way to associate morphological changes (such as component ejections or changes in magnetic field strength/orientation) close to the central black hole to activity in other wavebands, which is a powerful diagnostic of emission processes. Further, they have the potential to locate and constrain the size of emission regions in conjunction with high energy and intermediate energy observations. Key questions that can only be addressed with the help of VLBI monitoring include:

- Where are the gamma-rays produced in blazar jets? How big are these emission regions?
- How do gamma-ray variability patterns relate to intrinsic jet parameters?
- Are radio and gamma-ray emission beamed with the same Lorentz factor?
- What are jets made of? Are they purely leptonic or do they have a hadronic component?
- Are gamma-ray flares accompanied by jet-component ejections?
- Why are some blazars bright in gamma-rays while others are not?
- Under what conditions can non-blazar AGN become bright gamma-ray emitters?

When VLBI arrays are capable of measuring polarization, Very Long Baseline Polarimetry (VLBP) yields the magnitude and orientation of jet magnetic fields which enable studies of the collimation and acceleration of jets, jet composition and kinetic luminosity. VLBP can also trace magnetic fields external to jets e.g. in sheaths and boundary layers. Further, as a tracer of jet hydrodynamics VLBP is extremely useful for the identification of shocks, shears and aberration. VLBP observations have already shown that linear polarization is low (just a few percent) in the core region but is typically higher (in the tens of percent) in jets. There are extended regions in jets where the magnetic field is orthogonal to the jet direction and remains so even when the jet bends significantly. There are now several examples of jets showing a 'spine-sheath' structure. Faraday rotation gradients across jets appear to be common and there is some evidence for the existence of helical magnetic field structures that have been proposed by some theorists [44], [45].

Major VLBI monitoring programs include MOJAVE, the Boston University blazar group, VIPS and TANAMI. Though there is some overlap, these programs are largely complementary and together they effectively wield the power of VLBI monitoring. We will take a quick look at each program and then summarize the key findings so far.

With its origins in the VLBA 2cm Survey (1994-2002; [46]), the MOJAVE (Monitoring Of Jets in Active galactic nuclei with VLBA Experiments; [47]) program is probably the most comprehensive VLBI monitoring program and it includes many of the longest time baseline, well sampled VLBI observations available. MOJAVE makes full polarization observations of about 300 sources using the ten telescopes of the VLBA (Very Long Baseline Array; [48]) at a frequency of 15 GHz. MOJAVE observes a flux limited gamma-ray selected sample based on the *Fermi* 11 month list with a matching radio sample with $S_{5GHz} > 1.5$ Jy during the 11 month period. Its goal it to study the structure and kinematics of AGN jets and their connection to data at other wavebands including high energy emission. MOJAVE's large sample is particularly suitable for statistical studies.

The Boston University blazar group uses millimeter wavelength VLBI imaging to follow jet evolution particularly the motions of superluminal 'knots' in the jet. They also use the VLBA about once a month, to make full polarization observations of 34 blazars at 43 GHz [49]. They associate optical, X-ray and gamma-ray flares with a superluminal knot if a flare is coincident with the passage of the knot through the 43 GHz core which is several parsecs downstream of the central black hole. This group also has an optical polarization monitoring program and during a flare it attempts to match the optical polarization angle to a VLBI feature with a similar value of polarization angle. They are attempting to use the timelag between high energy variations and changes in optical and millimeter band fluxes to determine the location of X-ray and gamma-ray emission sites.

The VIPS (VLBA Imaging and Polarimetry Survey; [50]) is a high dynamic range, full polarization survey of a large sample of about 1100 sources at 5 and 15 GHz using the VLBA. The sources have a flux limit $S > 85$ mJy at 8 GHz, are located at declination > 20 and $|b| > 10$ in the SDSS (Sloan Digital Sky Survey) northern cap to make identification and redshift information easier to obtain. With a flux limit almost an order of magnitude below that of MOJAVE and other surveys, it is able to study the large population of relatively faint radio sources finding, for example, that at lower flux levels the radio and gamma-ray flux densities are not directly correlated. Aside from characterizing *Fermi* sources, VIPS aims to understand polarization properties of different AGN classes, study AGN environments and to find close binary black hole systems.

The TANAMI program [43] makes dual frequency (8 and 22 GHz) observations about every two months, measuring spectral indices of jet features and their time evolution. It is the only program observing sources in the southern third of the sky which not only significantly improves the statistics for jet kinematics and flare-ejection studies but allows access to some of the most interesting individual sources in the sky e.g. the nearby radio galaxy Centaurus A. TANAMI uses the five telescopes of the Australian Long Baseline Array (LBA; e.g. [51]) augmented by two German telescopes (at Concepcion, Chile and O'Higgins, Antarctica), a telescope in Hartebeesthoek, South Africa and a telescope at NASA's Deep Space Network in Tidbinbilla, Australia. The array has recently been expanded with the addition of three new telescopes in Warkworth, New Zealand, in Yarragadee, Western Australia and in Katherine, in the Northern Territory. The TANAMI sample currently includes about 80 sources and is being expanded to include new *Fermi/LAT* detections. TANAMI recently published the highest resolution images ever made of Centaurus A [52] as well as the dual-frequency spectral index distribution along its jet (see Fig. 1). The jet of Centaurus A appears to be already well collimated on scales of 10^{15} m and multiple possible sites for the production of high energy radiation can be identified.

As the large number of published results testify, VLBI monitoring is rapidly improving our understanding of AGN but many of the deepest questions require a longer timeline of observations. However, it is already clear that apparent superluminal motion is quite common with the Lorentz factor distribution going up to $\Gamma = 50$ or perhaps even higher [47], [53]. Acceleration, both parallel and perpendicular to the motion of jet components, is also common [54]. In general, it is clear that gamma-ray loudness is directly related to radio properties. Gamma-ray loud jets generally have higher apparent speeds, higher Doppler factors, higher brightness temperatures, are oriented closer to the line of sight and have smaller opening angles. Jets with varying gamma-ray emission have higher apparent speeds than those with steady emission and gamma-ray bright jets are in an active radio state. There is evidence that only the energized portion of a broader underlying jet is visible [33], [55], [43], [56], [57]. The median fractional polarization (at 15 GHz) of the unresolved cores of gamma-ray emitting jets is significantly higher than for gamma-ray quiet jets [58].

5.1. High resolution followup of flaring sources

It is important to obtain high resolution observations of sources when they are flaring at high energies. As VLBI monitoring programs only observe periodically, this means Target of Opportunity type of observations are needed. This has proved to be relatively straightforward in the northern hemisphere as the VLBA observes throughout the year. For the southern hemisphere the Australian LBA only observes for a few weeks every year which limits its ability to observe flares. To provide rapid followup to a gamma-ray flare with milliarcsecond resolution, the TANAMI program has set up a 1700 km single baseline interferometer between a 30 m dish at Ceduna, South Australia and a 26 m dish in Hobart, Tasmania. Called the Ceduna Hobart Interferometer (CHI; [59]), it can observe at most centimeter bands up to 22 GHz and provide a measure of the brightness temperature when the full TANAMI array is not available.

6. Summary

Radio observations of AGN are critical to understanding high energy emission and indeed the basic physics of these objects. Virtually all high energy AGN are radio loud and thus accessible to observations in this band. The unparalleled angular resolution afforded by interferometric techniques and the ability to probe a range of time scales means radio observations play a special, indeed unique, role in the suite of multiwavelength observations that are set to transform our understanding of AGN. This somewhat unlikely but strong symbiosis between observations at opposite ends of the electromagnetic spectrum is central to efforts to answer the fundamental outstanding questions about these objects.

Acknowledgments

The author thanks Matthias Kadler for numerous very helpful discussions. He also acknowledges the generous help and advice of Denise Gabuzda, Marco Giroletti, Dan Homan, Svetlana Jorstad, Alan Marscher and Greg Taylor. Responsibility for any shortcomings, however, lies solely with the author. This research was funded in part by NASA through Fermi Guest Investigator grant NNH09ZDA001N. This research was supported by an appointment to the NASA Postdoctoral Program at the Goddard Space Flight Center, administered by Oak Ridge Associated Universities through a contract with NASA.

References

[1] Peterson B M 1997 *An introduction to Active Galactic Nuclei* (Cambridge, UK: Cambridge University Press)
[2] Kembhavi A K & Narlikar J V 1999 *Quasars and Active Galactic Nuclei*, (Cambridge, UK: Cambridge University Press)
[3] Krolik J H 1999 *Active Galactic Nuclei: From the Central Black Hole to the Galactic Environment* (Princeton, NJ: Princeton University Press)
[4] Dondi L & Ghisellini G 1995 *MNRAS* **273** 583
[5] Atwood W B et al. 2009 *ApJ* **697** 1071
[6] Gehrels N et al. 2004 *ApJ* **611** 1005
[7] Cohen M H et al. 2007 *ApJ* **658** 232
[8] Agudo I et al. 2011 *ApJ* **735** L10
[9] Abdo A A et al. 2010 *ApJS* **188** 405
[10] Böttcher M 2007 *Ap&SS* **309** 95
[11] Blandford R D & Königl A 1979 *ApJ* **232** 34
[12] Falcke H & Biermann P L 1995 *A&A* **293** 665
[13] Königl A 1981 *ApJ* **243** 700
[14] Sokolov A, Marscher A P & McHardy I M 2004 *ApJ* **613** 725
[15] Sikora M, Blazejowski M, Begelman M C & Moderski R 2001 *ApJ* **554** 1
[16] Mannheim 1995 *APh* **3** 295
[17] Becker J K & Biermann P L 2009 *APh* **31** 138
[18] Begelman M C, Blandford R D & Rees M J 1980 *Nature* **287** 307
[19] Marscher A P & Gear W K 1985 *ApJ* **298** 114
[20] Türler M 2010 *MmSAI* **82** 104
[21] Spada M, Ghisellini G, Lazzati D & Celotti A. 2001 *MNRAS* **325** 1559
[22] Camenzind M & Krockenberger, M. 1992 *A&A* **255** 59
[23] Ghisellini G, Tavecchio F, Bodo G & Celotti A 2009 *MNRAS* **393** L16
[24] Abdo A A et al. 2010 *Nature* **463** 919

[25] Abdo A A et al. 2010 *ApJ* **716** 30
[26] Abdo A A et al. 2011 *ApJ* **726** 43
[27] Abdo A A et al. 2011 *ApJ* **727** 129
[28] Abdo A A et al. 2011 *ApJ* **730** 101
[29] Jorstad S G et al. 2001 *ApJ* **556** 738
[30] Lähteenmäki A & Valtaoja E 2003 *ApJ* **590** 95
[31] Valtaoja E & Teräsranta H 1995 *A&A* **297** L13
[32] Sbarrato T, Foschini L, Ghisellini G & Tavecchio F 2011 *AdSpR* **48** 998
[33] Lister M L et al. 2011 *arXiv* **1107.4977**
[34] Jorstad S & Marscher A 2010 *cosp* **38** 2265
[35] Marscher A. P. et al. 2010 *ApJ* **710** L126
[36] Ackermann M et al. 2011 *ApJ* **741** 30
[37] Aller M F, Aller H D & Hughes P A 2011 *JApA* **32** 5
[38] Teräsranta H, Wiren S, Koivisto P, Saarinen V & Hovatta T 2005 *A&A* **440** 409
[39] Kovalev Y A & Kovalev Y Y 2006 *ASPC* **360** 137
[40] Angelakis E K et al. 2010 *ASPC* **427** 289
[41] Richards J L et al. 2011 *ApJS* **194** 29
[42] Ojha R, Kadler M, Tingay S & Lovell J 2008 *AIPC* **1053** 395
[43] Ojha R et al. 2010 *A&A* **519** A45
[44] Homan D C 2005 *ASPC* **340** 133
[45] Gabuzda D C, Vitrishchak V M, Mahmud M & O'Sullivan S 2008 *ASPC* **386** 444
[46] Kellermann K I, Vermeulen R C, Zensus J A & Cohen M H 1998 *AJ* **115** 1295
[47] Lister M L et al. 2009 *AJ* **137** 3718
[48] Napier P J, Bagri D S, Clark B G, Rogers A E E, Romney J D, Thompson A R & Walker R C 1994 *IEEEP* **82** 658
[49] Marscher A P, Jorstad S G, Larionov V M, Aller M F & Lähteenmäki A 2011 *JApA* **32** 233
[50] Taylor G B et al. 2007 *ApJ* **671** 1355
[51] Ojha R et al. 2004 *AJ* **127** 3609
[52] Müller C et al. 2011 *A&A* **530** 11
[53] Jorstad S et al. 2005 *AJ* **130** 1418
[54] Homan D C, Kadler M, Kellermann K I, Kovalev Y Y, Lister M L, Ros E, Savolainen T & Zensus J A 2009 *ApJ* **706** 1253
[55] Linford J et al. 2011 *ApJ* **726** 16
[56] Pushkarev A B, Kovalev Y Y, Lister M & Savolainen T 2009 *A&A* **507** 33
[57] Kovalev Y Y et al. 2009 *ApJ* **696** 17
[58] Hovatta T, Lister M, Kovalev Y Y, Pushkarev A B & Savolainen T 2010 *IJMPD* **19** 943
[59] Blanchard J, Lovell, J, Ojha R, Kadler M, Dickey J & Edwards P 2011 *A&A* **Submitted** http://pulsar.sternwarte.uni-erlangen.de/tanami/pubs/chipaper.pdf

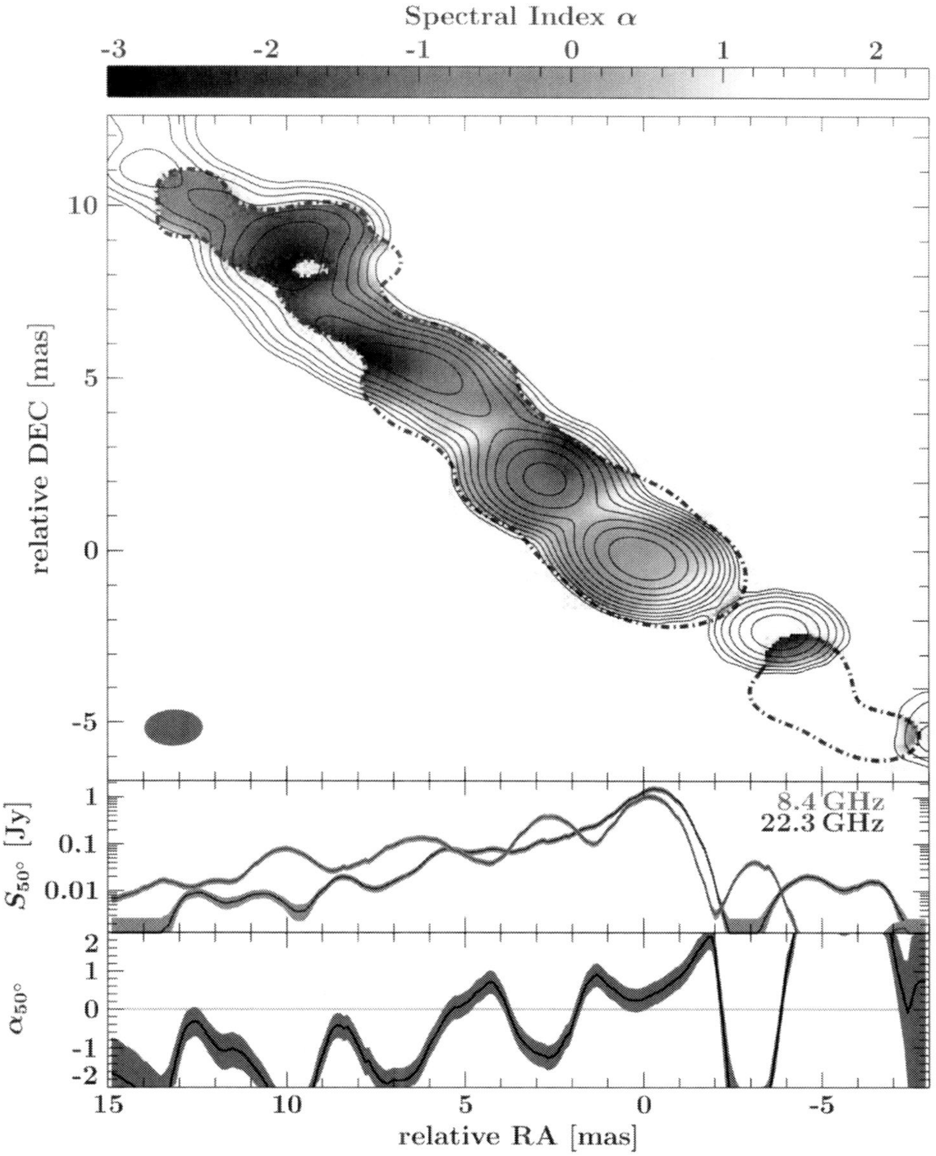

Figure 1: An example of what dual frequency VLBI imaging can reveal. Top: Spectral index map of the nearby radio galaxy Centaurus A, derived from the flux densities at 8.4 GHz and 22.3 GHz. The overlying contours show the flux density distribution at 8.4 GHz (solid black) and at 22.3 GHz (dashed gray) restored with a beam of 1.61×1.02 mas ($\theta = 88°$) represented by the gray ellipse in the lower left corner. Note that 1 mas corresponds to just 0.018 parsec at this distance. Middle: Flux density distribution at 8 GHz (red) and 22.3 GHz (blue) for a narrow strip of 0.1 mas along the jet at a P.A.=50° corrected for the shift of $\Delta\alpha \approx -0.25$ mas and $\Delta\delta \approx -0.2$ mas. Bottom: Spectral index given by the flux densities at P.A.=50°. Uncertainties corresponding to the absolute calibration uncertainties of $\sim 15\,\%$ and considering the image rms at both frequencies are shown as shaded regions around the best-fit distributions (solid lines). From [52].

Journal of Physics: Conference Series **355** (2012) 012007 doi:10.1088/1742-6596/355/1/012007

The OVRO blazar monitoring program

O G King[1,2], **W Max-Moerbeck**[1], **J L Richards**[1], **V Pavlidou**[1,3], **T J Pearson**[1] **and A C S Readhead**[1]

[1] Cahill Center for Astronomy and Astrophysics, California Institute of Technology, 1200 E California Blvd, Pasadena, CA 91125, USA
[2] Keck Institute Fellow
[3] Einstein Fellow

E-mail: ogk@astro.caltech.edu

Abstract. The OVRO 40 m Telescope monitoring program carries out twice-weekly measurements of the 15 GHz flux density of nearly 1600 blazars and other AGN, including all those associated with northern (declination $> -20°$) *Fermi* Large Area Telescope (LAT) detections and a preselected sample ideal for statistical studies. We present some results from the program and describe the statistical method we have developed to assess the intrinsic radio variability of the sources in our sample. We also present a method for assessing the significance of correlations between radio and gamma-ray light curves.

1. Introduction

The launch of the *Fermi* Gamma-ray Space Telescope in June 2008 provides an unprecedented opportunity for the systematic study of blazar jets [1]. Its Large Area Telescope (LAT) observes the sky at energies between 100 MeV and a few hundred GeV. In this energy range relativistic particles can be probed through their inverse Compton emission in the case of electron/positron jets [2, 3, 4], or a combination of pionic emission from primaries and inverse Compton emission from cascade-produced leptonic secondaries in the case of hadronic jets [5].

Blazars comprise the most numerous class of extragalactic GeV sources associated with lower-energy counterparts: the first-year *Fermi* point source catalog (1FGL) contains 1451 sources, of which 596 have been associated with blazars in the first AGN catalog (1LAC) [6, 7]. The exact location of the gamma-ray emission region in blazars and its proximity to the central black hole remain subjects of debate. Two possible models of the GeV emission region are that this emission comes from a "gamma-sphere" close to the base of the jet [4], or that it comes from the same shocked regions that are responsible for the radio emission seen in VLBI observations much further out in the jet [8].

The blazar monitoring program we discuss here is focused on the regular monitoring of a preselected, statistically complete sample of likely gamma-ray-bright objects, independent of their gamma-ray activity state. In anticipation of the unique opportunities offered by the *Fermi*-LAT sky monitoring at gamma-ray energies, three years ago we began the bi-weekly monitoring of a large sample including 1158 likely gamma-ray loud blazars, preselected according to uniform criteria, with the Owens Valley Radio Observatory (OVRO) 40-m telescope at 15 GHz.

The core sample for our monitoring program consists of the 1158 CGRaBS [9] sources north of declination $-20°$. However, we have since *added* (and we continue to add) to our monitoring

Published under licence by IOP Publishing Ltd

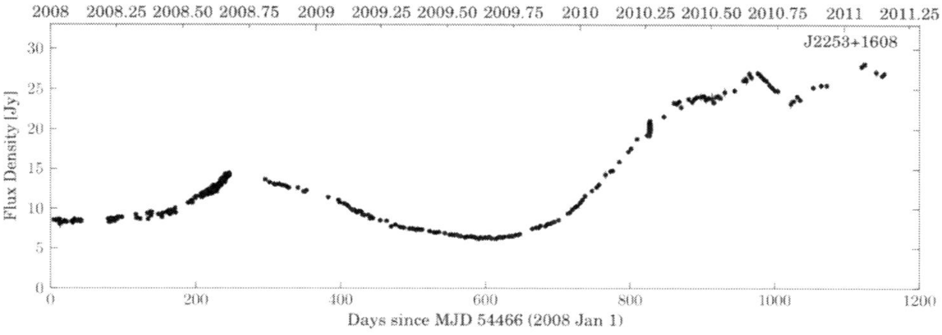

Figure 1. 15 GHz light curve for 3C454.3. Light curves for all the sources in our sample can be found in the program paper [10] and are available for download from the program website[1].

program all new *Fermi*-LAT blazars north of $-20°$ declination that are not CGRaBS members. As published, our subset of the CGRaBS sample contains 812 FSRQs, 111 BL Lacs, 53 radio galaxies, and 182 objects without spectroscopic identification. In our analysis we use redshifts from the CGRaBS publication, which covered 81% of the sample (100% of FSRQs, 49% of BL Lacs).

Our sample is statistically well defined and large enough to allow for statistical analyses and comparisons of sub-samples. We achieve a radio lightcurve sampling cadence comparable to those typically achieved by *Fermi*-LAT for bright gamma-ray blazars. The combination of sample size and cadence is unprecedented, making this by far the largest monitoring survey of radio sources that has been undertaken to date. Such a systematic study of radio and radio/gamma-ray population properties allows us to address a series of long-standing questions on the physical properties of blazar jets, including the location, structure, and radiative properties of the gamma-ray emission region, and the collimation, composition, particle acceleration, and emission mechanisms in blazar jets.

We discuss here the details of the monitoring program and an analysis of the radio variability of blazars [10, 11]. We also present a method for analyzing the significance of cross-correlations between radio and gamma-ray light curves [12].

2. Blazar monitoring program

Our average effective cadence for CGRaBS sources is about 128 measurements per source in the first two years of the program. The efficiency compared to our nominal cadence is 62%.

2.1. Light curves

A typical light curve for the CGRaBS program sources is shown in Figure 1. Regular updates to the data set, including data for sources outside the core sample, are available from the program website.[1]

2.2. Source variability

The questions of the variability amplitude of a source and the confidence with which this can be measured are complex ones and have been traditionally addressed using a variety of measures and tests, such as structure function analysis [13, 14];the variability index [15]; the fluctuation

[1] http://www.astro.caltech.edu/ovroblazars

index [16]; the modulation index [17]; the fractional variability amplitude [18, 19]; and χ^2 tests of a null hypothesis of non-variability. Each of these tools provides different insights to the variability properties of sources and is sensitive to different uncertainties, biases, and systematic errors.

We propose a new index for characterizing source variability: the *intrinsic modulation index* $\overline{m}$, which is the intrinsic standard deviation of the distribution of source flux densities in time, σ_0, measured in units of the intrinsic source mean flux density, S_0. Here the term "intrinsic" is used to denote flux densities and variations as would be observed with perfectly uniform sampling of adequate cadence and zero observational error:

$$\overline{m} = \frac{\sigma_0}{S_0}\,. \tag{1}$$

In this way, $\overline{m}$ is a measure of the true amplitude of variations in the source, rather than a convolution of true variability, observational uncertainties, and effects of finite sampling. Observational uncertainties and finite sampling will, of course, affect the accuracy with which $\overline{m}$ can be measured. We have developed a likelihood analysis, described in detail in [10], which estimates the intrinsic modulation index $\overline{m}$ and intrinsic mean flux density S_0, along with strictly defined $1, 2$ and 3σ uncertainties.

2.2.1. Variability analysis—Results In Figure 2(a) we plot the intrinsic modulation index $\overline{m}$ and associated 1-σ uncertainty against the intrinsic, maximum-likelihood average flux density, S_0, for all our CGRaBS and calibrator sources. The error bar on S_0 corresponds to the 1-σ uncertainty in mean flux density, calculated from the joint likelihood marginalized over $\overline{m}$. CGRaBS sources are shown as black or magenta points or blue triangles for upper limits, while calibrators are shown as green points.

2.2.2. Variability analysis—Population studies—Results We now address the question of whether the intrinsic variability amplitude at 15 GHz, as quantified by $\overline{m}$, correlates with the physical properties of the sources in our sample. To this end, we will determine the *distribution* of intrinsic variability indices $\overline{m}$ for various subsets of our monitoring sample, and we will examine whether the various subsets are consistent with being drawn from the same distribution.

We again use a likelihood analysis and assume that the distribution of $\overline{m}$ in any subset is an exponential distribution of the form:

$$f(m)dm = \frac{1}{m_0}\exp\left[-\frac{m}{m_0}\right]dm \tag{2}$$

with mean m_0 and variance m_0^2. Since distributions of this family are uniquely described by the value of the mean, m_0, we determine m_0, or rather the probability distribution of possible m_0 values, in any specific subset.

Figure 2(b) shows the probability density of m_0 for two populations. The set of sources that are included in 1LAC is depicted by a solid line, while the set of sources that are not in 1LAC is depicted by a dashed line. The two are not consistent with each other at a confidence level of 6σ, with a maximum-likelihood difference of 5.7 percentage points, with gamma-ray-loud blazars exhibiting, on average, a higher variability amplitude *by almost a factor of 2* versus non gamma-ray-loud blazars. This agrees with earlier work which found that EGRET-detected AGN are more variable at radio wavelengths than gamma-ray-quiet blazars [20].

We analyze the subsets of CGRaBS BL Lacs and FSRQs. The probability densities for the mean m_0 of the two subsets are shown in Figure 3(a). The results for BL Lacs (FSRQs) are plotted as a solid (dashed) line. The two curves are not consistent with each other—the BL Lacs

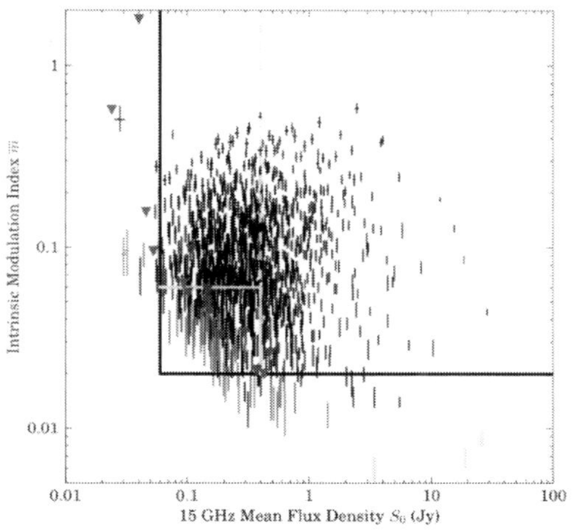

(a)

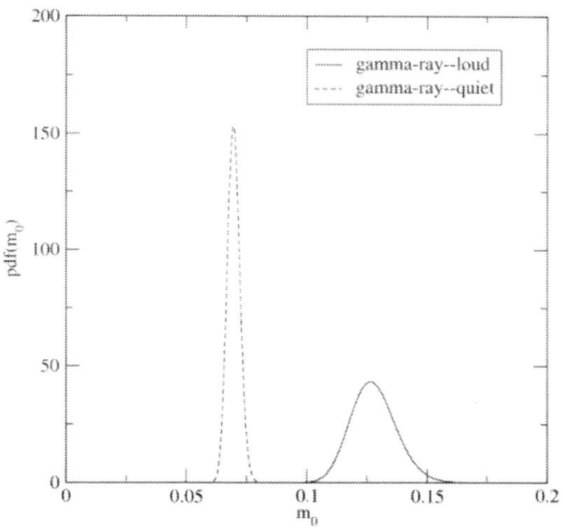

(b)

Figure 2. (a) Intrinsic modulation index $\overline{m}$ and associated 1-σ uncertainty, plotted against intrinsic maximum-likelihood average flux density, S_0. Black points: CGRaBS sources found to be variable with 3-σ confidence by χ^2 test; magenta points: CGRaBS sources found consistent with non-variable by χ^2 test; green points: calibrators 3C 286, DR 21, and 3C 274; blue triangles: 3-σ upper limits for CGRaBS sources for which variability could not be established at $\geq 3\sigma$ confidence level. The error bar on S_0 corresponds to the 1-σ uncertainty in mean flux density, calculated from the joint likelihood marginalized over $\overline{m}$. Data, except for upper limits, outside the yellow and cyan shaded areas are used in the population studies. (b) Probability density of m_0 for CGRaBS blazars in our monitoring sample that are (solid line, maximum-likelihood value and 1-σ error $m_0 = 0.127^{+0.010}_{-0.009}$) and are not (dashed line, maximum-likelihood value and 1-σ error $m_0 = 0.070 \pm 0.003$) included in 1LAC. The two distributions are *not* consistent with a single value.

appear to have, on average, higher variability amplitude than the FSRQs. A similar result was found using the much longer time series data of the Metsähovi and University of Michigan Radio Observatory data sets by [21], though their result was biased by including only the BL Lacs with distinct variability features in their study.

Finally, we examine the dependence of variability amplitude on redshift. In Figure 3(b) we plot the mean $\overline{m}$ (as calculated by a simple average rather than the likelihood analysis) in redshift bins of $\Delta z = 0.5$ for bright ($S \geq 0.4$ Jy) FSRQs with known redshifts in our monitoring sample. We exclude BL Lacs from this analysis so as not to bias the result, as BL Lacs with known redshifts are located at low z, and we have also already shown that they have a higher mean $\overline{m}$ compared to FSRQs. Although the errors are large, there is a hint that variability amplitude decreases with increasing redshift.

3. Radio/gamma-ray time lags

The existence of correlated variability can be investigated using the cross-correlation between the radio and gamma-ray light curves. This has been done using the method of Edelson and Krolik [22], which is applicable to unevenly sampled time series and also provides error estimates on the cross-correlation for each time lag. Blazar or AGN light curves can be modelled as red-noise processes that frequently show flare-like features. These flare-like features can produce spurious

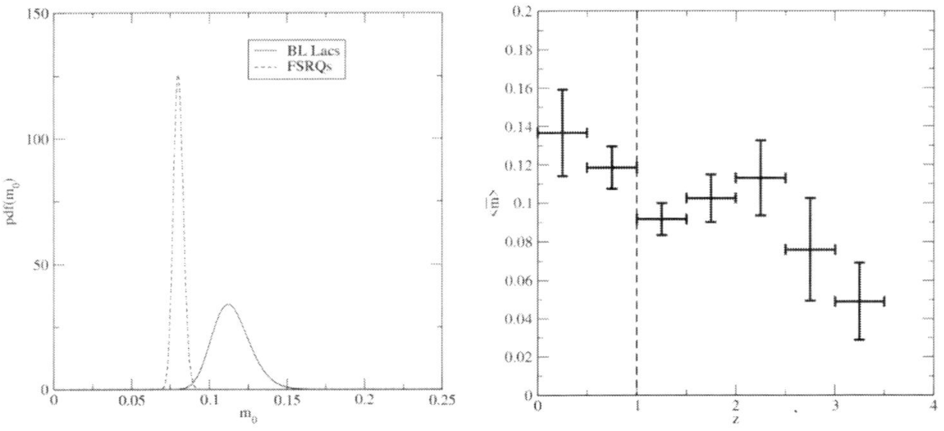

(a) (b)

Figure 3. (a) Probability density of m_0 for BL Lac (solid line, maximum-likelihood value and 1-σ error $m_0 = 0.112^{+0.013}_{-0.011}$) and FSRQ (dashed line, maximum-likelihood value and 1-σ error $m_0 = 0.080 \pm 0.003$) CGRaBS blazars in our monitoring sample. The two distributions are *not* consistent with a single value. (b) Mean $\overline{m}$ in redshift bins of 0.5 for bright ($S > 0.5$ Jy) FSRQs in our monitoring sample.

cross-correlations in unrelated time series. The distribution of these chance associations needs to be investigated in order to attach statistical significance to the cross-correlation results. An example of our data is shown in Figure 4(a).

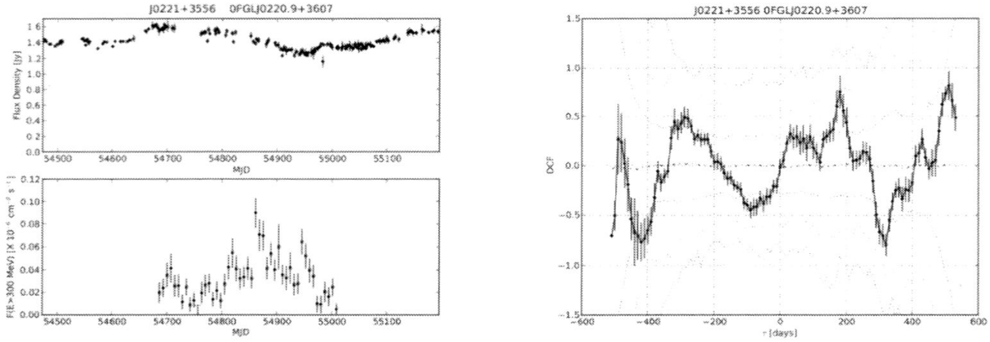

(a) (b)

Figure 4. (a) Example radio (top panel) and gamma-ray (bottom panel) light curves. (b) Discrete cross-correlation function and its 1-,2-, and 3-σ significance curves for the data in (a).

Due to the complexities of the sampling pattern and autocorrelation on the light curves, among others issues, the only practical way to measure the significance is through Monte Carlo simulations. A method to deal with these has been described in [23] and references therein, and will be discussed in detail in [12]. We assume the light curves can be modeled as red-noise processes with simple power-law power spectral density ($S(f) \propto f^{-\beta}$), in which the exponent β can in principle depend on the frequency band and source. A large number of light curve pairs are simulated following a particular model. The simulated light curves are sampled as the data are, and each data point is randomized according to the measurement error on the data. The resulting light curves are cross-correlated in the same way as the original data. Using the results for a large number of simulated light curve pairs the distribution of chance cross-correlations can be estimated at each time lag and used to estimate the probability of a chance cross-correlation. The results depend on the chosen power-law exponent and the particulars of the sampling. For the power-law exponent we use $\beta \sim 2 - 2.5$ for radio and $\beta \sim 1.5 - 2.0$ for gamma-rays. These values are consistent with those found in the literature [24, 25]. An example of the method is shown in Figure 4(b).

The significance of the cross-correlation peaks depends on the assumed model for the light curves; we characterize this dependence using a Monte Carlo method [26]. In this method different power spectral density models can be compared with the data allowing for the selection of the best fit. We fit simple power-law models and study the source-to-source variabilty for the radio light curves.

4. Conclusions

We have presented data from the largest monitoring survey of radio sources that has been undertaken to date. A statistically rigorous analysis of the variability of different source populations has been performed, in which we find that gamma-ray loud blazars are intrinsically more variable at radio wavelengths than gamma-ray quiet blazars, that BL Lacs are more variable than FSRQs, and that there is a trend towards decreasing variability amplitude with increasing redshift for FSRQs. We have presented a method for assessing the significance of cross-correlations between radio and gamma-ray light curves.

Acknowledgments

We acknowledge support from NASA grants NNX08AW31G and NNG06GG1G and NSF grant AST-0808050. OGK acknowledges the support of a Keck Institute for Space Studies Fellowship. WM acknowledges support from the US Department of State and CONICYT in Chile for a Fulbright-CONICYT scholarship. VP acknowledges support for this work provided by NASA through Einstein Postdoctoral Fellowship grant number PF8-90060.

References

[1] Atwood *et al* 2009 *ApJ*, **697** 1071
[2] Dermer C D, Schlickeiser R and Mastichiadis A 1992 *A&A* **256**, L27
[3] Sikora M, Begelman M C and Rees M J 1994 *ApJ* **421**, 153
[4] Blandford R D, and Levinson A 1995 *ApJ* **441**, 79
[5] Mannheim K 1993 *A&A* **269**, 67
[6] Abdo A A *et al* 2010 *ApJS* **188**, 405
[7] Abdo A A *et al* 2010 *ApJ* **715**, 429
[8] Jorstad S G, Marscher A P, Mattox J R, Aller M F, Aller H D, Wehrle A E and Bloom S D 2001 *ApJ* **556**, 738
[9] Healey S E *et al* 2008 *ApJ* **175**, 97
[10] Richards J L *et al* 2011 *ApJS*, **194** 29
[11] Pavlidou V *et al* 2011 in prep.
[12] Max-Moerbeck W *et al* 2011 in prep.
[13] Hughes P A, Aller H D and Aller M F 1992 *ApJ* **396**, 469
[14] Lainela M and Valtaoja E 1993 *ApJ* **416**, 485
[15] Aller M F, Aller H D and Hughes P A 1992 *ApJ* **399**, 16
[16] Aller M F, Aller H D and Hughes P A 2003 *ApJ* **586**, 33
[17] Kraus A *et al* 2003 *A&A* **401**, 161
[18] Edelson R, Turner T J, Pounds K, Vaughan S, Markowitz A, Marshall H, Dobbie P and Warwick R 2002 *ApJ* **568**, 610
[19] Soldi S *et al* 2008 *A&A* **486**, 411
[20] Lähteenmäki A and Valtaoja E 2003 *ApJ* **590**, 95

[21] Hovatta T *et al* 2008 *A&A* **485**, 51
[22] Edelson R A and Krolik J H 1998 *ApJ* **333**, 646
[23] Max-Moerbeck W *et al* 2010 *Proc. Workshop "Fermi meets Jansky: AGN in Radio and Gamma-Rays"* vol 1 (Bonn: Max-Planck-Institut für Radioastronomie) pp 77–80
[24] Abdo A A *et al* 2010 *arXiv* **1004.0348**
[25] Chatterjee R *et al* 2008 *ApJ* **689**, 79
[26] Uttley P, McHardy I M and Papadakis I E 2002 *MNRAS* **332**, 231

Journal of Physics: Conference Series **355** (2012) 012008 doi:10.1088/1742-6596/355/1/012008

Faraday rotation in the MOJAVE blazars: 3C 273 a case study

T Hovatta[1,2], M L Lister[1], M F Aller[3], H D Aller[3], D C Homan[4], Y Y Kovalev[5,6], A B Pushkarev[7,8] and T Savolainen[6]

[1] Department of Physics, Purdue University, 525 Northwestern Ave, West Lafayette, IN 47907, USA

[3] Department of Astronomy, University of Michigan, 817 Denison Building, Ann Arbor, MI 48109-1042, USA

[4] Department of Physics and Astronomy, Denison University, Granville, OH 43023, USA

[5] Astro Space Center of Lebedev Physical Institute, Profsoyuznaya 84/32, 117997 Moscow, Russia

[6] Max-Planck-Institut für Radioastronomie, Auf dem Hügel 69, 53121 Bonn, Germany

[7] Pulkovo Observatory, Pulkovskoe Chaussee 65/1, 196140 St. Petersburg, Russia

[8] Crimean Astrophysical Observatory, 98409 Nauchny, Crimea, Ukraine

E-mail: thovatta@caltech.edu

Abstract. Radio polarimetric observations of Active Galactic Nuclei can reveal the magnetic field structure in the parsec-scale jets of these sources. We have observed the γ-ray blazar 3C 273 as part of our multi-frequency survey with the Very Long Baseline Array to study Faraday rotation in a large sample of jets. Our observations re-confirm the transverse rotation measure gradient in 3C 273. For the first time the gradient is seen to cross zero which is further indication for a helical magnetic field and spine-sheath structure in the jet. We believe the difference to previous epochs is due to a different part of the jet being illuminated in our observations.

1. Introduction

According to current knowledge, jets of AGN are magnetically launched from a rotating black hole or accretion disk [1, 2, 3]. It is then natural to expect the magnetic field structure in the jet to be helical [4]. We do not know, however, if the helical magnetic field structure dominates down to parsec-scale jets or if the field becomes predominately tangled due to re-collimation shocks or interaction with the external medium [5]. Magnetic field structure of parsec-scale AGN jets can be studied by using polarimetric multifrequency observations. Evidence for helical magnetic field structure could be a detection of a Faraday rotation measure (RM) gradient transverse to the jet [6].

Faraday rotation occurs when polarized waves propagate through non-relativistic plasma within or external to the source [7]. It can be described with a linear dependence between the observed electric vector position angle (EVPA, $\chi_{\rm obs}$) and wavelength squared (λ^2) by the

[2] Present address: Cahill Laboratory of Astronomy and Astrophysics, California Institute of Technology, 1200 E California Blvd, Pasadena CA 91125

Journal of Physics: Conference Series **355** (2012) 012008 doi:10.1088/1742-6596/355/1/012008

following formula

$$\chi_{\rm obs} = \chi_0 + \frac{e^3\lambda^2}{8\pi^2\epsilon_0 m^2 c^3}\int n_e \mathbf{B}\cdot d\mathbf{l} = \chi_0 + {\rm RM}\lambda^2, \quad (1)$$

where χ_0 is the intrinsic EVPA and RM is the rotation measure, related to the electron density n_e and the magnetic field component **B** parallel to the line of sight. The sign of the RM then depends on the direction of the line of sight component of the magnetic field with RM being positive if the magnetic field points towards the observer and negative if it points away from the observer. Thus a helical magnetic field viewed directly from the side would have a RM gradient going from positive to negative and crossing zero in the middle of the jet. Due to relativistic aberration effects blazars that are observed at small viewing angles are viewed from the side in the rest frame of the jet. Recent simulations [8] show that launching the jet from a rotating black hole can indeed produce such structure.

The first observational evidence for a RM gradient across a jet was reported in quasar 3C 273 by Asada et al. [9], who detected a RM change from 450 to 250 rad/m^2 across the jet in observations between 4.6 and 8.6 GHz in 1995. The gradient was confirmed by Zavala & Taylor [10] who in their two observing epochs in 2000 detected a change from 2000 to 0 rad/m^2 across the jet of 3C 273 about 5 mas from the core. They attribute the large difference in the RM values in comparison to the previous observations [9] to the better resolution of their high frequency observations between 12.1 and 22.2 GHz. Follow-up observations on the gradient at frequencies between 4.6 and 8.6 GHz in 2002 reveal time variations on time scales of several years [11]. Since these observations were done at the same angular resolution as the first detection [9], the difference between the epochs cannot be due to observational effects. The total rotation they detect is larger than can be caused by internal Faraday rotation, as it would cause severe depolarization. They also rule out external clouds as the cause of the rotation as the variability time scales are too short to cause variations in foreground clouds. They conclude that the rotation is most likely caused by a sheath around the jet. RM gradients have been detected in other sources as well but the issue remains controversial due to the difficulty in assessing the reliability of a RM gradient [12]. Recently, RM gradients have been detected in the kpc-scale jets as well, see e.g. Gabuzda et al. (these proceedings).

In 2006 we conducted a large survey of Faraday rotation in 191 sources in the MOJAVE (Monitoring of Jets in Active galactic nuclei with VLBA Experiments) sample (Hovatta et al. 2011 in preparation). In this paper we report our observations of the RM gradient in 3C 273.

2. Observations and data reduction

We observed 3C 273 on March 9, 2006 and June 15, 2006 with the Very Long Baseline Array (VLBA) at 8.1, 8.4, 12.1 and 15.3 GHz. The initial data reduction and calibration were done in the standard manner as described in the AIPS cookbook[3]. The imaging and self calibration were done using the Difmap package [13]. For more details see Ref. [14]. The absolute EVPA calibration of the 15 GHz observations were done previously as part of the MOJAVE project using the stability of the D-term phases of some antennas and IFs [15, 16]. The 8 and 12 GHz bands were calibrated using nearby observations from the VLA/VLBA polarization calibration database[4], and single-dish observations from University of Michigan Radio Astronomy Observatory. We estimate the uncertainty in the EVPA calibration to be 3°, 2° and 4° in the 15, 12 and 8 GHz bands, respectively.

The (u,v)-range of all the bands was clipped to correspond to each other, and the images at all the frequencies were restored to the 8.1 GHz beam. Because the absolute position of the source is lost during our calibration process, we used 2D cross-correlation to properly align the images

[3] http://www.aips.nrao.edu

[4] http://www.aoc.nrao.edu/~smyers/calibration/

Journal of Physics: Conference Series **355** (2012) 012008 doi:10.1088/1742-6596/355/1/012008

at different frequencies. The images of the different bands were then combined to create the RM maps by performing a linear least-squares-fit of the EVPA vs. λ^2 in each pixel independently. Goodness of the fit was estimated by a χ^2 criterion. Details of the procedure will be reported in Hovatta et al. 2011 (in preparation).

3. RM maps of 3C 273

The RM maps for 3C 273 are shown in Fig. 1. The top panels show the RM map with color indicating the RM in rad/m^2 overlaid on top of total intensity contours at 15 GHz. The second panel from the top shows the error in RM, calculated from the variance-covariance matrix of the linear fit. The third panel from the top shows the EVPAs at 15 GHz which have been corrected for the Faraday rotation. In the bottom panel the sign of the RM is shown.

The maps are blanked for pixels with absolute RMs of over 20 000 rad/m^2. These are typically concentrated on the core or edge regions of the jets and they do not coincide between the two epochs. Therefore we do not think these extreme RM values are real but are artifacts that occur by chance due to random noise in the data and $n\pi$ ambiguities present in the analysis. We are currently investigating this in detail with simulations (Hovatta et al. 2011, in preparation), and our preliminary results show that it is possible to get these large RM values purely due to random noise. The errors in the RM are typically around 70 rad/m^2 rising to as high as 200 rad/m^2 in the jet edges where the signal-to-noise ratio in total intensity is typically smaller. The Faraday-corrected EVPAs are mostly perpendicular to the optically thin jet indicating that the magnetic field is predominantly parallel to the jet. The jet region about 7 mas from the core is an exception with clearly differing EVPA direction. This could be caused by a shock compressing the magnetic field which would also explain the enhanced total intensity and polarized emission from this region. The sign map of the RM shows clearly bilateral structure indicative of a helical magnetic field as shown by simulations [8].

4. Transverse RM gradient

Our observations in Figs. 2 and 3 show a clear gradient transverse to the jet and along the whole jet length. The maximum gradient is seen about 3-7 mas from the core where the RM changes from 500 to -600 rad/m^2. Moving down the jet the gradient becomes less pronounced as was also noted by Attridge et al.[17]. Our observations were done at resolution which is in between those by Refs. [9, 11] and Ref. [10] and that can explain some of the discrepancy in our values. The most notable difference is that our RM gradient goes from positive to negative RM values, as expected for a helical field viewed from the side, in contrast to Refs. [9, 11, 10] where the values are always positive. Our positive values also do not extend to as high values as in the maps of Ref. [10]. The most likely explanation is that we are seeing a different part of the jet being illuminated by components moving down the jet as our observations are several years apart from earlier observations [18]. The effect of seeing a different part of the jet at different epochs is clearly demonstrated in the stacked images of the MOJAVE program [14] which show how the jet at any given epoch may fill only a portion of the region visible in the stacked image. In fact, if we calculate the inner jet direction in 2000 and compare it to the jet direction in 2006, the difference is 10°. We are thus seeing a portion of the jet which is 10° to the South of the observations in Ref. [10]. This together with the slightly different resolution of our observations could explain the difference in the RM values. A similar effect is seen in the RM structure of the radio galaxy 3C 120 over several years time scales [19, 20].

Even though our observations are only three months apart we can still see differences in the RM between the two epochs. In particular, the core regions strongly differ. This is probably due to the complex sub-structure of the core region in 3C 273 [17, 21]. If we ignore the unreliable jet edges, the jet RM values are very similar in the two epochs and differences are likely due to observational effects such as signal to noise in the differing jet regions resulting in bad λ^2-fits in

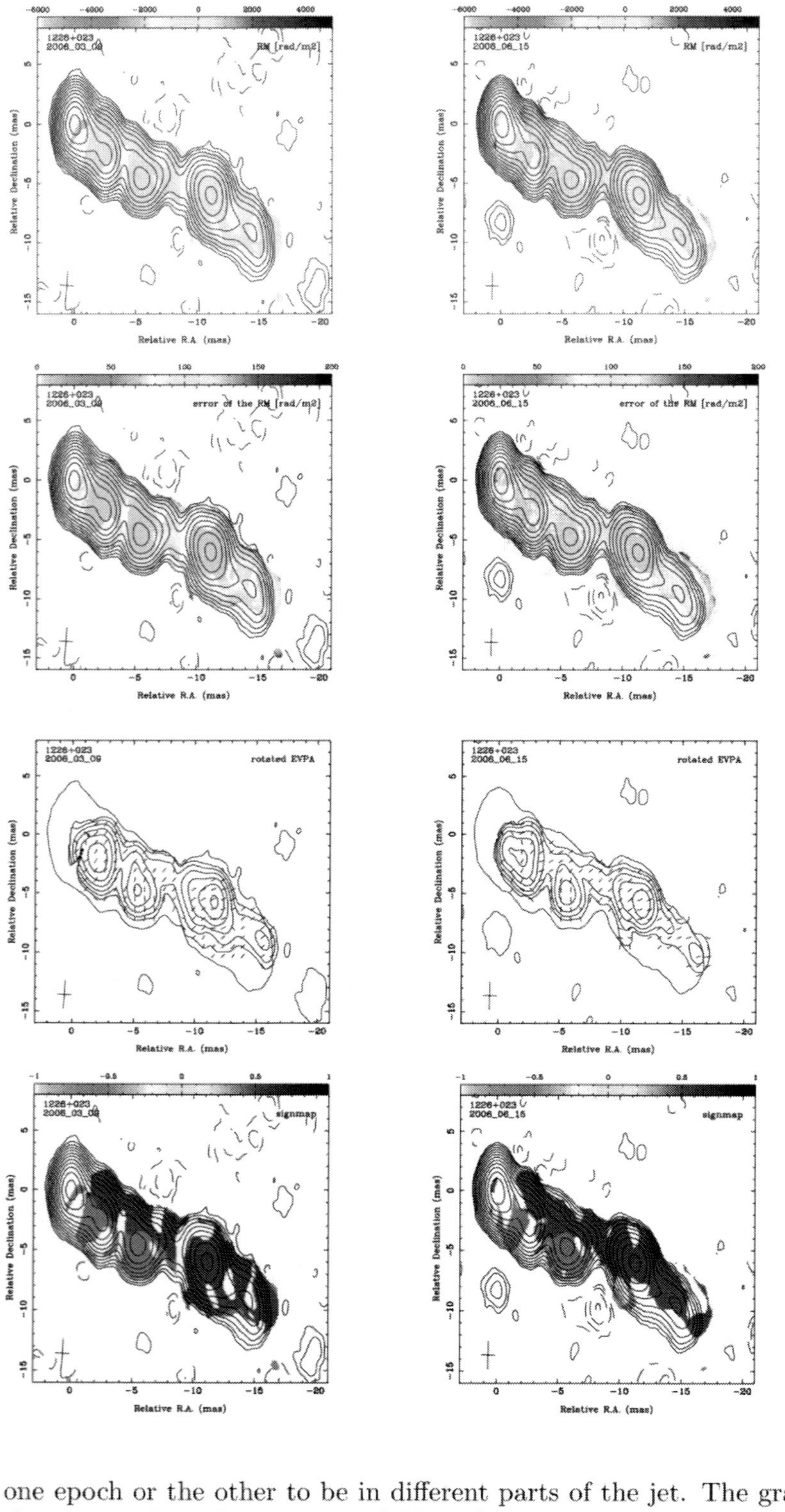

Figure 1. Left: Epoch 2006-03-09. Right: Epoch 2006-06-15. From top down: RM in color overlaid on 15 GHz total intensity contours, error of the RM, RM-corrected EVPA at 15 GHz overlaid on the 15 GHz polarized flux contours, sign of the RM. In the top panel the map has been blanked for very high absolute RM values.

one epoch or the other to be in different parts of the jet. The gradient looks very similar over the two epochs. The consistent RM gradient over the length of the whole jet suggests a scenario involving a sheath around the jet. Recent simulations show that observing a bilateral structure

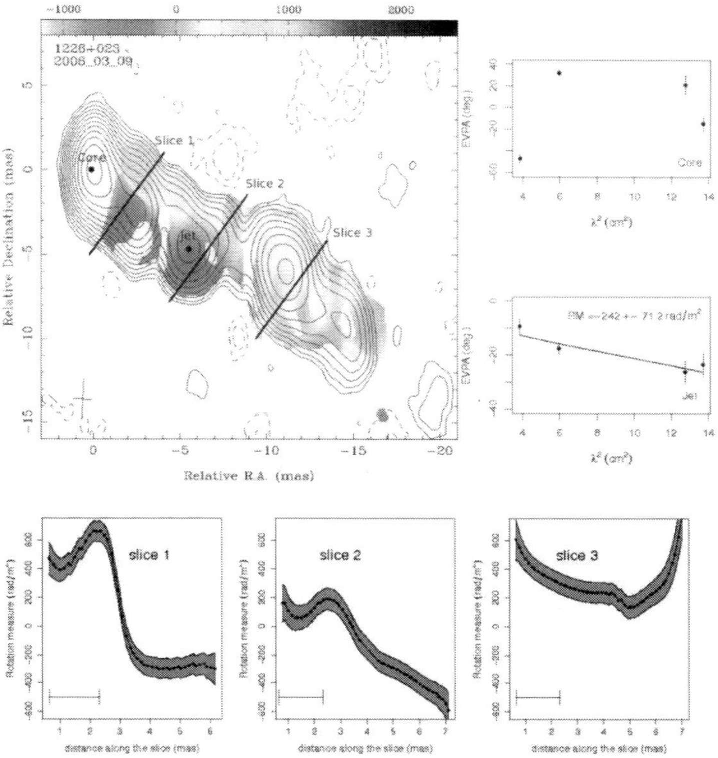

Figure 2. Left: RM map of 3C 273 in 2006-03-09 with additional high-RM pixels of over 4000 rad/m^2 blanked to show the gradient. Right: RM fit of the core and a jet component 7 mas from the core as marked by the dots in the RM map. Bottom: Transverse slices of the jet as marked by the lines in the RM map. The absolute EVPA calibration error has been subtracted from the error bars because it is the same in every pixel. The beam size along the slice is shown as a line in the bottom left corner.

in the sign map of the RM is a strong indication of helical magnetic field surrounding the jet which cannot be easily explained by a random foreground screen [8].

Detailed multi-epoch observations of the RM and polarization structure are needed to fully understand the Faraday-screen properties, as was recently done for 3C 120 [20]. Their observations showed that even though they detect a gradient and variations in the jet RM, these are not connected with the polarization changes. This suggests that the Faraday medium may be a distant external screen not connected with the jet.

5. Conclusions

We have observed 3C 273 with the VLBA as part of our large survey to study the Faraday rotation in AGN jets. Our observations confirm the rotation measure gradient transverse to the jet. For the first time the gradient is seen to cross zero which is further evidence for a helical magnetic field and a sheath around the jet.

Acknowledgments

The MOJAVE project is supported under National Science Foundation grant AST- 0807860 and NASA Fermi grant NNX08AV67G. Work at UMRAO has been supported by a series of grants from the NSF and NASA and by funds for operation from the University of Michigan.The VLBA is a facility of the National Science Foundation operated by the National Radio Astronomy Observatory under cooperative agreement with Associated Universities, Inc.

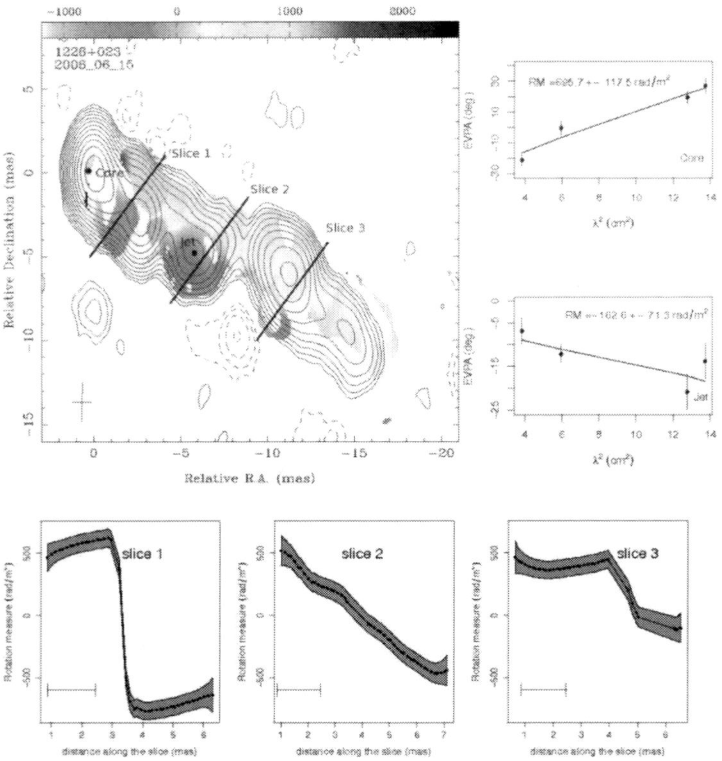

Figure 3. Same as Fig. 2 but for epoch 2006-06-15.

References

[1] Blandford R D and Znajek R L 1977 *MNRAS* **179** 433
[2] Meier D L, Koide S and Uchida Y 2001 *Science* **291** 84
[3] Vlahakis N and Königl A 2004 *ApJ* **605** 656
[4] McKinney J C and Narayan R 2007 *MNRAS* **375** 531
[5] Marscher A P et al. 2008 *Nature* **452** 966
[6] Blandford R 1993 *Astrophysical Jets, Astrophysics and Space Science Library* (Cambridge: Cambridge Univ. Press) Vol 103, 15
[7] Burn B J 1966 *MNRAS* **133** 67
[8] Broderick A E and McKinney J C 2010 *ApJ* **725** 750
[9] Asada K et al. 2002 *PASJ* **54** L39
[10] Zavala R T and Taylor G B 2005 *ApJ* **626** L73
[11] Asada K, Inoue M, Kameno S and Nagai H 2008 *ApJ* **675** 79
[12] Taylor G B and Zavala R T 2010 *ApJL* **722** L183
[13] Shepherd M C 1997 *in ASP Conf. Proc. 125, Astronomical Data Analysis Software and Systems VI* ed. Hunt G & Payne H E (San Francisco, CA: ASP) 77
[14] Lister M L et al. 2009 *AJ* **137** 3718
[15] Gomez J-L, Marscher A P, Alberdi A, Jorstad S G and Agudo I 2002 *VLBA Scientific Memo* No. 30
[16] Lister M L and Homan D C 2005 *AJ* **130** 1389
[17] Attridge J M, Wardle J F C and Homan D C 2005 *ApJ* **633** 85
[18] Savolainen T, Wiik K, Valtaoja E and Tornikoski M 2006 *A&A* **446** 71
[19] Gomez J-L, Marscher A P, Jorstad S G, Agudo I and Roca-Sogorb M 2008 *ApJ* **681** L69
[20] Gomez J-L, Roca-Sogorb M, Agudo I, Marscher A P and Jorstad S G 2011 *ApJ* **733** 11
[21] Savolainen T, Wiik K, Valtaoja E and Tornikoski M 2008 *ASPC* **386** 451

International Workshop on Beamed and Unbeamed Gamma-Rays from Galaxies IOP Publishing
Journal of Physics: Conference Series **355** (2012) 012009 doi:10.1088/1742-6596/355/1/012009

Investigating the Effects of Finite Resolution on Observed Transverse Jet Profiles

E. Murphy, D.C. Gabuzda

Physics Department, University College Cork, Cork, Ireland

E-mail: egm@student.ucc.ie

Abstract. Both the emission properties and evolution of Active Galactic Nuclei (AGN) radio jets are dependent on the magnetic fields that thread them. Faraday Rotation gradients are a very important way of investigating these magnetic fields, and can provide information on the orientation and structure of the magnetic field in the immediate vicinity of the jet; for example, a toroidal or helical field component should give rise to a systematic gradient in the observed Faraday rotation across the jet, as well as characteristic intensity and polarization profiles. However, real observed radio images have finite resolution, usually expressed via convolution with a Gaussian beam whose size corresponds to the central lobe of the point source response function. This will tend to blur transverse structure in the jet profile, raising the question of how well resolved a jet must be in the transverse direction in order to reliably detect transverse structure associated with a helical jet magnetic field. We present results of simulated intensity, polarization and Faraday rotation images designed to directly and empirically investigate the effect of finite resolution on observed transverse jet structures.

1. Introduction

At radio wavelengths the jets of active galaxies emit synchrotron radiation, which is characterised by appreciable linear polarization, with the plane of the polarization perpendicular to the plane of the jet magnetic field in the optically thin region. The polarization structure provides information about the structure of the magnetic fields threading these jets, which influence the evolution and emission properties of the jets, as well as their stability. In addition, an understanding of the magnetic field structure is important if we wish to correctly infer the intrinsic jet structure and physical processes occurring in the jet from the observed radio images. Yet despite much observational effort the nature of the magnetic field structures of AGN remain incompletely understood. Several types of observational results suggest that the magnetic field threading the jet may have a significant helical component on parsec scales.

First, helical magnetic fields in a jet with perfect circular symmetry yield projections of the magnetic fields onto the plane of the sky which are either parallel or perpendicular to the jet axis, as is frequently observed [1,2]. Second, many parsec-scale jets show obvious antisymmetries in total intensity and polarization profiles that are reminiscent of those revealed in the helical field simulations of Laing [2,3]. Third, the presence of transverse Faraday rotation gradients across a number of VLBI jets can also be explained by helical fields in the immediate vicinity of the jets [4].

These observational results are all affected by the resolution of the observations used to detect them. In order to use these as identifiers for potential helical magnetic fields threading the jets

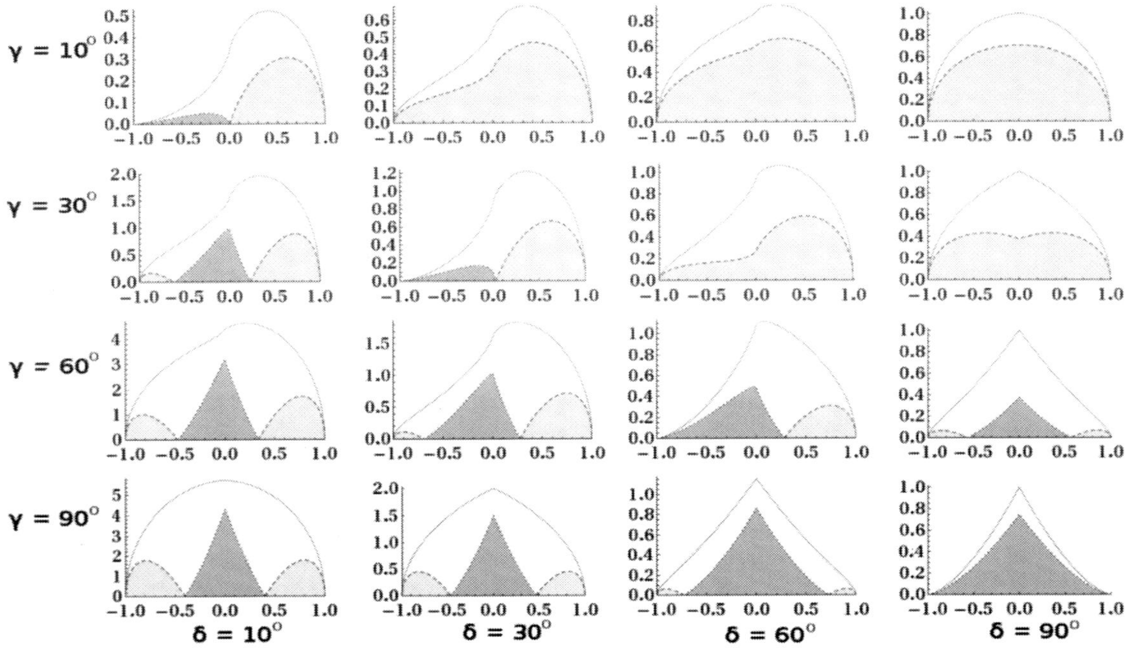

Figure 1. Transverse structure produced by the helical field model, for various viewing angles, δ, and helix pitch angles, γ. Solid lines correspond to total intensity, purple regions to longitudinal polarization and green regions to transverse polarization. The vertical scaling is arbitrary; the centre of the jet is at 0.0 on the horizontal axis.

of AGN, it is very important to understand the effects of finite resolution on observed transverse jet profiles.

2. Total Intensity and Polarized Intensity Profiles

To investigate the effects of finite resolution on transverse jet profiles, infinite resolution intensity (I) and polarization (P) profiles where generated using a simple helical field model and then convolved with various beams. The model assumes a helical B field of constant pitch angle and uniform flux density threading a cylindrical jet, and can be used to predict the I and P distributions across a jet as functions of the helical pitch angle (γ) and viewing angle (δ) [2,3,5]. Both γ and δ are in the rest frame of the jet. Fig. 1 shows the asymmetric intensity and polarization structure expected for various values of δ and γ.

An example of the effects of convolution with a Gaussian beam on the I and P transverse profiles can be seen in Figs. 2 and 3. The infinite resolution I and P profiles are shown in Fig. 2 (left). Convolution with a Gaussian beam (Fig. 3) rapidly reduces the asymmetry in transverse I profiles, but reduces the asymmetry expected in P profiles much more slowly. One of the major effects convolution can have on transverse P profiles is changing the observed polarization configuration. In this example, the original polarization structure was 'spine and sheath', a spine of longitudinal polarization with a sheath of transverse polarization (Fig. 2). When convolved with a beam whose width is 0.2 times the intrinsic jet width, the intensity profile appears only slightly asymmetric, but the asymmetric 'spine and sheath' polarization structure is clearly visible (Fig. 3, top left). After convolution with a beam of width equal to 0.6 times the intrinsic jet width, the intensity profile has become very symmetric and one sheath is no longer observed

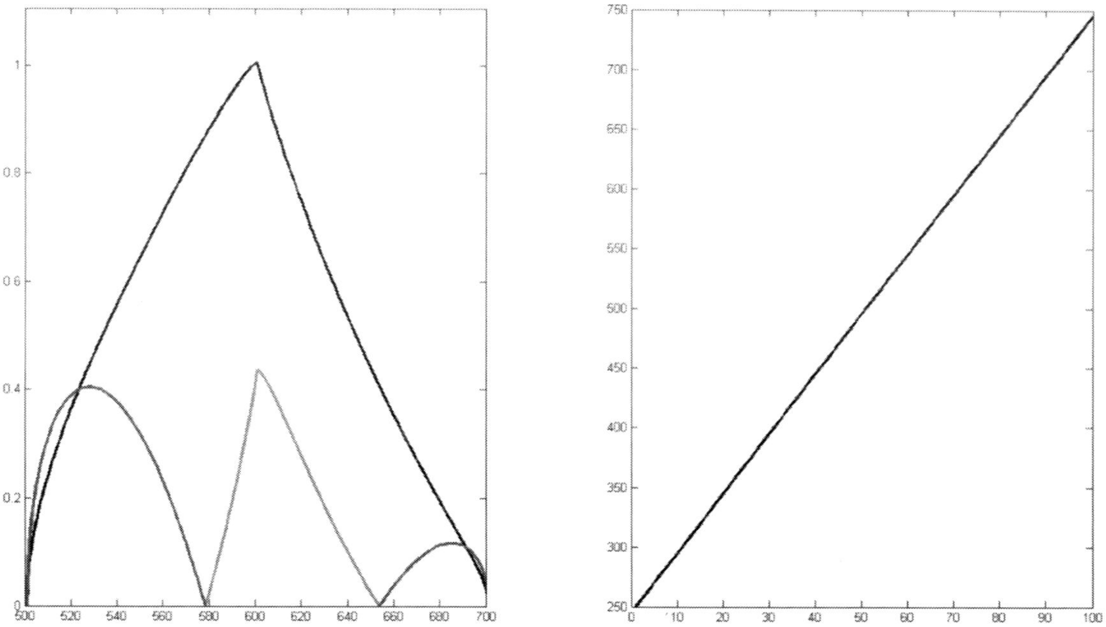

Figure 2. Plots of the infinite resolution I and P (left) and Rotation Measure profiles (right) used in analysis. Transverse profiles of I and P were generated for a jet with $\delta = 80°$ and $\gamma = 53°$. The black, blue and red lines correspond to total intensity, transverse polarization and longitudinal polarization, respectively. The I and P values are normalized to the peak I value. The polarization has in addition been scaled by a factor of two in order to increase visibility. The vertical axis in the RM plot is in rad/m^2. The horizontal scales are in arbitrary units: the intrinsic jet width is 200, the centre of the jet is at 600 in the left panel and at 50 in the right panel.

(Fig. 3, top right). The P profile for this case is now longitudinal on one side, transverse on the other. After convolution with a beam of width equal to the intrinsic jet width, the longitudinal 'spine' becomes very weak (Fig. 3, bottom left). Finally, after convolution with a beam width equal to 1.4 times the intrinsic jet width, only longitudinal polarization offset to one side of the jet is observable (Fig. 3, bottom right). However, even after convolution with a very large beam such as this last one, the position of the maximum polarization intensity is still clearly offset from the position of maximum total intensity. Thus, the observed transverse P structure can still reflect the presence of helical magnetic fields threading the jets of AGN even if the beam used in the observations is large with respect to the size of the jet, although it will not be possible in this case to determine accurate values of γ and δ based on the observed images.

3. Faraday Rotation

Faraday Rotation occurs when an electromagnetic wave propagates through a region with charged plasma and a magnetic field. Faraday Rotation rotates the polarization of the electromagnetic wave because the left-hand circularly polarized component of the EM wave has different refractive index than the right-hand circularly polarized component, given by

$$\chi = \chi_0 + RM\lambda^2 \quad (1)$$

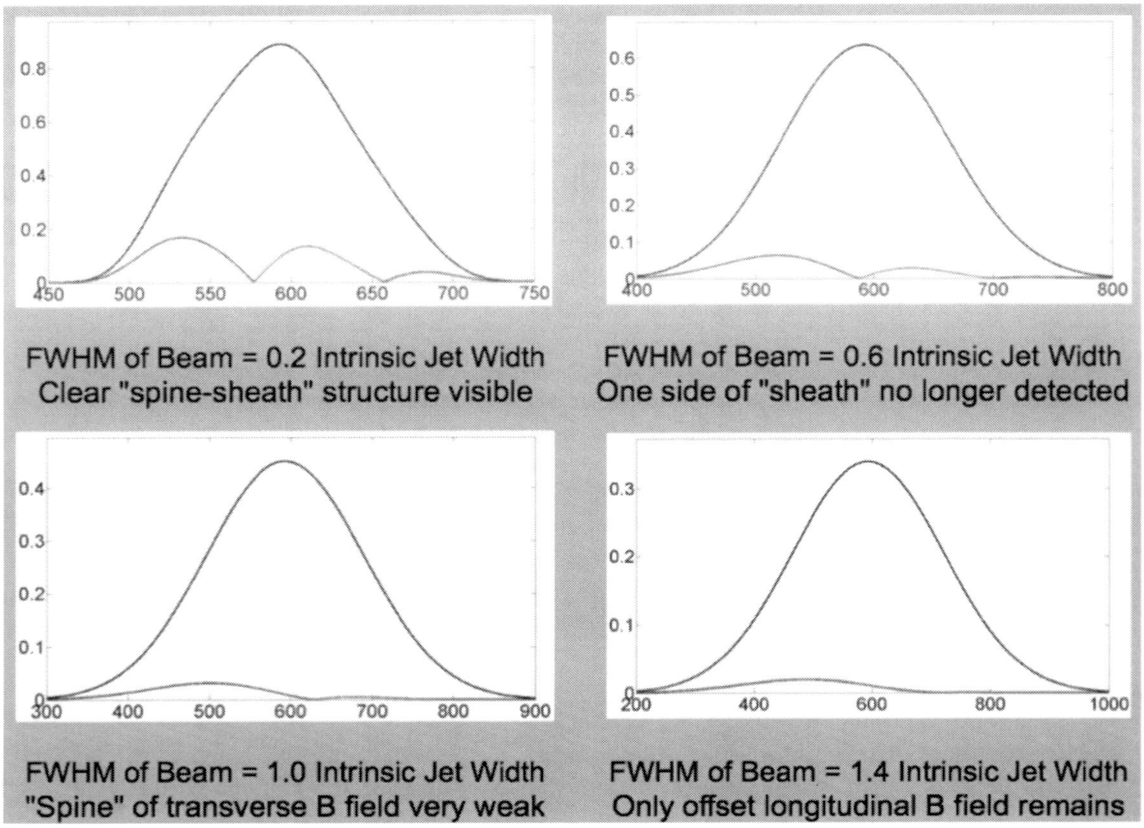

Figure 3. Transverse profiles of I and P generated for a jet with $\delta = 80°$ and $\gamma = 53°$ after convolution with Gaussian beams of various sizes. The black, blue and red lines correspond to total intensity, transverse polarization and longitudinal polarization, respectively. The I and P values are normalized to the peak I value. The polarization has in addition been scaled by a factor of two in order to increase visibility. The horizontal scale is in arbitrary units, with an intrinsic jet width of 200 with a centre at 600.

where χ is the observed polarization angle, χ_0 is the emitted polarization angle, λ is the wavelength and RM is the Rotation Measure. The Rotation Measure is given by

$$RM = \frac{e^3}{8\pi^2 \epsilon_0 m_e^2 c^3} \int n_e \vec{B} \bullet \vec{dl} \tag{2}$$

where e is the elementary charge, ϵ_0 is the permittivity of free space, m_e is the mass of the electron, n_e is the number density of electrons in the plasma and B is the magnetic field strength in the plasma. We would expect to observe a transverse RM gradient if a helical B field threads the jet, due to the systematic change in $\vec{B} \bullet \vec{dl}$ across the jet[4].

In order to analyse the effects of finite resolution on observations of a RM gradient, the effects of a calculated RM gradient produced by the helical field model were applied to the Stokes Q and U profiles for an infinite resolution jet. The resulting Q and U profiles were then convolved with various beams, and the RM profile was determined from these convolved Q and U profiles.

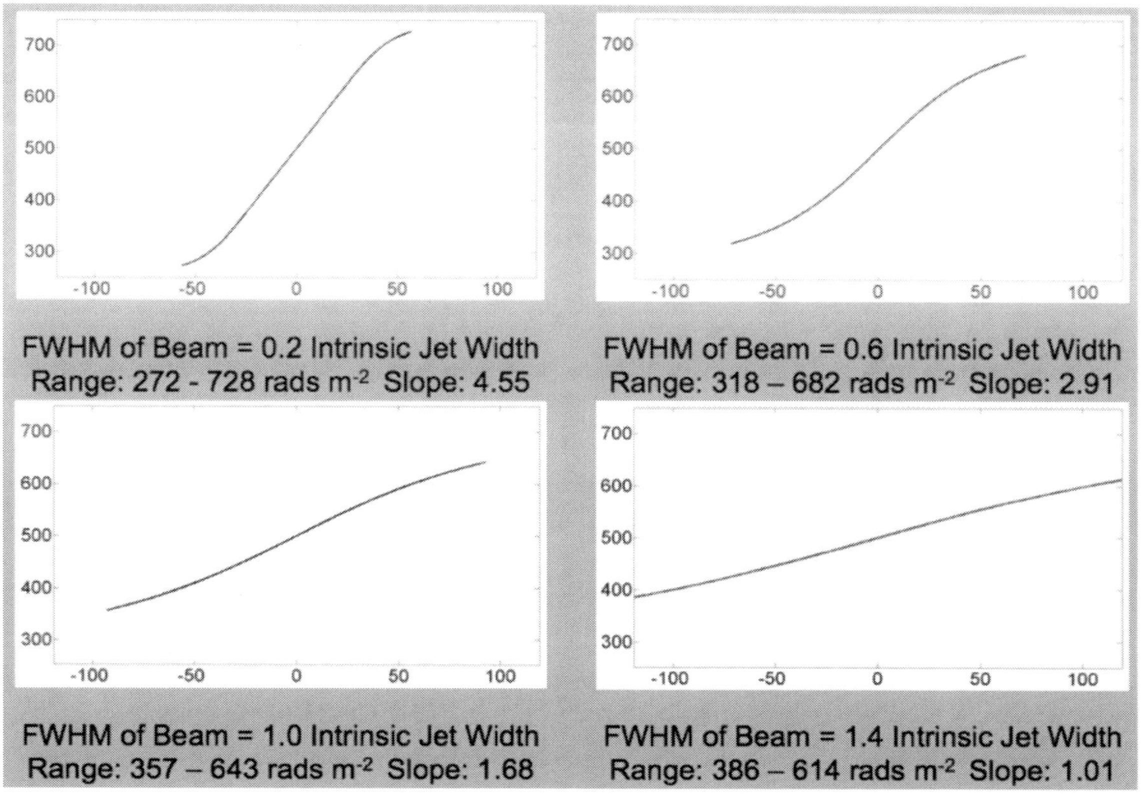

Figure 4. Transverse profiles of Faraday Rotation Measure after convolution with Gaussian beams of various sizes. The vertical axes are in rad/m^2. The values given for the slopes correspond to the slope of the best fit line fitted to the gradients. The horizontal scale is arbitrary, with a intrinsic jet width of 100 and the jet centre at 0.

The original RM gradient applied to the Q and U profiles ranged from 250 to 750 rad/m^2 and had a slope of 5 (Fig. 2). Figure 4 shows the observed RM gradients after convolution with the same beams used in Figure 3. Convolution significantly changes the values of the observed Faraday RMs but does not destroy the RM gradients produced by the helical field, essentially independent of the size of the beam. Transverse RM gradients can potentially remain visible, even with beams significantly larger than the width of the jet.

4. Conclusion

The presence of a helical magnetic field threading an AGN jet can be inferred from characteristic transverse structures in the intensity, polarized intensity and rotation measure profiles. The detection of this transverse structure is affected by the resolution of the observations used. Convolution with a Gaussian beam rapidly reduces the asymmetry in the I profiles, while weakening the P structure much less severely. Appreciable transverse P structure remains visible even when the I profile has become completely symmetrical, and even when the beam is appreciably larger than the intrinsic jet width. Thus, it may be possible to clearly detect transverse structure consistent with a helical magnetic field, but not reliably deduce the parameters of the field if the beam is large. Convolution also significantly changes the values of the Faraday RMs observed across AGN jets, but does not destroy the RM gradients produced by the helical field, even in the case of large beams. Here as well, the presence of the RM gradient associated with a helical B field is robust, but it is not possible to accurately deduce the parameters of the field if the beam is large compared to the jet. This suggests that the proposed criterion of [6] that a transverse RM gradient must span three beamwidths to be reliable is inappropriate and overly conservative. The important criteria are the others proposed in [6]: that the candidate RM gradients be monotonic and smooth, that the RM change by at least three times the error across the gradient, and that the region sampled be predominantly optically thin at all the frequencies used to make the RM distribution. These three criteria have, in fact, been satisfied by the majority of published RM analyses reporting the presence of transverse RM gradients [7–12].

5. Acknowledgments

Funding for this research was provided by the Irish Research Council for Science Engineering and Technology (IRCSET).

References

[1] Lyutikov M., Pariev V.I. and Gabuzda D.C. 2005, *MNRAS* **360** 869
[2] Papageorgiou A. 2005 *PhD. Thesis* University of Central Lancashire
[3] Laing R.A. 1981 *ApJ* **248** 87
[4] Blandford R.D., Livio M. and O'Dea C.P. 1993 *Astrophysical Jets. Cambridge University Press* p. 26
[5] Murphy E., Gabuzda D.C. and Cawthorne T. 2010, Proceedings of Science, http://pos.sissa.it/archive/conferences/125/041/10th
[6] Taylor G.B. and Zavola R. 2010 *Ap J* **722** L183
[7] Gabuzda D.C., Murray E. and Cronin P. 2004, *MNRAS* **351**, L89
[8] Asada K., Inoue M., Nakamura M., Kameno S. and Nagai H. 2008, *Ap J*, **682**, 798
[9] Gabuzda D.C., Vitrishchak V.M., Mahmud M. and O'Sullivan S.P. 2008, *MNRAS*, **384**, 444
[10] Mahmud M., Gabuzda D.C. and Bezrukovs V. 2009, *MNRAS*, **400**, 2
[11] Asada K., Nakamura M., Inoue M., Kameno S. and Nagai H. 2010, *Ap J*, **720**, 41
[12] Croke S.M., O'Sullivan S.P. and Gabuzda D.C. 2010, *MNRAS*, **402**, 259

International Workshop on Beamed and Unbeamed Gamma-Rays from Galaxies IOP Publishing
Journal of Physics: Conference Series **355** (2012) 012010 doi:10.1088/1742-6596/355/1/012010

On leptonic models for blazars in the Fermi era

Charles Dermer[1] & **Benoit Lott**[2]

[1]Code 7653, Space Science Division, U.S. Naval Research Laboratory, 4555 Overlook Ave., SW, Washington, DC 20375-5352

[2]Centre d'Études Nucléaires Bordeaux Gradignan, Université de Bordeaux, CNRS/IN2P3, UMR 5797, Gradignan, 33175, France

E-mail: charles.dermer@nrl.navy.mil

Abstract. Some questions raised by *Fermi*-LAT data about blazars are summarized, along with attempts at solutions within the context of leptonic models. These include both spectral and statistical questions, including the origin of the GeV breaks in low-synchrotron peaked blazars, the location of the gamma-ray emission sites, the correlations in the spectral energy distributions with luminosity, and the difficulty of synchrotron/SSC models to fit the spectra of some TeV blazars.

1. Introduction

The 11 June 2008 launch of the Gamma-ray Large Area Space Telescope, subsequently renamed the Fermi Gamma ray Space Telescope, opened a new era into studies of radio galaxies and blazars. The Large Area Telescope (LAT) on Fermi, with a peak on-axis effective of ≈ 8000 cm^2 and a single photon angular resolution for 68% containment at 1 GeV of $\approx 0.8°$ [1], represents an improvement by $1-2$ orders of magnitude in sensitivity over its predecessors EGRET on the Compton Gamma-ray Observatory and AGILE [2]. In the Third EGRET Catalog of Gamma Ray Sources [3], 66 high-confidence ($\gtrsim 5\sigma$) detections of blazars and one probable radio galaxy, Centaurus A were reported. Ten high-latitude sources at high galactic latitude ($|b| > 10°$) were detected with significance $> 10\sigma$. Now, just over three years into the Fermi mission, three important lists of *Fermi*-LAT AGNs have been or are about to be published:

(i) The LAT Bright AGN Sample (LBAS) [4]. Based on the first three months of scientific operations of the Fermi satellite, the LBAS consists of 106 high-latitude sources with $TS \gtrsim 100$, or significance $\gtrsim 10\sigma$. It consists of 58 flat spectrum radio quasars (FSRQs), 42 BL Lac objects, 2 radio galaxies, namely, Centaurus A and NGC 1275 (3C 84), and 4 blazars with unknown classification.

(ii) The First LAT AGN catalog (1LAC) [5]. This catalog contains 671 sources that are associated at high confidence with 709 AGNs (some γ-ray sources have multiple associations). The associations are based on counterpart catalogs, such as the flat-spectrum 8.4 GHz CRATES (Combined Radio All-Sky Targeted Eight GHz Survey) Catalog [6] and the Roma BZCAT blazar catalog [7]. The 1LAC clean sample is a subset of the high-confidence associations, and consists of 599 AGNs with no multiple associations or other analysis flags. These sources subdivide into 275 BL Lac objects, 248 FSRQs, 50 blazars of unknown type (either because a source lacks an optical spectrum, or the optical spectrum is inadequate to determine if it is a BL Lac or FSRQ), and 26 other AGNs. These "other" AGNs include 7 radio galaxies/misaligned AGNs, 4 radio-loud narrow line Seyfert 1 (RL-NLSy1) galaxies, 4 narrow line radio galaxies, 3 starburst galaxies (which of course are not actually AGNs), and 9 other sources.

(iii) The Second LAT AGN catalog (2LAC) [8]. The 2LAC contains 1016 sources that are associated at high confidence with AGNs. The 2LAC clean sample consists of 885 sources (46% larger than the 1LAC clean sample), consisting of 395 BL Lacs, 310 FSRQs, 156 blazars of unknown type, 8 misaligned AGNs, 4 RL-NLSy1 galaxies, 8 AGNs of other types, and 2 starburst galaxies (NGC 4945 has fallen out of the list).

Important questions that are impacted by or arise from the *Fermi*-LAT observations of blazars and radio galaxies, and interpretation that relates to these questions within the context of leptonic models, will occupy the remainder of this short review.

2. Important questions

After summarizing the standard leptonic model for blazars, we consider some important questions raised by the *Fermi*-LAT data, including

(i) Limitations of one-zone synchrotron/SSC model for TeV blazars
(ii) Relation between aligned and misaligned jet sources
(iii) Validity and meaning of the blazar sequence
(iv) GeV spectral cutoffs of LSP and ISP blazars
(v) Super-Eddington luminosities of blazars

2.1. Leptonic Models for blazar SEDs

The general outlines of blazar leptonic models are now well established for the standard one-zone model within the context of blazar unification [9]. In this paradigm, the basic ingredients of a blazar are an accretion disk, regions of broad and narrow emission-line gas, and a dusty IR torus. Radio-loudness is a consequence of collimated relativistic outflows of plasma directed along the polar axis of the accretion disk. When observing at small angles to the plane of the accretion disk, the torus gas conceals the broad-line region formed close to the black hole, so that only narrow emission lines are detected [10]. These are the Type II Seyfert galaxies or, if radio-loud, narrow line radio galaxies. Observations at larger angles to the plane of the accretion disk, where the torus no longer obscures the central region, reveal the broad emission lines. These are the Type I Seyferts or, if radio-loud, the broad-line radio galaxies.

When viewed at very small angles with respect to the jet axis, the Doppler-boosted nonthermal radiation overwhelms the accretion-disk radiation and makes the familiar two-humped broadband blazar continuum [11]. The featureless continuum is made by nonthermal particles accelerated at shocks formed, for example, by colliding shells [12] or through sheared flows in structured jets [13]. In the standard blazar model for flaring activity, a compact plasma zone filled with nonthermal electrons and positrons radiate synchrotron photons that form the highly polarized radio continuum that extends to optical frequencies in flat spectrum radio quasars (FSRQs), and to X-ray frequencies in the high-synchrotron peaked (HSP; with peak synchrotron frequency $\nu_{syn}^{pk} > 10^{15}$ Hz) BL Lac objects. Accompanying the synchrotron emission component is the synchrotron self-Compton (SSC) γ-ray component. This simple one-zone model successfully fits the SEDs of HSP BL Lac objects like Mrk 421 [14] or Mrk 501 [15]. The synchrotron/SSC model for these objects implies Doppler factors $\delta_{\rm D} \sim 10 - 50$, comoving magnetic fields $B' \sim 0.01 - 0.1$ G, and absolute powers $\approx 10^{44}$ erg s^{-1}. The outflow Lorentz factors are compatible with values determined by superluminal radio observations [16], and the inferred jet powers are far below the Eddington luminosity for black holes with mass $\approx 10^9 M_\odot$ that are thought to power these objects.

The simple one-zone synchrotron/SSC model has difficulty to explain the SEDs of FRSQs and intermediate (ISP; 10^{14} Hz $< \nu^{pk}_{syn} < 10^{15}$ Hz) BL Lacs like BL Lac itself [17], which have dominant γ-ray components that would require, for a synchrotron/SSC model, parameters far from equipartition, which would consequently place large demands on jet power. External radiation fields, either from the disk [18], the broad-line region (BLR), or the IR torus [19], are then invoked to provide seed photons that are Compton-scattered to form an additional γ-ray emission component. Within this framework, precise detailed fits are made to sources like 3C 279 [11]. Blazar modeling seemed somewhat satisfactorily explained at the close of the EGRET era with synchrotron/SSC models for X-ray selected BL Lac objects, and with the addition external Compton scattering components for FSRQs and radio-selected BL Lac objects.

2.2. Difficulties of simple blazar models

Even before the launch of *Fermi*, cracks in our understanding of the blazar phenomenon became apparent from observations with TeV telescopes. The July 2006 flaring data of PKS 2155-305 observed with HESS [20] and Swift required $\delta_D \gtrsim 100$ for good synchrotron/SSC model fits [21]. Very high energy (VHE; $\gtrsim 100$ GeV) emission from the FSRQ 3C 279 observed with the MAGIC telescope [22] was found to pose severe difficulties to either synchrotron/SSC or external Compton scattering models [23]. Unexpectedly low extragalactic background light (EBL) intensities at the level of that implied by galaxy counts were required to make sense of the unusually hard deabsorbed TeV spectrum of H 2356-309 and 1ES 1101-232 [24]. Furthermore, doubts about the γ-ray emission site being located within the BLR were raised by radio/γ-ray correlations [25] and multwavelength observations of BL Lacertae, suggesting emission in an inner acceleration zone and at an outer recollimation zone [26].

These difficulties have been compounded with the advent of Fermi and long baseline GeV data correlated with targeted campaigns at radio, optical, X-ray, and VHE energies. Broadband modeling [27] of a large sample of TeV BL Lacs found that most are well fit with a synchrotron/SSC model with $B' \sim 0.01 - 1$ G and $\delta_D \sim 20 - 50$, but with a second group of blazars displaying unusually low magnetic fields and large break electron Lorentz factors. These latter sources include 1ES 1101-232, 1ES 0229+200, and 1ES 0347-121, and are peculiar for BL Lac objects by being very weakly variable or nonvariable, and displaying very weak GeV fluxes. They are now being used to make inferences about the level of the intergalactic magnetic field (IGMF) [28], insofar as the absorbed primary TeV radiation make a large flux of secondary electrons and positrons from attenuated γ rays. These pairs would make GeV radiation by Compton scattering photons of the CMBR and would overproduce the measured Fermi flux unless the pairs are deflected by a sufficiently strong intergalactic magnetic field. The implied IGMF ranges from $\approx 10^{-18}$ G to $\approx 10^{-15}$ G [29, 30, 31] depending on the IGMF correlation length and duration of the blazar engine.

If the IGMF is $\approx 10^{-16} - 10^{-18}$ G, then a separate, quiescent GeV emission component will be present in the SED of HSP BL Lac objects. Indeed, this could account for the slowly varying GeV flux in sources like Mrk 501 [32]. The situation is further complicated if TeV blazars accelerate ultra-high energy cosmic rays (UHECRs) and produce a separate high-energy radiation component due to photopair production by UHECR protons while traveling through intergalactic space [33, 34]. This could affect inferences of the EBL based on the assumption that the highest energy photons are made at the source. This ambiguity can be resolved by observations of multi-TeV radiation by CTA and HAWC from sources like 1ES 0229+200, which can be used to discriminate between a photon-induced and UHECR proton-induced origin of the high-energy radiation and determine whether TeV blazars are sites of UHECR acceleration [35].

For FSRQs such as PKS 1510-089 [36] and 3C279 [37], coherent optical polarization position-angle swings over timescales of many weeks with late-time flares coinciding with strong γ-ray activity seem to be most simply explained if the emission is formed in a helical jet on multi-pc scales. It is not clear, however, how this relates to indications in PKS 1510-089 that the γ-ray flux precedes the optical flux by $\approx$ 13 d [38]. Moreover, the interpretation of distant emission regions may collide with the inference about sharp GeV cutoffs in the SEDs of these sources, as described in the next section.

2.3. GeV spectral cutoffs of LSP and ISP blazars

Already by the time of the LBAS and the first paper on 3C 454.3 [39] it was evident that the γ-ray spectra of FSRQs, which are nearly all low synchrotron-peaked (LSP) sources with νF_ν peak synchrotron frequency $\nu_{pk}^{syn} < 10^{14}$ Hz, as well as ISP objects such as AO 0235+164, display remarkably sharp cutoffs between 1 and 10 GeV. This feature is particularly evident in the compilation of spectra of Fermi bright blazars [40], where a distribution of source-frame break energies E_{br}^* as a function of apparent isotropic γ-ray luminosity L_γ is displayed. Essentially all sources have E_{br}^* within 3σ of 5 GeV. Two other important aspects of this result should be mentioned: (1) The break energy changes only slightly with flux state. In the case of 3C 454.3, E_{br}^* increases by a factor of 2 when the flux varies by more than a factor of 20, as indicated by spectral analysis of the bright GeV blazar 3C 454.3 [41]. (2) The change in spectral index below and above the break is typically larger than 1 unit, ruling out naive radiative cooling models that predict a spectral break of 0.5 units.

This has generated considerable theoretical interest. A hybrid scattering scenario [42] involving both accretion-disk and BLR photons can reproduce the SED of 3C454.3, with the added feature that the spectral break energy is insensitive to location if the BLR has a wind-like profile, because the energy densities of both the accretion-disk and BLR radiation fields then scale with the same dependence on radius. Poutanen and Stern argue in print [43, 44] and at this conference that He II recombination line and continuum radiation attenuate a continuum GeV emission component through $\gamma\gamma$ pair production. For He II recombination radiation at energy $\approx 4 \times 13.6 = 54.4$ eV or $\epsilon_s \cong 54.4/511000 = 1.06 \times 10^{-4}$ (in m_ec^2 units), photons with energy $m_ec^2\epsilon_1 > 2/\epsilon \cong 9.6$ GeV, close to E_{br}^*, can be strongly absorbed. The γ-ray emitting region must be found deep within the BLR for this model to work, and the excessive target photon energy density may introduce difficulties to fit the multi-band spectral data of 3C 454.3.

A further possibility is that the breaks are due to Klein-Nishina effects on Compton-scattered Ly α photons [41]. GALEX observations of 3C 454.3, as reported by Bonnoli and coworkers [45], reveal the presence of a strong Ly α radiation field in the BLR of 3C 454.3, amounting to a line luminosity of $(2-4)\times10^{45}$ erg s^{-1}. The onset of the Klein-Nishina decline takes place when $4\gamma\epsilon \cong 1$ for electrons with Lorentz factor γ scattering Ly α target photons with dimensionless energy $\epsilon = 10.2/511000 = 2 \times 10^{-5}$. Photons are Compton-scattered to energies $\epsilon_s = (3/2)\gamma^2\epsilon$, implying $\epsilon\epsilon_s = 3/32$, i.e., $m_ec^2\epsilon_s = 2.4$ GeV, consistent with the measured break energy. Spectral calculations employing a power-law electron spectrum produce a break at the right energy, but with an insufficiently sharp break [41]. Combined Compton and synchrotron losses will harden the electron spectrum and reduce the magnitude of the break [46, 47], but a curved electron spectrum described by a log parabola function [48], formed when acceleration dominates over cooling, might resolve this difficulty.

2.4. Blazar sequence

The search for an organizing principle in blazar studies has centered on the so-called blazar sequence, which is the well-known anti-correlation of the νF_ν synchrotron and Compton peak frequencies with luminosity [50, 51]. It is reflected in the Fermi data by plotting the blazar spectral index Γ_γ versus the γ-ray luminosity. This figure reveals a clear separation between

Journal of Physics: Conference Series **355** (2012) 012010 doi:10.1088/1742-6596/355/1/012010

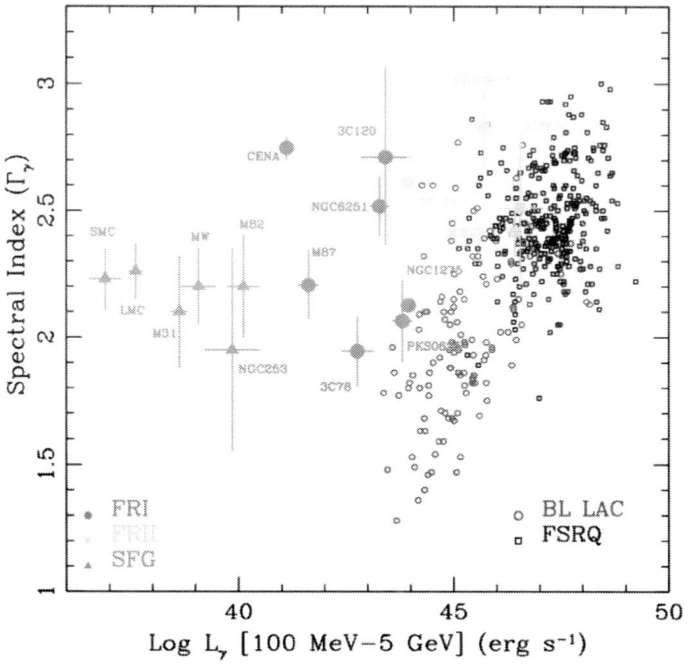

Figure 1. Gamma-ray spectral slopes Γ_γ of BL Lac objects (open blue circles), FSRQs (open black squares), FR1 radio galaxies (red circles), FR2 radio sources (green squares), and star-forming galaxies (magenta diamonds), are plotted as a function of their 100 MeV - 5 GeV γ-ray luminosity [49]. The γ-ray emitting misaligned AGNs are 1LAC sources associated with objects in the Third Cambridge Revised (3CRR) and Molonglo radio catalogs. The γ-ray emission from star-forming galaxies are powered by cosmic-ray processes rather than AGNs.

BL Lac objects and FSRQs—a blazar divide [52]—that is attributed to a difference in accretion regime. As plotted in Figure 1, one sees that all FSRQs have $\Gamma_\gamma > 2$. Inasmuch as the presence of strong emission lines means that there has to be a significant surrounding scattered radiation field, then a consistent interpretation of FSRQ SEDs is to attribute the γ-ray emission to external Compton processes [51], with the trend in the SEDs from FSRQ to BL Lac objects resulting from the reduction in the intensity of the surrounding radiation field [53]. This is reasonable provided that the emission region of FSRQs is found within the BLR. Indeed, this seems to be the most natural explanation for rapid variability [54]. Yet this expectation collides with detection of VHE γ rays from FSRQs like 3C 279 [22] and, most recently and most spectacularly, 4C +21.35, which exhibits a 10 minute flaring episode of 70 – 400 GeV radiation [55] from a source with very strong IR radiation [56] that would strongly attenuate high-energy radiation made within the pc scale. Two-zone models have been advocated to deal with these questions [23, 57].

The blazar sequence and blazar divide are not, however, immune to criticisms, most notably that a large fraction of BL Lac objects lack redshifts. This fraction amounts to more than 50% in the 2LAC. Because strongly beamed nonthermal radiation would conceal emission lines, then the highest redshift, highest luminosity BL Lac objects would be selectively excluded from the blazar sequence diagram, and these would also likely be the HSP BL Lacs where the nonthermal radiation is strongest in the optical/UV range. Besides outliers to the blazar sequence, there is evidence for a population of HSP FSRQs [58]. A further concern is that the BL Lac objects are primarily identified from and biased towards the radio and X-ray surveys from which they are selected [59].

In any case, the simplest characterization of the blazar sequence must consider the contribution of slightly off-axis jetted sources and misaligned radio galaxies. When observing away from the beaming axis, the apparent luminosities from the smaller Doppler boosting should appear at lower peak frequencies, opposite to the blazar sequence correlation. Consequently radio galaxies like NGC 6251 [60] will fill in a distinct portion of a diagram of the blazar sequence, turning it rather into an envelope [61].

Journal of Physics: Conference Series **355** (2012) 012010 doi:10.1088/1742-6596/355/1/012010

2.5. Relation between aligned and misaligned jet sources

A curiosity revealed by Fig. 1 has to do with the spread in luminosities of the radio galaxies compared with the aligned blazar jet sources. The misaligned AGN (MAGN) population detected with Fermi consists of 11 sources [49]. Theses includes 7 FRI and 2 FRII radio galaxies, which are thought to be the parent population of BL Lac objects and FSRQs, respectively, in addition to 2 steep spectrum radio quasars that are believed to be slightly misaligned FSRQs. The MAGN sources in the 3CRR catalog have large core dominance parameters compared to the general 3CRR source population, implying that the beamed component makes an appreciable contribution to the γ-ray flux. This is furthermore supported by the fact that three of these sources—3C 78, 3C 111, and 3C 120—do not appear in 2LAC, evidently due to variability.

Why the ratio of measured γ-ray luminosities of FRI galaxies and BL Lac objects span a much larger range than the comparable ratio for FRII galaxies and FSRQs is an open question, if in fact it is not due to statistics. The simplest possibility is that there are a lot more nearby FRIs, and limitations on Fermi sensitivity will therefore favor detection of these nearby sources. At $z \ll 1$, the luminosity distance $d_L \approx 4200z(1+z)$ Mpc. The *Fermi*-LAT reaches a limiting energy-flux sensitivity of $\approx 5 \times 10^{-12}$ erg cm^{-2} s^{-1} for two years of observations that is, unlike integral photon flux, only weakly dependent on source spectral index [8]. Fermi can thus only detect sources with luminosity $L_\gamma \gtrsim 10^{46} z^2(1+2z)$ erg s^{-1}. The number of nearby FRI and FRII galaxies depends on the space density of these objects which, in the study by Gendre, Best, and Wall [62], is inferred from 1.4 GHz NVSS-FIRST radio observations (recall that the Fanaroff-Riley dichotomy is based on 178 MHz luminosities $P_{178\ \rm MHz}$). From their Figure 12, FRI and FRII radio galaxies have a local ($z < 0.3$) space density of $\approx 35 \times 10^{-6}$ Mpc^{-3} and $\approx 2 \times 10^{-6}$ Mpc^{-3}, respectively, at $P_{1.4\ \rm GHz} \gtrsim 10^{22}$ W Hz^{-1} sr^{-1}, implying that the local space density of FRIs exceeds that of FRIIs by ≈ 20.

This implies that there are several hundred FRIs and FRIIs within $z \cong 0.1$, yet other than 3C 111, all LAT-detected radio galaxies within this volume are FRI sources (see Figure 3 of [63]). This is further remarkable in that FSRQs and their putative parent population of FRII radio galaxies are far more radio luminous than BL Lac objects and FRI galaxies. With only a few hundred randomly aligned sources within $z = 0.1$, it might be the case that a narrower γ-ray beaming cone in FSRQs, with a more rapid fall-off in off-axis flux, makes detection of these nearby sources far less likely than a broader γ-ray emission cone in BL Lac objects, as expected for the different beaming factors from SSC emission and external Compton processes [64, 65]. This could also reflect differences in jet structure between FSRQs and BL Lac objects [66], or extended jet or lobe emission in FRIs that is missing in FRII galaxies.

2.6. Apparent super-Eddington luminosities of blazars

We end this contribution on a speculative note. The time-averaged luminosities of FSRQs extend to values in excess of $L_\gamma \approx 10^{49}$ erg s^{-1} (Figure 1). In the extraordinary 2010 November flare of 3C 454.3 [67], L_γ reached apparent isotropic luminosity of $(2.1 \pm 0.2) \times 10^{50}$ erg s^{-1} over a period of a few hours, making it the most luminous blazar yet observed. Black-hole mass estimates for 3C 454.3 are in the range $0.5 \lesssim M_9 \lesssim 4$ [45], where $10^9 M_9 M_\odot$ is the mass of the black hole powering this AGN. For this range of masses, the Eddington luminosity therefore ranges from $\approx 6 \times 10^{46}$ erg s^{-1} to $\approx 5 \times 10^{47}$ erg s^{-1}. During this extreme outburst, the apparent luminosity of 3C 454.3 was more than a factor of ≈ 400 greater than its Eddington luminosity. Even its time-averaged luminosity of $L_\gamma \approx 10^{49}$ erg s^{-1} is super-Eddington by a factor of ≈ 20.

Assuming that the Eddington condition *does* limit accretion flow onto the black hole, which is likely to be the case for the long-term average luminosity if not for the flaring luminosity, then the absolute radiant luminosity is limited to a value of $L_{abs} \lesssim 5 \times 10^{47}$ erg s^{-1}. This is consistent with the large apparent luminosities if the emission is highly beamed. For a simple top-hat jet beaming factor, a jet opening angle θ_j implies a beaming factor $f_b = \theta_j^2/2$ for a two-sided jet with $\theta_j \ll 1$. A mechanism for collimation is, however, required. Should this arise from the Blandford-Znajek process, then we are still restricted to values of the absolute Blandford-Znajek power

$$\frac{d\mathcal{E}}{dt}|_{\rm BZ} \approx \pi c(\frac{a}{M})^2 r_+^2 B_+^{r2} \; , \tag{1}$$

where a/M is the angular momentum per unit mass, r_+ is the radius of the event horizon, and B_+^r is the radial component of the magnetic field threading the event horizon [68]. Scaling the energy density of the magnetic field to the energy density of accreted matter near the event horizon shows that the Blandford-Znajek power is likewise Eddington-limited. Making the hypothesis that the extraction of energy through black-hole rotation collimates the jet outflow with $\cos\theta_j \approx a/M$, then $f_b \cong 1-(a/M)$ and $\theta_j \cong \sqrt{2(1-a/M)}$, implying $a/M > 1-(L_{\rm Edd}/L_\gamma)$. If the jet opening angle is a consequence of the bulk Lorentz factor Γ of the outflow, then $\Gamma \gtrsim \sqrt{L_\gamma/2L_{\rm Edd}}$.

For the case of 3C 454.3 in its flaring state, when $L_\gamma/L_{\rm Edd} \gtrsim 10^3$, this hypothesis then implies that $a/M > 0.999$ and $\Gamma \gtrsim 23$, which is consistent with the value $\Gamma_{min} \approx 14$ from $\gamma\gamma$ opacity arguments [67]. In the most conservative case with $M_9 = 4$, $a/M \cong 0.998$ and $\Gamma \gtrsim 15$. This is marginally consistent with the limiting maximum value $a/M \cong 0.996$ suggested by Aschenbach from analyses of microquasars and the Galactic Center black hole [69]. The smaller black-hole mass estimate, $M_9 = 0.5$ (Bonnoli et al. 2010), implies a value of a/M that violates this limit by a large margin. Much work, both numerical and theoretical, has been devoted to jet formation from the Blandford-Znajek process, and it is unclear if jet collimation can be described by the guess that $\cos\theta_j \approx a/M$, but it is interesting to suggest this possibility, which leads to values of Γ consistent with separate inferences regarding the outflow Lorentz factor.

Acknowledgments

This work is supported by the Office of Naval Research and the Fermi Guest Investigator program. We thank Paola Grandi for preparing Fig. 1 and stressing the unusual properties of LAT-detected radio galaxies, and Maxim Barkov, Amir Levinson, Govind Menon, Juri Poutanen, Anita Reimer, and Boris Stern for discussions. CD would also like to thank the organizers, especially Robert Wagner, for the opportunity to visit Finland.

References

[1] Atwood W B *et al.* 2009 *Astrophys. J.* **697** 1071–1102 (*Preprint* 0902.1089)
[2] Tavani M *et al.* 2009 *Astron. & Astrophys.* **502** 995–1013 (*Preprint* 0807.4254)
[3] Hartman R C *et al.* 1999 *Astrophys. J. S.* **123** 79–202
[4] Abdo A A *et al.* 2009 *Astrophys. J.* **700** 597–622 (*Preprint* 0902.1559)
[5] Abdo A A *et al.* 2010 *Astrophys. J.* **715** 429–457 (*Preprint* 1002.0150)
[6] Healey S E, Romani R W, Taylor G B, Sadler E M, Ricci R, Murphy T, Ulvestad J S and Winn J N 2007 *Astrophys. J. S.* **171** 61–71 (*Preprint* arXiv:astro-ph/0702346)
[7] Massaro E, Giommi P, Leto C, Marchegiani P, Maselli A, Perri M, Piranomonte S and Sclavi S 2009 *Astron. & Astrophys.* **495** 691–696 (*Preprint* 0810.2206)
[8] Abdo A A *et al.* 2011 *Astrophys. J.*
[9] Urry C M and Padovani P 1995 *Publ. Astron. Soc. Pacific* **107** 803–+ (*Preprint* arXiv:astro-ph/9506063)
[10] Antonucci R 1993 *Ann. Rev. Astron. Astrophys.* **31** 473–521
[11] Böttcher M 2007 *Astrophys. & Space Sci.* **309** 95–104 (*Preprint* arXiv:astro-ph/0608713)
[12] Böttcher M and Dermer C D 2010 *Astrophys. J.* **711** 445–460 (*Preprint* 1001.1606)
[13] Aloy M A, Gómez J L, Ibáñez J M, Martí J M and Müller E 2000 *Astrophys. J. Lett.* **528** L85–L88 (*Preprint* arXiv:astro-ph/9911153)
[14] Abdo A A *et al.* 2011 *Astrophys. J.* **736** 131–+ (*Preprint* 1106.1348)
[15] Abdo A A *et al.* 2011 *Astrophys. J.* **727** 129–+ (*Preprint* 1011.5260)
[16] Jorstad S G, Marscher A P, Mattox J R, Wehrle A E, Bloom S D and Yurchenko A V 2001 *Astrophys. J. S.* **134** 181–240 (*Preprint* arXiv:astro-ph/0101570)

[17] Abdo A A *et al.* 2011 *Astrophys. J.* **730** 101–+ (*Preprint* 1101.5905)
[18] Dermer C D and Schlickeiser R 2002 *Astrophys. J.* **575** 667–686 (*Preprint* arXiv:astro-ph/0202280)
[19] Sikora M, Stawarz L, Moderski R, Nalewajko K and Madejski G M 2009 *Astrophys. J.* **704** 38–50 (*Preprint* 0904.1414)
[20] Aharonian F *et al.* 2007 *Astrophys. J. Lett.* **664** L71–L74 (*Preprint* 0706.0797)
[21] Finke J D, Dermer C D and Böttcher M 2008 *Astrophys. J.* **686** 181–194 (*Preprint* 0802.1529)
[22] MAGIC Collaboration, Albert J *et al.* 2008 *Science* **320** 1752– (*Preprint* 0807.2822)
[23] Böttcher M, Reimer A and Marscher A P 2009 *Astrophys. J.* **703** 1168–1175
[24] Aharonian F *et al.* 2006 *Nat.* **440** 1018–1021 (*Preprint* arXiv:astro-ph/0508073)
[25] Lähteenmäki A and Valtaoja E 2003 *Astrophys. J.* **590** 95–108
[26] Marscher A P *et al.* 2008 *Nat.* **452** 966–969
[27] Tavecchio F, Ghisellini G, Ghirlanda G, Foschini L and Maraschi L 2010 *Monthly Not. Roy. Astron. Soc.* **401** 1570–1586 (*Preprint* 0909.0651)
[28] Neronov A and Vovk I 2010 *Science* **328** 73– (*Preprint* 1006.3504)
[29] Dolag K, Kachelriess M, Ostapchenko S and Tomàs R 2011 *Astrophys. J. Lett.* **727** L4+ (*Preprint* 1009.1782)
[30] Dermer C D, Cavadini M, Razzaque S, Finke J D, Chiang J and Lott B 2011 *Astrophys. J. Lett.* **733** L21+ (*Preprint* 1011.6660)
[31] Tavecchio F, Ghisellini G, Bonnoli G and Foschini L 2011 *Monthly Not. Roy. Astron. Soc.* **414** 3566–3576 (*Preprint* 1009.1048)
[32] Neronov A, Semikoz D V and Taylor A M 2011 *ArXiv e-prints* (*Preprint* 1104.2801)
[33] Essey W and Kusenko A 2010 *Astroparticle Physics* **33** 81–85 (*Preprint* 0905.1162)
[34] Essey W, Kalashev O E, Kusenko A and Beacom J F 2010 *Physical Review Letters* **104** 141102–+ (*Preprint* 0912.3976)
[35] Murase K, Dermer C D, Takami H and Migliori G 2011 *ArXiv e-prints* (*Preprint* 1107.5576)
[36] Marscher A P *et al.* 2010 *Astrophys. J. Lett.* **710** L126–L131 (*Preprint* 1001.2574)
[37] Abdo A A *et al.* 2010 *Nat.* **463** 919–923
[38] Abdo A A *et al.* 2010 *Astrophys. J.* **721** 1425–1447 (*Preprint* 1007.1237)
[39] Abdo A A *et al.* 2009 *Astrophys. J.* **699** 817–823 (*Preprint* 0904.4280)
[40] Abdo A A *et al.* 2010 *Astrophys. J.* **710** 1271–1285
[41] Ackermann M *et al.* 2010 *Astrophys. J.* **721** 1383–1396
[42] Finke J D and Dermer C D 2010 *Astrophys. J. Lett.* **714** L303–L307
[43] Poutanen J and Stern B 2010 *Astrophys. J. Lett.* **717** L118–L121 (*Preprint* 1005.3792)
[44] Stern B E and Poutanen J 2011 *ArXiv e-prints* (*Preprint* 1105.2762)
[45] Bonnoli G, Ghisellini G, Foschini L, Tavecchio F and Ghirlanda G 2011 *Monthly Not. Roy. Astron. Soc.* **410** 368–380 (*Preprint* 1003.3476)
[46] Dermer C D and Atoyan A M 2002 *Astrophys. J. Lett.* **568** L81–L84 (*Preprint* arXiv:astro-ph/0202406)
[47] Moderski R, Sikora M and Błażejowski M 2003 *Astron. & Astrophys.* **406** 855–865 (*Preprint* arXiv:astro-ph/0205228)
[48] Massaro E, Perri M, Giommi P and Nesci R 2004 *Astron. & Astrophys.* **413** 489–503 (*Preprint* arXiv:astro-ph/0312260)
[49] Abdo A A *et al.* 2010 *Astrophys. J.* **720** 912–922
[50] Fossati G, Maraschi L, Celotti A, Comastri A and Ghisellini G 1998 *Monthly Not. Roy. Astron. Soc.* **299** 433–448 (*Preprint* arXiv:astro-ph/9804103)
[51] Ghisellini G, Celotti A, Fossati G, Maraschi L and Comastri A 1998 *Monthly Not. Roy. Astron. Soc.* **301** 451–468 (*Preprint* arXiv:astro-ph/9807317)
[52] Ghisellini G, Maraschi L and Tavecchio F 2009 *Monthly Not. Roy. Astron. Soc.* **396** L105–L109 (*Preprint* 0903.2043)
[53] Böttcher M and Dermer C D 2002 *Astrophys. J.* **564** 86–91 (*Preprint* arXiv:astro-ph/0106395)
[54] Foschini L, Ghisellini G, Tavecchio F, Bonnoli G and Stamerra A 2011 *Astron. & Astrophys.* **530** A77+ (*Preprint* 1101.1085)
[55] Aleksić J *et al.* 2011 *Astrophys. J. Lett.* **730** L8+ (*Preprint* 1101.4645)
[56] Malmrose M P, Marscher A P, Jorstad S G, Nikutta R and Elitzur M 2011 *Astrophys. J.* **732** 116–+ (*Preprint* 1103.1682)
[57] Tavecchio F, Becerra-Gonzales J, Ghisellini G, Stamerra A, Bonnoli G, Foschini L and Maraschi L 2011 *ArXiv e-prints* (*Preprint* 1104.0048)
[58] Padovani P, Perlman E S, Landt H, Giommi P and Perri M 2003 *Astrophys. J.* **588** 128–142 (*Preprint* arXiv:astro-ph/0301227)
[59] Padovani P 2007 *Astrophys. & Space Sci.* **309** 63–71 (*Preprint* arXiv:astro-ph/0610545)
[60] Migliori G, Grandi P, Torresi E, Dermer C, Finke J, Celotti A, Mukherjee R, Errando M, Gargano F,

Giordano F and Giroletti M 2011 *ArXiv e-prints* (*Preprint* 1107.4302)
[61] Meyer E T, Fossati G, Georganopoulos M and Lister M L 2011 *ArXiv e-prints* (*Preprint* 1107.5105)
[62] Gendre M A, Best P N and Wall J V 2010 *Monthly Not. Roy. Astron. Soc.* **404** 1719–1732
[63] Dermer C D and Razzaque S 2010 *Astrophys. J.* **724** 1366–1372 (*Preprint* 1004.4249)
[64] Dermer C D 1995 *Astrophys. J. Lett.* **446** L63+
[65] Georganopoulos M, Kirk J G and Mastichiadis A 2001 *Astrophys. J.* **561** 111–117 (*Preprint* arXiv:astro-ph/0107152)
[66] Chiaberge M, Celotti A, Capetti A and Ghisellini G 2000 *Astron. & Astrophys.* **358** 104–112 (*Preprint* arXiv:astro-ph/0003197)
[67] Abdo A A *et al.* 2011 *Astrophys. J. Lett.* **733** L26+
[68] Levinson A 2006 *Energy Extraction from a Kerr Black Hole - An Ultimate Power Source in the Universe?* (Nova Science Publishers) pp 119–+
[69] Aschenbach B 2004 *Astron. & Astrophys.* **425** 1075–1082 (*Preprint* arXiv:astro-ph/0406545)

On the Physics of Hadronic Blazar Emission Models

A. Reimer

Leopold-Franzens-Universität Innsbruck, Institut für Theoretische Physik, und Institut für Astro- und Teilchenphysik, 6020 Innsbruck, Austria

E-mail: `anita.reimer@uibk.ac.at`

Abstract. The material of relativistic jets of active galactic nuclei is composed of leptons and hadrons. If a significant fraction of the overall jet power is converted into energizing protons to relativistic energies, along with the leptons, hadronic interactions play a role in converting particle energy into radiation and neutrino production. I will discuss the physics and some properties of hadronic blazar emission models in the framework of recent developments in this field.

1. Introduction

Active Galactic Nuclei (AGN) are among the most luminous objects in the universe. Only a small percentage ($\sim 10\%$) of them turn out to possess prominent jets which continuously transport material from the innermost region of the AGN to kpc-, sometimes even Mpc-scale distances with relativistic speed. Following the unification scheme of such radio-loud AGN, flat spectrum radio-quasars (FSRQs) and BL Lac objects (commonly refered to as "blazars") are those which are observed under a small viewing angle with respect to the jet axis, whereas those observed at large viewing angles are classified as Fanaroff Riley I and II radio galaxies, hence commonly considered as the blazars' parent populations (e.g., [41]). Such jets of AGN are oberved as broadband photon sources, being detected from the radio band up to very-high energy (VHE) gamma-ray energies. The census of GeV-detected blazars has dramatically increased from less than 100 blazars as detected by the EGRET instrument onboard the *Compton Gamma-Ray Observatory* to nearly 10^3 LAT (=Large Area Telescope)-detected high-latitude objects that have been associated with AGN since the launch of the *Fermi* satellite about two years ago [4]. The (nearly permanent) survey observation mode of the LAT-instrument onboard *Fermi* has triggered many follow-up observations of selected AGN also with the permanently improving instrument capabilities of modern Cherenkov-telescopes (ACTs). As a result, blazars of all subtypes (FSRQs, all types of BL Lac objects: low-frequency peaked BL Lacs (LBLs), intermediate-frequency peaked BL Lacs (IBLs), high-frequency-peaked BL Lacs (HBLs)) have meanwhile been detected at VHEs, thereby nearly doubling the census of blazars detected in this extreme photon energy band during the last two years. Most of the photon spectral energy distribution (SED) of blazars is dominated by radiation from non-thermal particle distributions in the jet. This is because the jet radiation is strongly beamed in these objects, thereby overwhelming any possible thermal signatures from the accretion disk, dusty torus or host galaxy.

Among the most striking properties of blazars is their variability at all wavelengths with variability time scales from weeks to months in the radio domain, down to even minutes in a few cases at VHEs (e.g., PKS 2155-304 [7], Mkn 501 [9]. Such extreme behaviour occurs in general during an enhanced flux and variability stage. The greatly improved sensitivity of gamma-ray instruments, however, allows meanwhile to detect a number of blazars at various variability stages including a quiescent non-variable state (see e.g., [2] for a study of gamma-ray variability properties of blazars).

The SED of jetted AGN consists of two broad components. The low-energy component is in general attributed to synchrotron radiation from relativistic electrons (e^-) and positrons (e^+) in a relativistically moving "blob" in the jet. The origin of the high-energy hump is still a matter of debate, depending strongly on the overall jet composition (see below). Recently, a new classification scheme for a spectral classification of blazars has been proposed based on the synchrotron peak of their broadband SED [3], and independent of being a BL Lac object or a quasar. Here, low-synchrotron-peaked blazars (LSPs) have their low-energy peak at $\nu_{s,peak} < 10^{14}$Hz, intermediate-synchrotron-peaked blazars (ISPs) at $10^{14} \leq \nu_{s,peak} \leq 10^{15}$Hz, and high-synchrotron-peaked (HSPs) at $\nu_{s,peak} > 10^{15}$Hz.

Despite the rapid increase of photon data in recent years many questions on the jet physics of AGN are still unsolved, among them the composition of their relativistic particle populations and their properties.

Depending on this jet matter composition two types of emission models have emerged during the last decade. So-called "leptonic" models consider relativistic e^-e^+ as the emitting relativistic particle population, while in so-called "hadronic" emission models the relativistic jet material is composed of relativistic protons (p) and e^-. In both types of jet material cold pairs and/or protons may exist as well, allowing charge neutrality to be fulfilled. So-called "lepto-hadronic emission models" follow the same physics as "hadronic emission models". In the following I will therefore follow the historical path and use the term "hadronic model" for describing radiation from relativistic e^-p jet matter compositions.

The historical motivation to consider relativistic protons in AGN jets is linked with the observation of ultra-high energy cosmic rays (UHECRs) up to several tens of EeV impinching the Earth from all directions. Indeed, AGN jets are among the prime candidate sources for these extremely energetic events. They fulfill all required criteria (e.g., [17]): Luminous AGN jets are located within the GZK horizon ($\sim$ 50 Mpc), provide sufficient power to potentially accelerate the charged jet particles beyond EeV scales and contribute significantly to the overall observed cosmic ray energetics, and fulfill the so-called "Hillas criterium" [21]. Yet, deflections of the charged particles in the intergalactic magnetic field complicate a direct signature of the sources in the observed CR distributions.

Observational indications of the existence of protons in AGN jets are meanwhile accummulating from various directions: If the observed X-ray cavities emerging from radio AGN jets in galaxy clusters (e.g., Cyg A; [42, 20]) are in pressure equilibrium with the intercluster medium, a significant matter component in those bubbles beyond of what pairs can provide is required. The non-detection of the predicted soft X-ray excess radiation and X-ray precursors to blazar flares from bulk Comptonization of cold pairs in quasars with strong accretion disk radiation puts serious limits to the jet's overall pair content (e.g., [39]). The modeling of blazar SEDs with leptonic models reveals a total luminosity in pairs and magnetic field that is significantly less than the observed radiative power L_{rad} in the comoving frame, L_{rad}/Γ^2 (e.g., [16]), with Γ the bulk Lorentz factor. Energy conservation thus requires the existence of nucleons and/or nuclei in such blazar jets, a robust lower limit of the total jet power is given by $L_{jet} > L_{rad}/\Gamma^2$. The role of pairs as jet energy carrier is therefore negligible, making hadrons the energetically and dynamically dominating matter component in blazar jets. What remains unclear so far are the properties of this hadron component: Is this hadron population (partly) relativistic, or entirely cold? If the former, what is the fractional relativistic hadron to electron ratio, and what are the maximum energies reached? Where are those relativistic nucleons produced/accelerated and by which mechanism(s)? Where are the observational signatures of these relativistic hadrons in AGN jet spectra? The application of hadronic emission models to blazar SEDs is thus required to answer those questions at least partly.

2. Photon and particle production in hadronic blazar emission models

In hadronic jet emission models, both a relativistic electron and proton population exist in the emitting magnetized[1] "blob" which itself moves with relativistic speed βc along the jet axis. As a consequence, leptonic and hadronic[2] processes must be fully considered: synchrotron radiation and inverse Compton scattering of charged particles, photon-photon and Bethe-Heitler pair production, photomeson production, inelastic nucleon-nucleon interactions, neutron decay. In this sense, leptonic emission models constitute a part of hadronic emission models.

Hadronic models where nucleon-nucleon interactions play an integral part have been proposed occasionally over the last decades (e.g., [12, 31, 11]). In those senarios (star-jet interaction, relativistic blast wave, jet-red giant interaction, etc.) extremely dense target media for proton-proton interactions are considered. In the following I shall, however, focus on the more common hadronic emission models where particle-photon interactions dominate over particle-particle interactions (e.g., [23, 32, 13, 27, 29, 30, 10]).

Figure 1 shows the relevant energy loss rates of protons which is compared to the energy loss rates for electrons in Figure 2. Overall, the photon production rate from electron interactions is in general larger than from proton interactions, electrons loose their energy more rapidly in dense radiative environments than protons. This makes electrons efficient radiators, nucleons - irrespective of whether they are cold or relativistic - are excellent energy carriers. Both properties are required in AGN jets.

Because the interaction threshold for Bethe-Heitler pair production is located at lower energies (with center-of-momentum frame (CMF) energy $\sqrt{s} = (m_p + 2m_e)c^2)$) than for photomeson production (with CMF energy $\sqrt{s} = (m_p + m_\pi)c^2)$), energy loss rates below the pion photoproduction threshold are dominated by pair production. Above this latter threshold energy losses from photomeson production typically dominate over those from pair production due to their several orders of magnitude greater inelasticity, ranging between ~ 0.2 in steep power-law target photon fields to nearly ~ 0.5 in very flat target photon fields (see Figure 3; [37]). In contrast to pair production on protons, photomeson production allows proton to neutron conversion (and vice versa) via charge exchange (e.g. [28]) [3]. As a consequence collimated neutron beams may form [19, 10] which can be used to transport a significant portion of the initial energy to large distances from the black hole.

There are various target photon fields for particle-photon interactions in AGN jets: the internal jet synchrotron photon field [23, 29, 30] , and fields external to the jet such as direct accretion disk radiation [32, 13], jet or accretion disk radiation reprocessed in the BLR [10], or radiation by the dusty torus. The secondary (stable) particles from those are γ-rays, pairs and neutrinos (ν). For power-law target photon fields the resulting γ-ray to ν energy content E_γ/E_ν[4] lies around unity ($\sim 1 - 1.4$) for a wide spectral index range, with 10% (8%) for

[1] In the following I consider the magnetic field uniformly and isotropically distributed in the comoving "blob" frame for simplicity. Other scenarios may consider stochastic magnetic fields (e.g., [24] in the case of GRBs) giving rise to jitter radiation in the relativistic regime.

[2] The following restricts to proton interactions only. Heavy nuclei are generally more likely to photo-disintegrate into protons and neutrons in dense radiation fields rather than suffer photomeson production losses. Photo-disintegration has, however, recently been shown to fail providing an efficient process for VHE photon production [8], leaving protons as a more efficient photon producer.

[3] e.g., in hadronic pγ-interactions 30%-70% of the initial protons are converted into neutrons [28]

[4] $e^\pm$ are assumed to be 100% radiatively efficient.

Journal of Physics: Conference Series **355** (2012) 012011 doi:10.1088/1742-6596/355/1/012011

steep power-laws to 25% (20%) in case of flat power-laws of the initial proton energy stored into photons (neutrinos) (see Figure 4; [37]). The average photon and neutrino energy per injected nucleon energy is dependent on the CMF energy where the dominant part of photomeson production takes place. In steep target photon spectra this is rather around the Δ-resonance region while for hard target photon fields the multipion production region plays a significant role [25]. Consequently the average photon (neutrino) energy ranges from 6% (4%) the injected nucleon energy in steep target photon fields to a fraction of % in flat target photon spectra (see Figure 5; [37]).

Figure 1 compares also the energy loss time scales from synchrotron radiation of muons in typically proposed blazar magnetic field strengths[5] with their decay time scales. Obviously, for sufficient energetic muons, such as e.g., produced in photomeson production initiated by UHE protons, muon synchrotron losses prior to their decay can not be neglected (e.g., [34]). This applies similarly also to charged pions and kaons (e.g., [30]). Indeed, if field strengths are sufficiently high ($\geq$ several tens of Gauss) p-[27, 5] $\mu^{\pm}$- and $\pi^{\pm}$-synchrotron radiation may constitute an important high energy photon component.

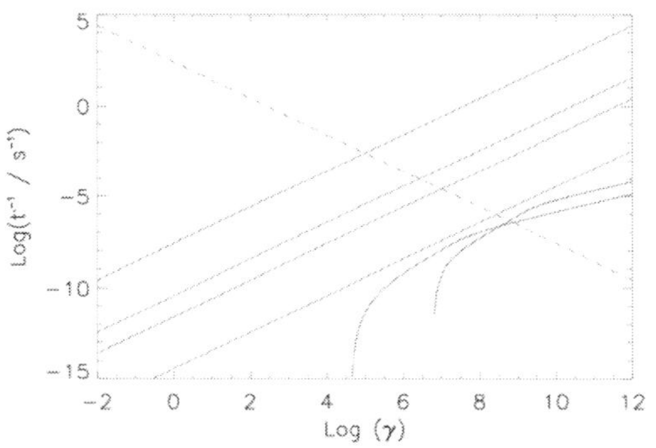

Figure 1. Energy loss rates (in the jet frame) for protons due to synchrotron radiation (red) in a 1 G (lower) and 100 G (upper curve) magnetic field, photomeson production (blue) and Bethe-Heitler pair production (green) in a broken power-law target photon field with photon index 1.5 below the break energy of 0.1 eV (jet frame) and 2.25 above this energy and with a total energy density $u_\gamma = 10^{11}\mathrm{eV/cm^3}$. The magenta solid line represents synchrotron losses of charged muons in a 1 G (lower) and 100 G (upper curve) magnetic field in comparison to their decay time scale (magenta dashed line).

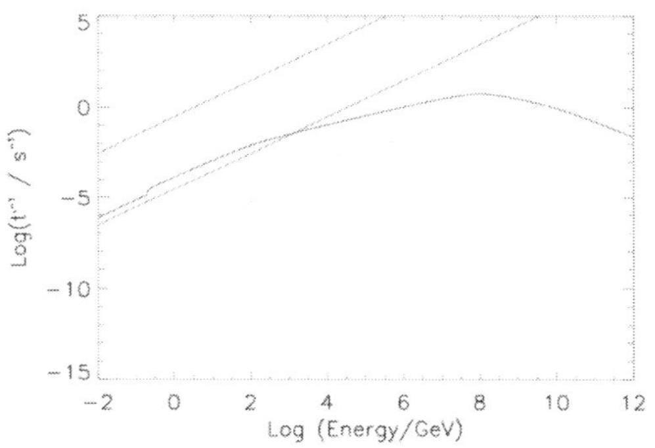

Figure 2. Energy loss rates (in the jet frame) for electrons due to synchrotron radiation (red) in a 1 G (lower) and 100 G (upper) magnetic field and inverse Compton scattering (dark blue) in comparison to photon-photon pair production (light blue) in a broken power-law target photon field with photon index 1.5 below the break energy of 0.1 eV (jet frame) and 2.25 above this energy and with a total energy density $u_\gamma = 10^{11}\mathrm{eV/cm^3}$.

Very high energy photons (produced, e.g., through decay of photomeson produced πs) may find the emission region optically thick (see Figure 2): a pair cascade then develops which

[5] except for one-zone SSC models where broadband SED modeling in general requires significantly lower field strengths

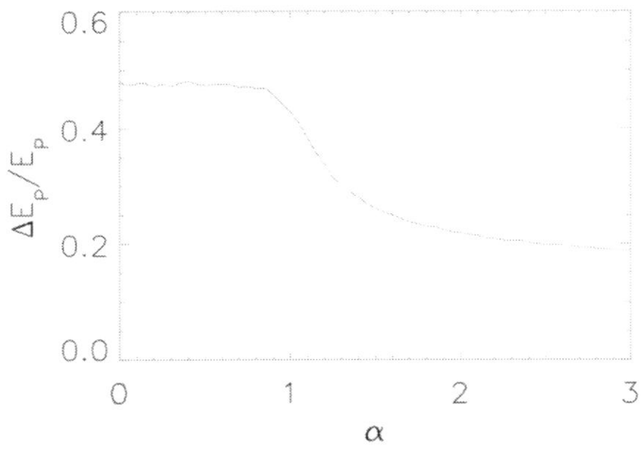

Figure 3. Inelasticity for photomeson production in power-law target photon fields with photon index α [37].

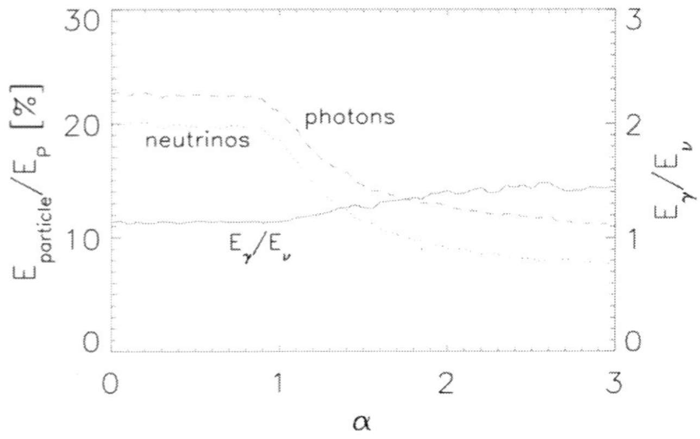

Figure 4. Photon and neutrino energy content per primary nucleon energy in photomeson production in power-law target photon fields with photon index α [37]. $e^{\pm}$ are assumed to be 100% radiatively efficient.

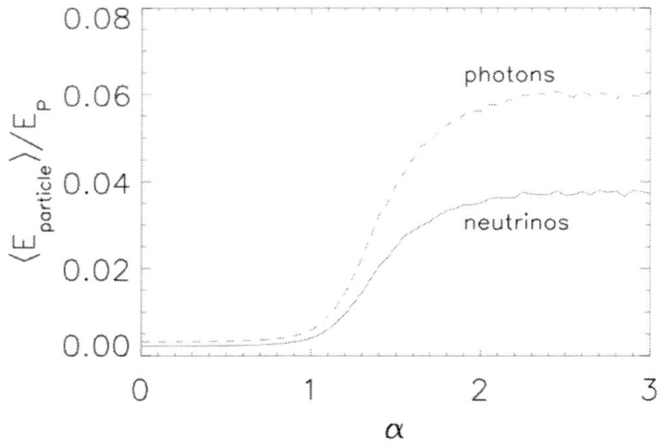

Figure 5. Average photon and neutrino energy per injected nucleon energy in photomeson production in power-law target photon fields with photon index α [37]. $e^{\pm}$ are assumed to be 100% radiatively efficient.

redistributes the power from very high to lower energies. Depending on the overall field to photon energy density in the emission region synchrotron- [23, 29, 30, 10] or inverse Compton supported pair cascades [10] may dominate (see Figure 2). Electromagnetic cascades can be initiated by photons from π^0-decay ("π^0-cascade"), electrons from the $\pi^{\pm} \to \mu^{\pm} \to e^{\pm}$ decay ("$\pi^{\pm}$-cascade"),

p-synchrotron photons ("p-synchrotron cascade"), and μ-, π- and K-synchrotron photons ("$\mu^\pm$-synchrotron cascade"). Ref. [29] and ref. [30] have shown that the "π^0 cascades" and "$\pi^\pm$ cascades" generate rather featureless γ-ray spectra, in comparison to "p-synchrotron cascades" and "$\mu^\pm$-synchrotron cascades" which produce distinct features in the high energy SED.

Generally, hadronic AGN emission models also consider all leptonic processes (including inverse Compton scattering off low energy photons) owing to the presence of ultrarelativistic e^- in the emission region. For sufficient high field strengths, any inverse Compton component is, however, strongly suppressed, leaving the proton-initiated radiation component as the dominating high energy emission mechanism.

3. Efficiencies and energetics

3.1. Proton radiation efficiency

The efficiency of producing photons from a proton of energy $\gamma_p m_p c^2$ within hadronic emission models is determined by the total photon production rate devided by the total proton loss rate r_{tot}:

$$\tilde{\xi}(\gamma_p) = \frac{\sum_i r_{i,photon}}{r_{tot}}$$

where loss rates include proton synchrotron, photomeson production, Bethe-Heitler pair production and adiabatic losses, and $\sum_i r_{i,photon}$ considers all processes that produce photons. By averaging over the input proton spectrum $\propto \gamma_p^{-p}$ the total radiation efficience of protons is derived:

$$\xi_p = \frac{\int \tilde{\xi}_p(\gamma_p)\gamma_p^{1-p} d\gamma_p}{\int \gamma_p^{1-p} d\gamma_p}$$

(e.g., [33]). The evaluation of this formula is not trivial in the case of large magnetic field strengths where parts of the neutrino producing channels are re-directed to photon producing channels via significant synchrotron losses of secondary charged particles produced in photomeson production (e.g., μ, π, K, etc.). Radiation efficiencies from photomeson production have been estimated in the past with simplifying assumptions (e.g., neglecting the synchrotron channel of secondary charged particles) giving, for example, values of 1-10% in typical GRB environments (e.g., [18]) and even lower values for photomeson production in external only radiation fields of blazars (e.g., [40]). The fractional rate of photon production during one dynamical time scale after instantaneous injection of a power-law proton spectrum $\propto E^{-2}$ into a power-law radiation field of given density can also be derived directly from a given blazar model taking into account all radiation processes self-consistently. Depending on the chosen parameter set of the model we find [38] the radiation efficiency of protons in hadronic emission models to range from percent (e.g., for parameter settings with an early cutoff of the proton spectrum, or very low magnetic field values) to several tens of percent (e.g., for the case of a strong contribution from proton synchrotron radiation). The observed bolometric blazar luminosity can therefore be estimated by $L_{rad} \sim \Gamma_{10}\xi_{p,10}\xi_{conv,10}L_{jet}$ with Γ the bulk Lorentz factor $\Gamma = 10\Gamma_{10}$, ξ_p the proton radiation efficiency $\xi_p = 0.1\xi_{p,10}$ ξ_{conv} the efficiency of converting jet power into relativistic particle energies $\xi_{conv} = 0.1\xi_{conv,10}$, and L_{jet} the jet power. For a sufficient large jet power and favorable parameter choices typically observed blazar luminosities can therefore be accommodated in hadronic emission models.

3.2. Neutrino Production Efficiency and Fluxes

An inevitable by-product of hadronic interactions is the production of neutrinos. The prime neutrino production channel is here the decay of charged pions ($\pi^\pm$) and muons ($\mu^\pm$). The neutrino production efficiency in hadronic emission models is calculated by analogy with the photon production efficiency. Because of proton losses such as, e.g. proton synchrotron radiation, competing with photomeson production and $\pi^\pm/\mu^\pm$-decay competing with synchrotron losses prior to its decay in highly magnetized environments it is clear that applying scaling relations to predict neutrino fluxes from gamma-ray observations can lead to neutrino flux upper limits only (e.g., Doert et al, this proceedings). We find [38] that in general the neutrino production efficiency is dependent on the model parameter values, reaching typically not more than $\sim 20\%$. It can be extremely low in the case of high field strengths and weak target photon fields (because of dominating proton synchrotron losses), and/or for low cutoff energies of the injected proton spectrum (threshold effect). Notably, in virtually no case is the neutrino production efficiency higher than the corresponding radiation efficiency.

Because of the potentially denser target photon field provided by powerful FSRQs as compared to BL Lacs, members of the former source class are considered a more promising neutrino source class as the close-by HSP-BL Lacs. Indeed, neutrino fluxes from typical HSP-BL Lacs can be several orders of magnitude lower than from FSRQs (e.g., 3C 279: [36]).

The predicted diffuse neutrino fluxes are also not only highly model-dependent but also suffer from large uncertainties (which can extend up to several orders of magnitude [30]) which originate from unknowns in the model parameters, uncertainties of the class luminosity functions and evolution, etc.

Current instruments are meanwhile capable of limiting the neutrino fluxes to not significantly above the "Waxman-Bahcall limit" (e.g., [1]) thereby ruling out hadronic neutrino production models with optimistically high predicted ν-fluxes.

4. Hadronic blazar modeling along the sequence HSP - ISP - LSP

There is overall consensus that synchrotron radiation from primary electrons dominates most parts of the low energy blazar SED in either model, leptonic or hadronic[6]. In the hadronic Synchrotron-Proton Blazar (SPB) model, e.g., the high energy bump stems mainly from direct proton, $\mu^\pm/\pi^\pm$ synchrotron radiation and their reprocessed components, and possibly contributions from the π-cascades [30]. Which of those processes dominate the SED at gamma-ray energies depends on the chosen parameter set. E.g., in environments with high field strengths but relatively weak low energy photon fields a dominance of proton synchrotron radiation is expected with a negligible cascading component. Such environments offer, e.g., the low-luminosity HSP-BL Lacs (e.g., PKS 2155-304: Fig. 6; [6]). By increasing the target photon density the probability of photomeson production and reprocessing increases. The resulting high energy SED appears smoother, often accompagnied with a higher level of X-ray/soft gamma-ray emission. Because of the higher photomeson production losses the injected proton spectrum is expected to cut off earlier, as does the corresponding gamma-ray spectrum, in comparison to the case of very low photon densities. Examples may be the ISP-BL 3C 66A where the high energy radiation is found to be dominated by a heavily reprocesses proton synchrotron component [35], and the LSP-BL BL Lacertae where pion production losses are found to dominate over proton synchrotron losses with a resulting dominance of strongly reprocessed μ/π-synchrotron radiation at X- to gamma-rays (see Fig. 7; [14]). In the case of FSRQs additional external target photon fields may play a role in forming the overall gamma-ray spectrum also in hadronic emission models, although satisfactorily data representations by hadronic emission models with only internal target photon fields may be found as well (e.g., [15] where modellings of the 2006 flare of 3C 279 have been presented; see Fig. 8). The large observed luminosity of FSRQs require

[6] Synchrotron radiation from secondary electrons may in some cases also contribute to the low energy hump in the SED .

then large jet powers in the framework of hadronic emission models (see Sect. 3). Because of the cascading of power from high to lower energies, a characteristic feature of hadronic emission models, gamma-ray spectra can naturally be extended to very high energies even with unusually hard spectral shapes (in the galaxy frame) also for FSRQs.

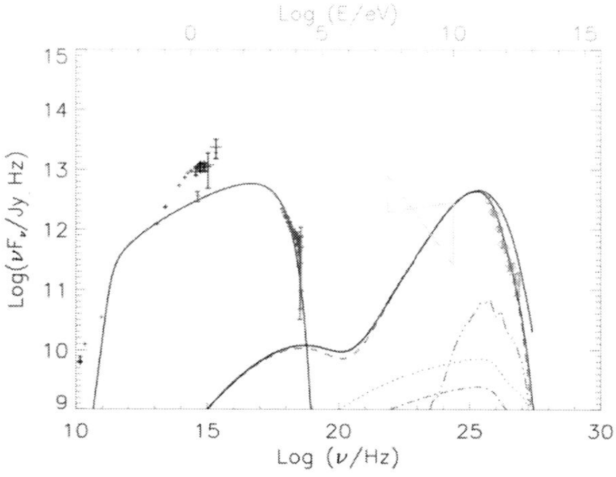

Figure 6. Hadronic SPB modeling of the quiescent state broadband SED in 2003 of the HBL PKS 2155-304 (see also [6]). Left solid line: synchrotron spectrum of primary electrons; right solid lines: emerging cascade spectrum and cascade spectrum corrected for absorption in the EBL. The dominant contribution to the high energy component is proton synchrotron radiation plus its reprocessed part (dashed line) while the contribution from μ synchrotron cascade (triple-dot-dashed line), π^0 cascade (dotted line), and $\pi^{\pm}$ cascade (dot-dashed line) is negligible.

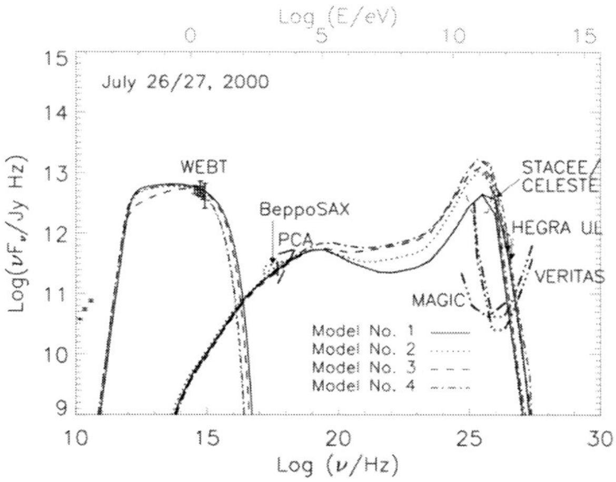

Figure 7. Various hadronic SPB model applications (corrected for absorption in the EBL) in comparison to the broadband SED data of July 2000 of the LBL BL Lacertae [14]. The μ synchrotron cascade contributes significantly to the high energy radiation component.

Acknowledgments

AR acknowledges support by Marie Curie IRG grant 248037 within the FP7 Program.

References

[1] Abbasi R *et al,* (The IceCube collaboration) 2011 *Preprint* arXiv:1104.5187
[2] Abdo A A *et al* (The Fermi-LAT collaboration) 2010a *ApJ* **722** 520

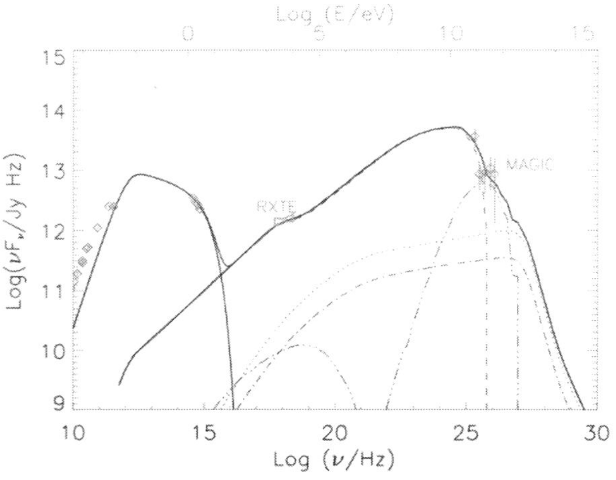

Figure 8. Hadronic SPB modeling of the broadband SED of 2006 of the FSRQ 3C 279 [15]. Left solid line: synchrotron spectrum of primary electrons; right solid line: emerging cascade spectrum (corrected for absorption in the EBL) as the sum of the proton synchrotron cascade (dashed line), μ synchrotron cascade (triple-dot-dashed line), π^0 cascade (dotted line), and $\pi^{\pm}$ cascade (dot-dashed line). No external photon fields were used here. The proton and μ synchrotron cascades dominate the high energy radiation. See [15] for a SPB model fit where external BLR photons serve as an additional seed photon field.

[3] Abdo A A *et al* 2010b (The Fermi-LAT collaboration) *ApJ* **715** 429
[4] Abdo A A *et al* 2011 (The Fermi-LAT collaboration) *ApJ* in preparation
[5] Aharonian F A 2000 *New Astr.* **5** 377
[6] Aharonian F A *et al* (HESS collaboration), 2005, *A&A* **430** 865
[7] Aharonian F A *et al* (HESS-collaboration) 2007 *ApJ Lett* **664** L71
[8] Aharonian F A and Taylor A M 2010 *APh* **34** 258
[9] Albert J *et al* (MAGIC-collaboration) 2007 *ApJ* **669** 862
[10] Atoyan A M and Dermer C D 2003 *ApJ* **586** 79
[11] Barkov M V, Aharonian F A and Bosch-Ramon V 2010 *ApJ* **724** 1517
[12] Bednarek W and Protheroe R J 1997 *MNRAS* **287** 9
[13] Bednarek W & Protheroe R J 1999 *MNRAS* **302** 373
[14] Böttcher M & Reimer A 2004 *ApJ* **609** 576
[15] Böttcher M, Reimer A, & Marscher A 2009 *ApJ* **703** 1168
[16] Celotti A and Ghisellini G 2008 *MNRAS* **385** 283
[17] Dermer C D, Razzaque S, Finke J D, Atoyan A 2009 *New Journal of Physics* **11** 065016
[18] Dermer C D 2010, Proc. of *"The Shocking Universe: Gamma Ray Bursts and High Energy Shock Phenomena"* Venice, Italy, September 2009
[19] Eichler D & Wiita P J 1978 *Nature* **274** 38
[20] Guo A and Mathews W G 2011 *ApJ* **728** 121
[21] Hillas A M 1984 *Ann.Rev. A&A* **22** 425
[22] Levinson A 2000 *PhRvL* **85** 912
[23] Mannheim K and Biermann, P L 1992 *A&A* **253** 21
[24] Medvedev M V 2000 *ApJ* **540** 704
[25] Mücke A, Rachen J P , Engel R , *et al.* 1999 *PASA* **16** 160
[26] Mücke A, Rachen J P , Engel R , *et al.* 2000 *NuPhS* **80** 08/10
[27] Mücke A & Protheroe R J 2000 *AIPC* **515** 149
[28] Mücke A, Engel R, Rachen J P , *et al.* 2000 *CoPhC* **124** 290
[29] Mücke A and Protheroe R J 2001 *APh* **15** 121
[30] Mücke A, Protheroe R J, Engel R 2003 *et al.*, *APh* **18** 593
[31] Pohl M and Schlickeiser R 2000 *A&A* **354** 395 & Erratum: **355** 829
[32] Protheroe R J 1997 *IAU Coll. 163* **Vol. 121** 585
[33] Protheroe R J & Mücke A 2001 ASP Conference Proceedings, 250, 113
[34] Rachen J P and Meszaros P 1998 *Phys Rev* D **58** 123005
[35] Reimer A, Joshi M, Böttcher M 2008 *AIP* **1085** 502
[36] Reimer A 2009 *Int. Journal of Modern Physics* D **18** 1511
[37] Reimer A and Böttcher 2012, in: *"Relativistic Jets from Active Galactic Nuclei"* ed M Böttcher, D Harris,

H Krawczynski (John Wiley & Sons) chapter 3
[38] Reimer A *et al* 2011, in preparation
[39] Sikora M and Madejski G M 2000 *ApJ* **534** 109
[40] Sikora M, Stawarz L, Moderski R, Nalewajko K and Madejski G 2009 *ApJ* **704** 38
[41] Urry C M and Padovani P 1995 *PASP* **107** 803
[42] Wilson A S, Smith David A, Young A J 2006 *ApJ* **644** 9

XMM-Newton and *Swift* observations of High Frequency Peaked BL Lacs in the TeV era

F. Massaro[1], A. Paggi[1,2], M. Elvis[1], A. Cavaliere[2]

[1] Harvard - Smithsonian Astrophysical Observatory, 60 Garden Street, Cambridge, MA 02138
[2] Dipartimento di Fisica, Università di Roma Tor Vergata, Via della Ricerca scientifica 1, I-00133 Roma, Italy

E-mail: fmassaro@head.cfa.harvard.edu

Abstract. Most of TeV detected extragalactic sources are BL Lac objects. They belong to the subclass of "high frequency peaked BL Lacs" (HBLs) exhibiting their spectral energy distributions with a lower energy peak in the X-ray band,interpreted as synchrotron emission. The X-ray spectra are generally curved, and well described in terms of a log-parabolic shape. In a previous investigation of TeV HBLs (TBLs) we found two correlations between their spectral parameters: (1) The synchrotron peak luminosity L_p increases with its peak energy E_p, (2) the curvature parameter b decreases as E_p increases. The first is consistent with the synchrotron scenario, while the second is expected with statistical/stochastic acceleration mechanisms for the emitting particles. Here, we present an extensive X-ray analysis of a sample of HBLs observed with XMM Newton and SWIFT but undetected at TeV energies (UBLs), to compare their spectral behavior with that of TBLs. We investigate the distributions of their spectral parameters, providing an interpretation of both the E_p and b distributions in terms of systematic and stochastic acceleration processes. We also compare the X-ray spectral behavior of UBLs with that in gamma-rays and propose a criterion to select the best HBLs candidates for future TeV observations.

1. Introduction

The great majority ($\geq 80\%$) of the extragalactic sources detected to April 2011 in γ rays at TeV energies are BL Lac objects. BL Lacs come in two flavors: the "high-frequency peaked BL Lacs" (HBLs) in which the low energy component of the SED peaks between the UV band and X-rays, and the "low-frequency peaked BL Lacs" (LBLs) when the spectral energy distribution (SED) peak falls in the IR-optical range [22]. It is widely agreed that this low-energy component is produced by synchrotron radiation of ultrarelativistic electrons accelerated in the jets, while the high energy component is likely due to inverse-Compton scattering of the synchrotron photons by the same electron population (Synchrotron Self-Compton, SSC, see e.g. [12,15].

In the following, we distinguish the HBLs detected at TeV energies from those not yet detected; we refer to the former as TBLs, and to the latter as UBLs.

A useful phenomenological description of the BL Lac X-ray spectra was introduced by [13] in terms of a *log-parabolic* (LP) model (i.e., a parabolic shape in a double-log plot); subsequently, this model has been frequently adopted for the low energy bump, e.g., by [16],[26] and other authors. Recently, the high energy component at TeV energies has also been successfully modeled with the same spectral shape [1,2,3,5,17].

The LP model has been used also to describe the SED of other classes of jet-dominated sources: plerions [6], high frequency peaked (HFPs) radio sources [14], and, recently, Solar Flares [10] and Gamma-Ray Bursts (GRBs) [19,20].

Adopting the LP model, the X-ray SED of HBLs is described in terms of 3 parameters: (1) the peak energy, E_p, in νF_ν space, (2) the maximum height of the SED, S_p, evaluated at E_p (or the corresponding peak luminosity $L_p \simeq 4\pi D_L^2 S_p$, with D_L being the luminosity distance), and (3) the spectral curvature, b, around E_p [18,27].

Extensive investigations of the TBLs, based on all the X-ray observations available in the *BeppoSAX*, *XMM-Newton* and *Swift* archives between 1997 and 2007, have shown that several TBLs trace two correlations in the (E_p, L_p, b) parameter space: (1) the peak luminosity L_p increases with E_p, as expected in the synchrotron scenario, (2) the curvature parameter b decreases as E_p increases [18] as expected in a stochastic acceleration scenario (e.g., [18,28]).

As a result, TBLs cover a well-constrained region in the $E_p - b$ plane. The correlation between b and E_p is evident for the 16 TBLs in [18], whilst no clear trend in the $E_p - L_p$ plane has been found for the whole sample.

Many HBLs have been targeted at TeV energies by HESS, Magic and VERITAS, but by no means all of them have been detected. It is striking that 19 out of 24 TBLs (to 2010, August 1st) belong to the Einstein Slew Survey Sample of BL Lacertae Objects (1ES, [8,24]), which includes only the brightest X-ray extragalactic sources at $\sim$ 1 keV. The remaining TBLs belong to three different samples, namely: 1) The ROSAT All-Sky Survey-Green Bank BL Lac catalog (RGB, []30); 2) The sedentary survey of extreme high energy peaked BL Lacs (SHBL[1],[9]); 3) The Hubble Space Telescope Survey of BL Lacertae Objects; (HST, [25,29]). Consequently, we selected all the UBLs in the above four samples to search for possible differences between these sources and the TBLs.

In this conference proceeding, we report only the main results of our analysis, the details regarding the sample selection criteria, the data reduction and data analysis procedures adopted to perform our investigation are presented in [21].

2. HBLs Detectable at TeV Energies

From comparing the distribution of the X-ray spectral curvature and the GeV *Fermi* LAT detections, we propose criteria to predict which UBLs are more likely to be detectable at TeV energies. TeV energies lie beyond the inverse Compton peak of the HBL SEDs. Hence to be detectable they need both a high GeV flux level and a small GeV - TeV spectral curvature. In the SSC scenario, the X-ray spectral curvature, b, of HBLs, evaluated at the synchrotron SED peak, E_p, is a good predictor of the curvature of the inverse Compton peak at GeV - TeV energies, although they are not always identical [17]. We can define three levels of confidence (i.e., TeV classes) in the prediction of TeV detectability (see Table 1, Col. 12):

Class 1: the best candidates for the future TeV detections are provided by UBLs with a GeV *Fermi* LAT detection and a curvature, b, lower than the characteristic value b_*=0.55 in all the X-ray observations. We found that four UBLs satisfy both conditions and so are the most likely new TeV detectable extragalactic sources: BZB J0326+0225, BZB J0442-0018, BZB J0744+7433 and BZB J1743+1953. Spectral variability could limit this prediction but UBLs appear to be less variable in the X-ray band than TBLs (see Section 6.2)

Class 2: six more UBLs have some X-ray observations with $b < b_*$, and are also detected by *Fermi* LAT and so are still TeV candidates: BZB J0208+3523, BZB J1136+6737, BZB J1417+2543, BZB J1442+1200, BZB J1728+5013, BZB J2250+3824 The variability of b leads us to expect the discovery of other new TBLs when their X-ray spectrum has $b \leq b_*$.

Class 3: UBLs with $b \leq b_*$ in at least one X-ray observations and $F_X \geq 10^{-11}$ erg s^{-1} cm^{-2}

[1] http://www.asdc.asi.it/sedentary/

Table 1. UBLs selected.

(1) BZCAT Name	(2) z	(3) D_L [Mpc]	(4) $N_{H,Gal}$ [$10^{20} cm^{-2}$]	(5) Φ_{XR}	(6) *Swift*	(7) *XMM*	(8) *Fermi*	(9) TeV class
BZB J0013-1854	0.094	420.1	2.13	2.24	4	—	—	3
BZB J0123+3420	0.272	1359.7	5.20	5.74	17	1	—	3
BZB J0201+0034	0.298	1511.2	2.23	2.71	1	—	—	-
BZB J0208+3523	0.318	1629.9	6.27	5.76	—	2	y	2
BZB J0214+5144	0.049	212.0	14.4	0.16	3	—	—	3
BZB J0227+0202	0.456	2499.3	2.67	5.05	2	—	—	-
BZB J0325-1646	0.291	1470.1	3.27	10.1	3	—	y	-
BZB J0326+0225	0.147	681.2	7.87	1.77	3	1	y	1
BZB J0441+1504	0.109	492.3	14.0	7.28	1	1	—	-
BZB J0442-0018	0.449	2453.2	4.83	3.35	4	—	y	1
BZB J0621-3411	0.529	2991.5	4.08	2.30	1	—	—	-
BZB J0744+7433	0.314	1606.0	3.28	2.74	—	2	y	1
BZB J0751+1730	0.185	878.4	4.93	1.78	1	—	—	-
BZB J0753+2921	0.161	752.9	3.44	0.28	1	—	—	-
BZB J0847+1133	0.199	953.2	3.17	3.34	1	—	y	-
BZB J0916+5238	0.190	905.0	1.43	0.53	1	—	—	-
BZB J0930+4950	0.187	889.1	1.38	7.94	1	—	—	-
BZB J0952+7502	0.179	846.8	2.23	2.07	2	—	—	-
BZB J1010-3119	0.143	660.9	8.48	1.37	2	—	—	3
BZB J1022+5124	0.142	655.9	1.02	6.88	1	—	—	-
BZB J1053+4929	0.140	645.8	1.50	0.13	1	—	y	-
BZB J1056+0252	0.236	1155.7	3.82	17.3	1	—	—	-
BZB J1111+3452	0.212	1023.5	1.64	5.71	1	—	—	-
BZB J1117+2014	0.139	640.7	1.35	3.26	1	—	y	-
BZB J1136+6737	0.136	625.7	1.09	3.28	5	—	y	2
BZB J1145-0340	0.167	784.0	2.22	2.28	2	—	—	-
BZB J1154-0010	0.254	1256.9	2.06	2.75	1	—	—	-
BZB J1231+6414	0.163	763.3	2.12	0.43	—	1	—	-
BZB J1237+6258	0.297	1505.3	0.97	1.90	13	2	—	-
BZB J1253-3931	0.179	846.8	7.66	1.47	1	—	—	3
BZB J1257+2412	0.141	650.8	1.25	5.16	1	1	—	-
BZB J1341+3959	0.172	810.0	0.80	1.26	4	—	y	-
BZB J1417+2543	0.237	1161.3	1.54	1.72	5	—	y	2
BZB J1439+3932	0.344	1787.1	1.14	2.64	2	—	y	-
BZB J1442+1200	0.163	763.3	1.58	1.13	4	—	y	2
BZB J1510+3335	0.114	516.7	1.54	3.16	—	1	—	-
BZB J1534+3715	0.143	660.9	1.33	0.10	1	—	—	-
BZB J1605+5421	0.212	1023.5	0.89	5.53	1	—	—	-
BZB J1626+3513	0.497	2773.1	1.36	0.35	—	2	—	-
BZB J1728+5013	0.055	239.0	2.35	1.01	4	—	y	2
BZB J1743+1935	0.080	354.0	7.36	0.14	3	—	y	1
BZB J2131-0915	0.449	2453.2	3.62	1.74	1	—	y	-
BZB J2201-1707	0.169	794.4	2.91	5.92	2	—	—	-
BZB J2250+3824	0.119	541.2	10.4	0.24	16	—	y	2
BZB J2308-2219	0.137	630.7	1.86	7.12	1	—	—	-
BZB J2322+3436	0.098	439.3	6.83	0.11	2	—	y	-
BZB J2343+3439	0.366	1922.6	6.75	1.60	2	—	y	-

Col. (1) ROMA BZCAT source names. Col. (2) the name in the selected sample. Cols.(3,4) the right ascension and declination, respectively. Col. (4) gives the redshift (from ROMA BZCAT). Col. (5) reports the luminosity distance. Cols. (6) the Galactic column density along the line of sight [11]. Col. (8) the X-ray to radio flux ratio Φ_{XR} (see Section 2). Cols. (9,10) report the number of X-ray observations per satellite. Col. (11) indicates if the source has been detected in the *Fermi* LAT 1st year catalog, wile Col. (12) the TeV candidate class derived from our analysis (see Section 6)

in the 0.5-10 keV energy range, but no LAT detection, make up our third class. The lower GeV normalization makes these less likely TeV candidates. However, in the single zone SSC scenario [23], the X-ray flux is similar to the detection threshold of 1yr *Fermi* LAT γ-ray flux [4] and the curvature is as broad as that of TBLs, we suggest that such UBLs can be detected at TeV energies. Five more UBLs fit class 3: BZB J0013-1854, BZB J0123+3420, BZB J0214+5144, BZB J1010-3119 and BZB J1253-3931.

Our source selection was concluded at the beginning of August 2010. Since then, of the 15 total candidates, the sources BZB J1442+1200 and BZB J2250+3824 from our class 2 and BZB J0013-1854 and BZB J1010-3119 from class 3 have been detected at TeV energies (see the

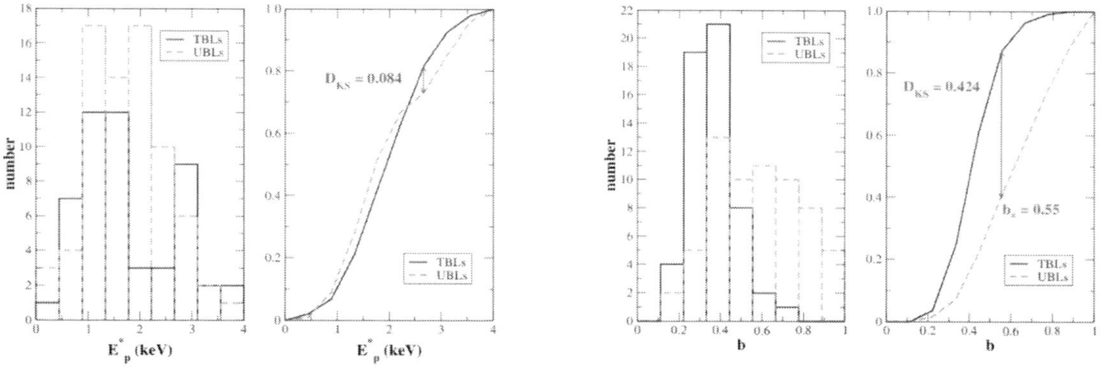

Figure 1. Left panel) The X-ray E_p distribution of UBLs (red) and TBLs (black). The sample of TBLs considered here does not include Mrk 421 and PKS 2155-304 and giant flares of Mrk 501 and 1H 1426+421, as described in Section 5. The maximum separation D_{KS}, of the two cumulative distributions, corresponding to the variable of the KS test, is also shown on the plot. Right panel) The X-ray curvature b distribution of UBLs (red) and TBLs (black). The sample of TBLs considered here does not include Mrk 421, PKS 2155-304 and the giant flares of Mrk 501 and 1H 1426+421, as described in Section 5. The maximum separation, D_{KS}, of the two cumulative distributions (i.e., the variable used for the KS test) and the corresponding boundary value of the curvature b_* are also shown on the plot.

TeVCAT[2] for new announced TeV sources).

3. Summary

We have carried out an extensive X-ray spectral analysis of HBLs to compare the spectral behavior of those not detected at TeV energies (UBLs) with those already known as TeV emitters (TBLs). We analyzed all 135 X-ray observations of a sample of 47 UBLs present in the *XMM-Newton* and *Swift* archives up to August 2010.

We found that the E_p distributions of UBLs and TBLs are similar, and symmetric around a value of a few keV for both subclasses. Instead the X-ray spectral curvature, b, of UBLs, is systematically lower than in TBLs, implying that the UBL X-ray spectra are *n*arrower. In addition, in the first year *Fermi* catalog, we found that the UBL and TBL MeV-GeV γ-ray spectral behavior is similar, yet only $\sim$ 40% of our selected UBLs have been detected in the *Fermi* LAT energy range vs 80% of TBLs.

On the basis of our analysis, we have developed criteria to predict likely TBLs. We present three lists with different levels of confidence for TeV detectability based on MeV-GeV flux level and keV spectral curvature, comprising a total of 15 TeV candidates. By December 2010, four of our candidates have already been detected at TeV energies, landing support to our selection criteria.

A crucial check for our TeV candidate criteria will be provided by X-ray monitoring of candidates from the different TeV classes, with simultaneous GeV and TeV observations, to investigate the variability timescales of the spectral curvature.

[2] http://tevcat.uchicago.edu/

4. Acknowledgments

We are grateful to A. Beardmore for discussions regarding the *Swift* calibration. F. Massaro thanks D. Harris for helpful suggestions that improve the presentation and A. Siemiginowska and B. Refsdal for their useful advices on the use of the Sherpa software package. The work at SAO was supported by the NASA grant NNX10AD50G.
F. Massaro acknowledges the Foundation BLANCEFLOR Boncompagni-Ludovisi, née Bildt for the grant awarded him in 2010 to support his research. Part of this work is based on archival data, software or on-line services provided by the ASI Science Data Center (ASDC). This research has made use of data obtained through the High Energy Astrophysics Science Archive Research Center Online Service, provided by the NASA/Goddard Space Flight Center.

5. References

[1] Abdo A. A. et al. 2011 ApJ submitted arXiv: 1106.1348
[2] Aleksic, J. et al. 2011 A&A submitted arXiv: 1106.1589
[3] Acciari, V. A. et al. 2011 A&A submitted arXiv: 1106.1210
[4] Atwood, W. B. et al. 2009, ApJ, 697, 1071
[5] Aharonian, F. et al. 2009, A&A, 502, 749
[6] Campana, R., Massaro, E., Mineo, T., 2009, A&A, 499, 847
[7] Dunkley, J., 2009 ApJ, 701, 1804
[8] Elvis, M., Plummer, D., Schachter, J., Fabbiano, G. 1992 ApJS, 80, 257
[9] Giommi, P.; Piranomonte, S.; Perri, M.; Padovani, P., 2005, A&A, 434, 385
[10] Grigis, P. G. & Benz A. O. 2008 ApJ, 683, 1180
[11] Kalberla, P.M.W., Burton, W.B., Hartmann, D., 2005, A&A, 440, 775
[12] Inoue, S., Takahara F., 1996, ApJ, 463, 555
[13] Landau, R., Golish, B., Jones, T. J., et al. 1986, ApJ, 308, L78
[14] Maselli, A. & Massaro, E. 2009, AN, 330, 295
[15] Marscher, A. P., Gear, W. K. 1985, ApJ, 298, 114
[16] Massaro, E., Perri, M., Giommi, P., et al. 2004, A&A, 422, 103
[17] Massaro, E., Tramacere, A., Perri, M., Giommi, P., Tosti, G., 2006, A&A, 448, 861
[18] Massaro, F., Tramacere A., Cavaliere A., et al. A&A 2008a, 478, 395 (M08)
[19] Massaro, F., Grindlay, J. E., Paggi, A. 2010a, ApJL, 714, 299
[20] Massaro, F., & Grindlay, J. E. 2011a ApJ, 727L, 1
[21] Massaro, F., A. Paggi, Elvis, M., & A. Cavaliere, 2011b ApJ in press
[22] Padovani, P., & Giommi, P., 1995, MNRAS, 277, 1477
[23] Paggi, A., Massaro, F., Vittorini, V. et al. 2009 A&A, 504, 821
[24] Perlman, E. S. et al. 1996 ApJS, 104, 251
[25] Scarpa, R. et al. 1999 ApJ, 521, 134
[26] Tanihata, C., Kataoka, J., Takahashi, T., et al. 2004, ApJ, 601, 759,
[27] Tramacere, A., Massaro, F., Cavaliere, A., 2007, A&A, 466, 521
[28] Tramacere, A., Giommi, P., Perri, M. et al. 2009 A&A, 501, 879
[29] Urry, C. M. et al. 2000 ApJ, 532, 816
[30] Laurent-Muehleisen S. A. et al. 1999 ApJ, 525, 127

Connection Between Optical and VHE Gamma-ray Emission in Blazar Jets

Riho Reinthal*, Elina J. Lindfors*, Daniel Mazin†, Kari Nilsson*,◇, Leo O. Takalo*, Aimo Sillanpää* and Andrei Berdyugin*,◇ on behalf of the MAGIC collaboration

***Tuorla Observatory, Department of Physics and Astronomy, University of Turku, 21500 Piikkiö, Finland**
†Institut de Fisica d'Altes Energies (IFAE), E-08193 Bellaterra (Barcelona), Spain
◇Finnish Centre for Astronomy (FINCA), University of Turku, Finland

Väisäläntie 20, FI-21500 Piikkiö, Finland

E-mail: rirein@utu.fi

Abstract. MAGIC has been performing optically triggered Target of Opportunity observations of flaring blazars since the beginning of its scientific operations. The alerts of flaring blazars originate from Tuorla Blazar Monitoring Programme, which started the optical monitoring of candidtate TeV blazars in 2002 and has now collected up to eight years of data on more than 60 blazars. These ToO observations have resulted in the discovery of five new VHE γ-ray emitting blazars (S5 0716+714, 1ES 1011+496, Mrk 180, ON 325 and B3 2247+381). In addition part of the discovery of BL Lac and the discovery of 3C 279 was made during a high optical state. In this contribution we present a detailed analysis of the optical light curves which are then compared to MAGIC observations of the same sources. We aim to answer the question: "Is there a connection between optical and VHE γ-ray high states in blazars or have we just been lucky?"

Keywords: blazars, observations, multi-wavelength

1. Introduction

Blazars are the most extreme case of Active Galactic Nuclei (AGN) in which the relativistic jet originating from close to the central black hole is oriented close to our line of sight. In these objects the dominant radiation component comes from the jet and is non-thermal in origin. Their Spectral Energy Distribution (SED) is characterised by two broad peaks of which the lower energy one is believed to originate from synchrotron emission of the charged particles in the jet. The higher energy one is most commonly explained by inverse Compton scattering of either the synchrotron (Synchrotron Self Compton – SSC, see e.g. [1]) or external (External Compton – EC, [2], [3]) seed photons by the electrons and positrons in the jet, but hadronic models (see e.g. [4], [5]) are also considered. Blazars are divided into flat spectrum radio quasars and BL Lacertae objects. The latter are further subdivided into low-, intermediate- and high-frequency peaked BL Lac (LBL, IBL and LBL, respectively) objects according to the frequency of the first peak which typically ranges from near-infrared in LBLs to UV or soft X-rays in HBLs.

Blazars show a variable flux in all wave bands from radio to Very High Energy (VHE, here defined as E>100 GeV) γ-rays on time scales ranging from months to a few minutes. The relationship between the variability at different wave bands appears to be rather complicated. In the simplest case both SSC and EC mechanisms predict a connection between optical and γ-ray flares. However, there could be time lags between the flares in the different wave bands due to several mechanisms (see e.g. [6], [7] for further discussion) with the optical leading the one in γ-rays or vice versa. Also, in case the synchrotron peak is located far from the optical like in the case of HBLs the variability in the optical band could be very small and the synchrotron flares should be best visible in other wavelengths, usually X-rays. This seems to be the case for the best studied TeV blazars Mrk 421 [8] and Mrk 501 [9]. In principle, hadronic models can also explain correlated variability and cannot be excluded as the emission mechanisms of γ-rays.

MAGIC is performing ToO observations of sources in a high flux state in other wave bands (e.g. optical and/or X-rays and/or high energy γ-rays). Alerts of optical high states are sent by the Tuorla Blazar Monitoring Program[1] [10]. Observations triggered by optical high states have so far led to the discovery of 5 new sources at VHE γ-rays. These are, in order of discovery: Mrk 180 [11], 1ES 1011+496 [12], S5 0716+714 [13], B3 2247+381 [14] and ON 325 [15]. In addition the discovery of 3C 279 was made and part of the data in the discovery of BL Lac were collected during an optical high state. Also, several γ-ray flares of known VHE emitters (e.g. the 2007 flare of 3C 279 [16] and the 2011 flare of 1ES 0806+524 [17]) have been detected during optically triggered observations. In this contribution we present an analysis of the optical light curves and compare them to simultaneous MAGIC observations. We discuss the implications of the results to the optical to VHE γ-ray connection.

2. Analysis of the optical light curves

The observations are performed using the KVA 35 cm telescope, which is located on the Canary island of La Palma, Spain and remotely operated from Tuorla Observatory, and the Tuorla 1 m telescope. Observations are done in the R-band and the magnitudes are measured using differential photometry with calibration stars in the same CCD frame as the object (see [19] for a more detailed overview). The light curves are available on the project web page[1] and are updated on a daily basis.

The monitoring started in 2002 and the original list consisted of 24 candidate TeV blazars chosen from [18] (with $\delta > 20°$ for them to be observable from Tuorla). Over time the number of monitored sources has grown to over 60 including many candidates detected at high energy (HE) γ-rays by the Fermi telescope. The goal of the monitoring is to study the optical variability of these sources and to determine their baseline flux. We have also studied the host galaxy's contribution to most objects' overall brightness [19]. This way we can get a direct measurement of the AGN's core brightness by subtracting the host galaxy contribution. For this study we chose 31 of the best sampled light curves including all 24 from the original sample.

Historically a source was said to be flaring in the optical if its core flux increased by $\geq 50\%$ compared to its quiescent level. For determining the quiescent level a general rule that a source would spend $\sim$20% of the time at or below this level was followed. Both the quiescent states and the flares were identified by visual inspection of the light curves. One of the purposes of this study was to calculate exact values for the quiescent states according to some common principles and to automate the identification of flares. As quiescent state we used the weighted average calculated from all the data points except for those belonging to 'flares'. For the purposes of these calculations 'flares' were all data points that were typically more than three (the exact

[1] http://users.utu.fi/kani/1m

value varied slightly from source to source) standard deviations above the average. As an initial value for the calculations we used the historical value of quiescent state i.e. the flux of which 20% of the whole data set was of lower level. In most cases the quiescent level ended up around the value where we would have put it by eye and also usually close to the 20% value used previously. The relative increase in core flux required for a brightening to be constituted as a flare was derived for each source independently depending on the amplitude of their variability – the larger the amplitude the higher increase in flux was required: $FlareCond \sim Qstate * \frac{F_{max}}{F_{min}}$ where $FlareCond$ is the flare condition, $Qstate$ the quiescent state and F_{max} and F_{min} are the highest and lowest measured core flux respectively. A total of 132 flares were counted since the beginning of the monitoring of which 108 occurred after MAGIC started regular observations. These numbers are slightly higher than the number of flares identified by eye from the light curves, because: a) if there is a long flare but with a gap in the data, software identifies it as two distinct flares and b) some sources (e.g. OJ 287) show "flaring states" where the flux can drop below the flaring condition briefly but immediately increase again which the software counts as separate flares. We also calculated the rise and decay times of the flares and flare durations where possible (refer to Fig. 1 for explanation of these terms). A summary of all the flares of all the sources can be seen in Table 1.

3. Optical to VHE γ-ray comparison

MAGIC (Major Atmospheric Gamma-ray Imaging Cherenkov) is a system of two imaging atmospheric Cherenkov telescopes (IACTs) located on the Canary island of La Palma, Spain at ~2200 m asl. The first telescope started observations in 2004 and since November 2009 the two telescopes have been observing in stereoscopic mode. Both telescopes have 17 m mirror dishes – the largest of any IACTs – allowing them to reach lower energy threshold than other instruments of their type. In the stereoscopic mode MAGIC reaches an energy threshold of ~50 GeV (~25 GeV with a special trigger), sensitivity of ~0.8% of Crab Nebula flux (5 σ in 50 h >250 GeV), an energy resolution of ~15-20% and an angular resolution of 0.07 degrees [20] making it the best instrument of its kind, particularly for energies below 200 GeV.

Because MAGIC is performing ToO observations of blazars that are in a high state in the optical there is simultaneous or quasi-simultaneous optical and VHE data for quite many flares. Of the 108 optical flares that occurred after the beginning of Cycle 1 in MAGIC (June 2005) for 56 there is some simultaneous MAGIC data but for only 49 of those there is more than 3 hours of good quality data. This is mostly constrained by the weather and moon conditions as well as conflicts in the schedule. On 27 occasions the source was detected at VHE (on 3 occasions in a low state) – a total of 14 different sources were seen (note that for Mrk 501 the VHE flare

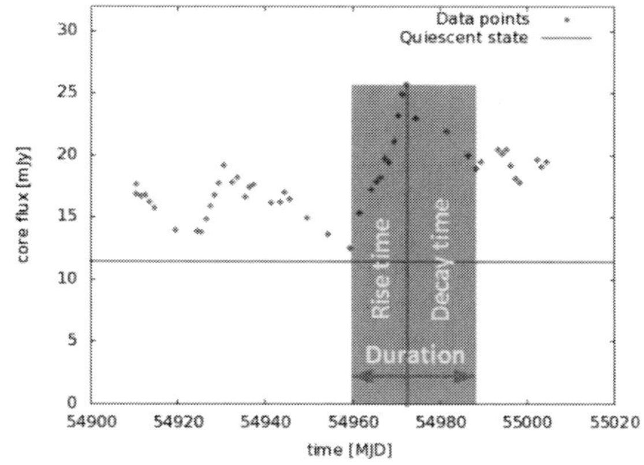

Figure 1. An example of the parts of the flare: peak of the flare is the point in the flare with the highest flux; beginning and end of the flare are the points with the lowest flux prior or after exceeding the flare condition respectively; rise time is the time from the beginning of the flare to the peak and decay time the time from the peak to the end.

Source	# flares[1]	Qstate[2]	below_qstate[3]	avg_rise_t[4]	avg_decay_t[5]	avg_dur[6]	t_betw_fl[7]	duty_cycle[8]
1ES 0033+595	3	0.18	15.7	11.06	18.98	23.71	73.6	0.07
1ES 0120+340	1	0.11	20.5	10.96	2.99	13.95	-	0.02
RGB 0136+391	2	1.77	18.1	3.96	55.95	57.93	-	0.09
RGB 0214+517	0	0.42	21.2	-	-	-	-	0
3C 66A	10	5.46	19.4	20.96	16.74	36.02	36.14	0.22
1ES 0647+250	11	0.93	10.3	15.3	24.23	35.71	42.84	0.38
S5 0716+714	9	12.83	23.6	11.49	10.3	20.51	45.56	0.16
1ES 0806+524	3	1.72	31.3	33.41	6.02	37.42	19.73	0.09
OJ 287	20	2.4	12.2	12.08	14.72	24.72	47.61	0.33
1ES 1011+496	6	2.05	17.7	21.8	20.65	42.45	20.98	0.22
1ES 1028+511	3	0.43	22.3	29.49	26.35	46.01	23.97	0.13
Mkn 421	10	11.41	12.5	16.17	13.19	29.36	42.73	0.21
RGB 1117+202	1	0.58	23.9	45.97	10.01	55.99	-	0.06
Mkn 180	2	1.13	21	20.65	3.69	24.34	-	0.04
RGB 1136+676	0	0.2	29.3	-	-	-	-	0
ON 325	7	2.27	12.8	14.7	11.4	24	42.12	0.22
1ES 1218+304	3	0.83	28.9	29.14	11.32	30.75	19.68	0.13
ON 231	4	2.81	8.2	34.31	22.3	42.46	50.25	0.35
3C 279	3	1.45	12.8	19.95	32.21	45.51	-	0.16
RGB 1417+257	0	0.38	20.9	-	-	-	-	0
1ES 1426+428	1	0.41	24.9	14.99	4.02	19	-	0.03
1ES 1544+820	0	0.59	18.6	-	-	-	-	0
PG 1553+113	4	9.31	19.7	34.12	30.12	64.24	-	0.33
Mkn 501	4	3.78	19.5	13.76	8.18	21.94	55.07	0.06
H 1722+119	7	1.73	16	24.67	13.96	36.64	40.98	0.49
OT 546	1	0.82	21.7	29.96	31.81	61.78	-	0.07
1ES 1959+650	3	2.98	39	32.84	32.78	65.62	-	0.13
PKS 2155-304	5	14.81	18.2	15.38	15.14	26.97	-	0.38
BL Lac	6	3.56	19.2	10.45	10.34	20.79	69.87	0.09
B3 2247+381	3	0.47	18.6	40.7	12.32	49.93	36.44	0.26
1ES 2344+514	0	0.53	26	-	-	-	-	0

Table 1. Results of the light curve analysis source-by-source.
[1]Absolute value of flares identified since the beginning of the programme.
[2]Quiescent state flux in mJy.
[3]Percentage of data points below quiescent state.
[4]Average rise time in days.
[5]Average decay time in days.
[6]Average flare duration in days.
[7]Average time between two flares in the same observation period (the ~6 months when the source is visible).
[8]Estimate of the duty cycle (fraction of time the source spends above flaring condition).
Note that for the calculations of the averages only values >0 were used. For the calculation of the duty cycles all data gaps longer than three weeks were excluded.

occurred ~1 month after the optical), 7 of them for the first time by MAGIC. In 19 cases the observations resulted in upper limit only and for 8 flares information was not available or the analysis is ongoing. The remaining two observations resulted in the detection of VHE γ-rays from the 3C 66A/B region [21]. The results of the comparison of the optical and VHE data are summarised in Table 2.

There are only two sources, Mrk 421 and Mrk 501, that can be reliably detected at low flux states in reasonable time scales and that have good enough VHE coverage that allows us to study the correlation between optical and VHE γ-ray states. The long-term monitoring of these sources by MAGIC is summarised in [22]. From the Mrk 421 data set they did not find any correlation between the optical and VHE wave bands whereas Donnarumma *et al* [23]

source	source type	redshift	VHE det?[1]	# of optical flares[2]	MAGIC observations[3]	supports connection?
1ES 0033+595	HBL	0.086*	NO	3	1: ~3 h, no detection	inconclusive
1ES 0120+340	HBL	0.272	NO	1	no observations	no info
RGB 0136+391	HBL	>0.3	NO	2	no observations	no info
3C 66A	IBL	0.444*	YES'	6	5: 1 non-det, 2 det (1 of which A/B), 2 analysis ongoing	inconclusive
1ES 0647+250	HBL	0.203*	NO	9	3: analysis ongoing	inconclusive
S5 0716+714	LBL	0.31*	YES'	7	4: 2 det, 2 analysis ongoing	yes
1ES 0806+524	HBL	0.138	YES'	3 (more like 1 long)	1: detection	yes
OJ 287	LBL	0.306	NO	18	2: no detection	no
1ES 1011+496	HBL	0.212	YES'	6	3: all detections	yes
1ES 1028+511	HBL	0.239	NO	3	3: no detections, 2 obs > 3 h	no
Mkn 421	HBL	0.031	YES'	7	3: twice in high state, once in low	inconclusive
RGB 1117+202	HBL	0.139	NO	1	no observations	no info
Mkn 180	HBL	0.045	YES'	2	2: 1 detection, only 1 obs > 3h	inconclusive
ON 325	LBL	0.130	YES'	6	3: 2 obs > 3 h, 2 non-det, 1 det	inconclusive
1ES 1218+304	HBL	0.182	YES	2	no observations	no info
ON 231	IBL	0.102	YES	4	2: both obs <3 h, no detections	inconclusive
3C 279	FSRQ	0.536	YES'	2	2: both detections	yes
1ES 1426+428	HBL	0.129	YES	1	1: poor quality data, no detection	inconclusive
PG 1553+113	HBL	>0.3	YES'	4	2: both showed increased VHE flux	yes
Mkn 501	HBL	0.034	YES'	1	1: gamma-ray flare detected ~1 month after optical	inconclusive
H 1722+119	HBL	0.018	NO	6	4: no detections, 3 obs <= 4 h	no
OT 546	HBL	0.055	NO	1	1: no detection, strong moonlight	inconclusive
1ES 1959+650	HBL	0.048	YES'	3	3: 1 obs < 2 h – no detection, twice in low state	no
PKS 2155-304	HBL	0.116	YES'	4	1: high state	yes
BL Lac	LBL	0.069	YES'	3	3: 1 det, 2 non-det	inconclusive
B3 2247+381	HBL	0.119	YES'	3 (more like 1 long)	1: detection	yes

Table 2. Summary of optical flares detected since MAGIC started observations and their comparison to contemporaneous VHE data. The redshifts marked with an asterisk are tentative values.
[1]Flag whether the source has been detected at VHE γ-rays by MAGIC or any other IACT. The ones seen by MAGIC are marked with a cross.
[2]Number of optical flares since the beginning of Cycle 1 in MAGIC.
[3]The number and short summary of the contemporaneous MAGIC observations.

saw a hint of correlation between the two during the 2008 flare. For both Mrk 421 and 501 extensive data sets have been collected (see e.g. [24], [25]) allowing to study the correlation between optical and VHE γ-rays in unparalleled detail. However, both of these sources are so-called extreme blazars [26] so they are not really representative of the whole blazar or the HBL classes. The long-term monitoring of 1ES 1959+650 by MAGIC has always detected the source in a low state [22] whereas in the optical the source has shown three flares since the beginning of the monitoring. Although difficult to observe from the Northern hemisphere due to large zenith angles, PKS 2155-304 can still be reliably detected by MAGIC during strong outbursts. One such was reported by H.E.S.S in July 2006 (coinciding with an optical flare) triggering MAGIC observations and leading to a strong detection of the source (see [27] and [28]). Also, H.E.S.S. saw a strong correlation between the optical and VHE bands during the multi-wavelength observations of PKS 2155-304 in August-September 2008 during a low VHE state of the source (see [29]) but this correlation does not always seem to be there. The results of five years of observations of PG 1553+113 by MAGIC reported in [30] suggest a correlation between the optical and VHE wave bands for this source. Only for a handful of the other sources we have MAGIC observations in both high and low flux states (e.g. S5 0716+714, 1ES 0806+524 and 3C 279) and in most cases there seems to be a connection between activities in the two wavebands but it is difficult to prove with the low statistics at VHE.

4. Discussion and conclusions

Optically triggered ToO observations have resulted in the discovery of five new VHE γ-ray emitting blazars by MAGIC so far. Additionally several other discoveries and γ-ray flares have coincided with optical high states. If the synchrotron emission and the IC emission arise from the same emission region, a connection between optical and VHE γ-ray states would be expected.

However, in case the synchrotron peak is located far from the optical waveband (which can be the case for HBLs) such connection would be difficult if not impossible to observe due to low amplitude of variability in the optical. And in case of different emission regions or mechanisms there might be no direct connection. It seems that for most sources the optical and VHE γ-ray states are somewhat connected. However there is at least one example (1ES 1959+650) where this does not seem to be the case which indicates that at least in some sources there are different emission mechanisms at work or that the two emissions originate from different regions.

In order to study the problem more in detail correlation studies between the optical and VHE γ-ray states are required. This would require more complete VHE light curves with reliable detections of the sources also at lower flux states. Unfortunately with the current generation of IACTs these kinds of studies are only feasible for a few sources and would require a significant fraction of the available observation time to be dedicated to monitoring. Otherwise we have to rely on future instruments to shed more light into this matter.

Acknowledgements: The work of RR and EL has been supported by grant 127740 of the Academy of Finland. We would also like to thank the Instituto de Astrofísica de Canarias for the excellent working conditions at the Observatorio del Roque de los Muchachos on La Palma. The support of the German BMBF and MPG, the Italian INFN, the Swiss National Fund SNF, and the Spanish MICINN is gratefully acknowledged. This work was also supported by the Marie Curie program, by the CPAN CSD2007-00042 and MultiDark CSD2009-00064 projects of the Spanish Consolider-Ingenio 2010 programme, by grant DO02-353 of the Bulgarian NSF, by the YIP of the Helmholtz Gemeinschaft, by the DFG Cluster of Excellence "Origin and Structure of the Universe", by the DFG Collaborative Research Centers SFB823/C4 and SFB876/C3, and by the Polish MNiSzW grant 745/N-HESS-MAGIC/2010/0.

References

[1] Maraschi, L., Ghisellini, G., Gelotti, A., 1992, ApJ, 397: L5-L9
[2] Dermer, C. D., & Schlickeiser, R., 1993, ApJ, 416: 458-484
[3] Ghisellini, G., Tavecchio, F., Chiaberge, M., 2005, A&A, 432: 401-410
[4] Mannheim, K., 1993, A&A, 269: 67-76
[5] Mücke, A., *et al* , 2003, APh, 18: 593-613
[6] Sokolov, A., Marscher, A. P., McHardy, I. M., 2004, ApJ, 613: 725
[7] Chatterjee, R., Jorstad, S. G., Marscher, A. P. *et al* , 2008, ApJ, 689: 79
[8] Lichti, G., Bottacini, E., Ajello, M., *et al* , 2008, A&A, 486: 721
[9] Pian, E., Vacanti, G., Tagliaferri, G., *et al* , 1998, ApJ, 492: 17
[10] Takalo, L. O., Nilsson, K., Lindfors, E., *et al* , 2008, AIPC, 1085: 705
[11] Albert, J., *et al* (the MAGIC Collaboration), 2006, ApJ, 648: L105
[12] Albert, J., *et al* (the MAGIC collaboration), 2007, ApJ, 667: L21-L24
[13] Anderhub, H., *et al* (the MAGIC collaboration), 2009, ApJL, 704: L129-L133
[14] Mariotti, M., *et al* (the MAGIC collaboration), 2010, ATel #2910
[15] Mariotti, M., *et al* (the MAGIC collaboration), 2011, ATel #3100
[16] Aleksić, J. *et al* (the MAGIC collaboration), 2011, A&A, 530: A4
[17] Mariotti, M., *et al* (the MAGIC collaboration), 2011, ATel #3192
[18] Costamante, L. & Ghisellini, G., 2002, A&A, 384: 56-71
[19] Nilsson, K., Pasanen, M., Takalo, L. O. *et al* 2007, A&A, 475: 199
[20] Albert, J., *et al* (the MAGIC collaboration), 2008, ApJ, 674: 1037-1055
[21] Aliu, E., *et al* (the MAGIc collaboration), 2009, ApJL, 692: L29-L33
[22] Hsu, C. C., Satalecka, K., Thom, M. *et al* , 2009, arXiv e-prints (Preprint 0907.0893)
[23] Donnarumma, I. *et al* , 2009, ApJ, 691: L13-L19
[24] Abdo, A. A. *et al* (the Fermi-LAT collaboration) and Aleksić, J. *et al* (the MAGIC collaboration), GASP-WEBT consortium and partners, 2011, arXiv e-prints (Preprint 1106.1348)
[25] Abdo, A. A. *et al* (the Fermi-LAT collaboration) and Aleksić, J. *et al* (the MAGIC collaboration), Acciari, V. A. *et al* (the VERITAS collaboration) *et al* , 2011, ApJ, 727: 129
[26] Costamante, L., Ghisellini, G., Giommi, P. *et al* , 2001, A&A, 371: 512
[27] Hadasch, D., Bretz, T., Mazin, D. for the MAGIC collaboration, 2007, AN, Vol.328, 680
[28] Hadasch, D., Bretz, T., Mazin, D. for the MAGIC collaboration, 2009, arXiv E-print: 0907.0370
[29] Aharonian, F. *et al* (the H.E.S.S. collaboration), Abdo, A. A. *et al* (the Fermi-LAT collaboration), 2009, ApJL, 696: L150-L155
[30] Aleksić, J. *et al* (the MAGIC collaboration), 2011, arXiv E-print: 1101.2764

Journal of Physics: Conference Series **355** (2012) 012014 doi:10.1088/1742-6596/355/1/012014

Bottom-up modelling of gamma-ray blazars

Joni Tammi

Aalto University Metsähovi Radio Observatory, P.O. Box 13000, 00076 Aalto, Finland

E-mail: joni.tammi@iki.fi

Abstract. I discuss blazar modelling from the point of view of the radiating particles and emphasise the role of low-energy emission in high-energy variability studies. I address problems that can arise from focusing on the high-energy spectrum and variability without taking into account the low-frequency behaviour and data – especially if the low-frequency emission is the most direct signature of the particles producing the gamma flares (shock-in-jet models). I use recent *Planck* observations combined with long-term monitoring of AGNs as examples and for motivation for the "bottom-up" direction – starting from the particles and the low-energy emission – in modelling gamma-ray blazars.

1. Introduction

The last decade or so has seen a rapid development in blazar observations across the electromagnetic spectrum with Cherenkov telescopes and X- and gamma-ray satellites at one end, and VLBI and now the *Planck* satellite at the other. The observational leaps have also enabled advances in theoretical modelling, although these possibilities are still often neglected. Power-distributions in radiating particle energy have already replaced the initial attempts that assume a monoenergetic or thermal distribution, and broken power laws are adding more physics into the models, but still few models go beyond the next step and really pay attention to the origin and evolution of the particle energy during a flaring event.

There are several ways in which we can update our models to meet the standards our new observations deserve. In this paper I call attention to one of these, accurate modelling of the emitting matter and the primary emission. Synchrotron emission is the first and the most straightforward (meaning: requiring least additional assumptions) observable signature of the particles and the environment that produce the high-energy spectra and variability, but only a part of the observations are typically used to constrain models, and especially the low-frequency radio data are often completely neglected.

2. Open questions and why the low energies matter

In the co-moving frame of the radiating region the relativistic electrons do two things: they emit synchrotron radiation and up-scatter photons to very high energies. The big questions are the location of the gamma-ray flares and the origin of the up-scattered photons (whether they are the synchrotron photons or whether they originate from somewhere else, *e.g.* from the accretion disk or the dust torus). In the following I limit the discussion to leptonic models, as they are the most popular ones, although the possibility of hadronic emission is still somewhat open (see Anita Reimer's review elsewhere in this volume for hadronic and lepto-hadronic models, and [1] for a more general review of spectral models in blazars).

2.1. *Location of gamma flaring*
The origin of the up-scattered seed photons tells us about the properties of the particles and the jet as well as the ambient radiation field. Generally speaking, external Compton emission (hereafter EC[1]) requires intense radiation field and works best near the central engine and the broad line region (BLR), whereas farther away the synchrotron self-Compton (SSC) is more likely to be the dominating mechanism.

Traditionally, the innermost jet has been the strongest candidate for the origin of high-energy photons, and the violent, extreme conditions near the black hole are the likely culprits for the extreme radiation. Also rapid variability and the inferred small size-scales (of the order of light-minutes or light-hours) have been argued to imply a central origin. Similarly, observed high-energy cutoffs in some blazars have been suggested to be evidence of absorption due to photon-photon interactions within the BLR [2, and Poutanen, elsewhere in this volume], or otherwise near the central engine [3].

Then again, the size scale of the emission region needs not always be linked to its distance from the central engine, as a shock front can be light-minutes in size even if it is light-years away down the jet. Shock waves can create violent and extreme conditions far away from the central engine, and statistical studies have found strong gamma-ray flares often happening *after* the onset of radio flares [4; 5; 6], as well as after the ejection of new VLBI components [7; 8, and Agudo, this volume]. This makes the shocks in the jet and in the radio core (parsecs away from the black hole) a viable source for some gamma-ray flares [9; 10; 11]. Even the high-energy cutoff due to absorption within the BLR has been shown *not* to exist in all cases [*e.g.*, 12].

All things considered, it appears likely that the origin and the dominating mechanism – as the details of the jets – vary between sources, perhaps depending on the object class [13].

2.2. *The synchrotron spectrum*
Gamma-ray models typically assume a single homogeneous component of some size and velocity filled with a particle population following a power-law distribution, and tune the model details to make the IC spectrum fit the X-to-γ-ray datapoints. EC models also require the existence of a bright IR-to-optical spectral component from, *e.g.*, the accretion disk or the dust torus in order to provide the seed photons needed to make the EC fit the data, although in most sources there is not much direct obervational evidence of such components. This simple approach can often produce good fits for the high-energy part of the SED, but it has also caveats one needs to take into account.

Firstly, although the synchrotron spectrum is the primary signature of the underlying particle population, it is often largely neglected; typically only the IR-to-optical-to-UV part of the synchrotron component is compared to the observations. Even advanced models that otherwise take the particles and the synchrotron spectrum accurately into account [see, *e.g.*, 14], concentrate on the optical-UV part and only consider upper limits for the lower frequencies. For a single-zone model focusing on the high-energy variability and the upper end of the particle energy distribution this is understandable, but in going beyond the one-zone approach, the wealth of information in the radio-to-IR regime can provide significant insights into the shape of the particle spectra, cooling and acceleration processes, and the underlying jet [*e.g.*, 15, Tammi et al., in preparation].

Secondly, regardless of the exact mechanism behind the high-energy peak, young shocks in the jet can also contribute to the IC spectrum, and this contribution needs to be taken into account in the model fits. In some sources the shocks can produce a part or all of the observed X- and gamma-ray emission through the SSC mechanism [*e.g.*, 15; 16; 17], and, as reported by

[1] The abbreviations EC and SSC are used both for the physical processes and the emission they produce, but the meaning is always clear from the context.

Türler & Björnsson (elsewhere in this volume), higher-order SSC might be enough to account for the gamma-rays in some cases. On the other hand, models including both SSC and EC from various seed photon sources suggest that the highest-energy gamma-rays could be dominated by the EC process [*e.g.*, 14]. In particular, as discussed in Sec. 2.1, the GeV break observed in 3C 454.3 [18], for example, appears to be best explained by EC models [2; 3]. It should be noted that as the distance of the radiating region from the black hole increased, the previously dominating EC yields to first- and second-order SSC [14], highlighting the importance of finding the location of the gamma-ray flaring.

Furthermore, in many cases there are multiple jet components contributing to the total SED at the same time. As a component ages it goes through various phases and emits a different spectrum in each phase [19; 20], and by modelling the evolution of the components it is possible to trace the spectral variability of many sources in great detail from radio to optical [see, for example, 15]. The possibility of contribution from multiple components needs to be taken into account when testing the models, and long-term multifrequency lightcurve and VLBI monitoring are needed to limit the number of free parameters.

3. What a gamma-ray modeller should know about the *Planck* results

Although the *Planck* satellite's task is mainly cosmological, its results are also highly relevant for AGN modelling. [21] published the spectral energy distributions of a hundred radio-bright northern AGNs from the first *Planck* all-sky survey, taken between August 2009 and June 2010 covering nine frequency bands between 30 and 857 GHz, together with simultaneous observations across the electromagnetic spectrum. (Simultaneous here means that the accompanying radio observations were taken within two weeks, and optical and higher energies within five days of the *Planck* measurement.) These early results mostly include one or two observing epochs depending on the source, and for many sources the published SED is the average of the two. The final release of the data will provide us with the complete SEDs for individual epochs for all observed sources, but already now a couple of interesting implications can be pointed out. In the following, all quantitative results are taken from [21] where they are discussed in detail. α and s are used for the power-law indices of the photon spectrum $S_\nu \propto \nu^\alpha$ and the particle energy distribution $N(E) \propto E^{-s}$, correspondingly.

First of all, the *Planck* radio spectra are, in general, flatter than expected. At lower frequencies ($\leq$ 70 GHz, the limiting frequency between *Planck*'s two instruments) the spectral indices concentrate around $\alpha \approx 0$ and on higher frequencies ($>$ 70 GHz) the average index is $\alpha \approx -0.6$ both for the single-epoch cases and for the total sample that also included the time-averaged sources. For the standard optically thin synchrotron spectrum the spectral index depends on the particle energy spectral index according to $\alpha_{\rm thin} = (1 - s)/2$. Traditionally $s = 2.2 - 2.5$, corresponding to $\alpha = -0.6 - -0.75$, and most *Planck*'s high-frequency spectra are flatter than this.

Low-frequency flatness The lower-frequency flatness is usually taken to be due to either an unresolved, optically thick core or a combination of aging synchrotron jet components moving to lower frequencies. Multiple individual spectra with different ages and self-absorption turnover peak frequencies are known to create a generally flat but bumpy spectrum [22]. It is likely that also here the low-frequency flatness in many sources is due to superposition of multiple synchrotron components, although in some cases the spectrum is so smooth that the multi-component explanation seems unlikely.

High-frequency flatness The higher-frequency spectral flatness seen in many sources is even more interesting. Even though many of the spectra have the "expected" optically thin spectral index between -0.7 and -1.2 (synchrotron losses steepen the spectrum by an additional

$\Delta\alpha = -0.5$), most of them are harder that this. 15 sources even have a high-frequency spectrum flatter than -0.3. [21] concluded that even though the early release data are not sufficient enough to be definite, it is unlikely that the flatness of all, or even most of the sources, could be due to multiple shock spectra or additional IR components, although in some cases this might be the case.

In particular, [21] listed 10 sources that were observed only once (to exclude the flatness caused by averaging over multiple epochs) and whose high-frequency radio spectrum was flatter than $\alpha \gtrsim -0.5$. For seven of these they ruled out flatness due to multiple components and claimed the spectrum to be straight within the error bars. In these example cases the synchrotron spectrum could not be explained with standard acceleration scenario and particle spectra with $s \geq 2$. Instead, the observations are compatible with an electron index $s \approx 1.5$ and $\alpha_{\rm thin}$ having flattest values around -0.2 and the steepest around -0.7 (presumably *after* the steepening due to synchrotron losses). Similar suggestions have been made already earlier by, *e.g.*, [23; 24; 25].

3.1. *Examples: Preliminary case studies*

[21] discussed a few examples in order to illustrate the general features. For most of them no detailed numerical modelling was done, but only simple spectral components were fitted to the radio–optical spectra according to the shock-in-jet scenario. In addition, they also analysed the radio lightcurves to gain some information regarding the past activity of the sources. Here I describe four of their sources as examples; the spectra and the lightcurves are given in [21].

0234+285 A good example of a single-component spectrum with optically thin high-radio-frequency spectral index $\alpha \approx -0.61$ that after a steepening of $\Delta\alpha = -0.5$ joins the optical data. This is what one would expect from one relatively old dominating component (shock), and is in accordance with the previous major flare, which peaked at 37 GHz two years before the *Planck* snapshot.

0235+164 A much bumpier spectrum shows evidence of multiple components contributing to the total emission, and [21] fitted the source with two synchrotron components. The lack of smoothness due to multiple components is again understandable looking at the long-time monitoring lightcurve that shows at least four large outbursts during the previous five years, but with no flaring during the *Planck* snapshot. They also speculated that these shocks would also still produce significant amounts of SSC X-rays while the fact that *Fermi* only observed upper limits for the gamma-ray flux is expected because even the most recent shock was already two years old during the *Planck* observation.

1253-055 (3C 279) Also here the highest-peaking radio component, peaking around 50 GHz, with $\alpha_{\rm thin} = -0.6$, can be made to join the optical spectrum after strong steepening (by $\Delta\alpha \approx 1$) in the IR regime. Especially in this source there is little room for additional IR components in the spectrum regardless of whether the spectrum is modelled by one flat and strongly steepening component or two steeper components, setting upper limits for possible EC seed photon distributions.

2251+158 (3C 454.3) This was the only source modelled numerically (Tammi et al., in preparation). The spectrum shows two separate structures best modelled with jet emission at the lowest frequencies (see Fuhrmann *et al.*, elsewhere in this volume) and a synchrotron spectrum from a strong shock-in-jet component dominating the sub-millimetre-to-optical spectrum. The spectrum was obtained during the early stages of the strongest outburst ever seen in this source, and the radio-to-optical part was fitted with a very flat ($\alpha_{\rm thin} = -0.2$) synchrotron spectrum.

3.2. Implications
As said, the results are compatible with an electron index $s \approx 1.5$, significantly harder than the ~ 2.2–2.5 associated with the standard first-order Fermi acceleration scenario. As discussed in [21], it is possible to have harder spectra by making assumptions about the shock geometry or the particle scattering process, or when the turbulence effects across the shock are considered. The last alternative of these is an especially interesting one, because the combination of low matter density and relatively strong magnetic field — conditions likely to be found in early Poynting-flux-dominated AGN jets — can enhance the effective strength of the shock and lead to very hard particle spectra [26; 27, and the references therein].

Furthermore, in these conditions also the *second-order* Fermi acceleration can become dominating over some energy range and produce very hard power-law spectra as well as enable high injection energies for the first-order shock acceleration on time-scales comparable to the fastest gamma-ray flares [28; 29]. Although there are only a few models that include the effects of the second-order mechanism and time-dependent particle acceleration, the attempts so far have been successful and encouraging [*e.g.*, 16; 17].

4. Summary
The synchrotron emission originating in in the shocks in the jet is the primary signature of the relativistic particles that are responsible for the IC scattering. Together with the radio spectra and monitoring improved synchrotron modelling offers information and tests that could benefit also many high-energy models.

The recent *Planck* early results have increased the motivation and need for developing multi-zone and multi-component models and working our way from particle-level physics to primary emission and all the way to the observed EC and SSC spectra. In particular they have highlighted the importance of the radio-to-submillimetre observations and the synchrotron modelling even in blazar models concentrating on the highest energies.

Based on the early release data of about one hundred northern AGNs over *Planck* frequencies and simultaneous multifrequency observations, the following key points relevant to gamma-ray modelling can be made [21]:

- The low-frequency radio spectra often shows signs of multiple synchrotron components of different ages contributing to the total spectrum. Different emission sites mean different sets of parameters, making single one-zone models unsuitable for these sources. The number of components affecting the observations at a given frequency range can be estimated from long-term multifrequency lightcurves and VLBI monitoring.
- The high-frequency radio spectrum is in many cases too flat to be explained with the traditional scenario. Instead, particle spectral indices closer to $s = 1.5$ seem to be required.
- Many sources showed ongoing millimetre-to-submillimetre flaring and spectra that can be remarkably well modelled with one dominating synchrotron component. On the other hand, in sources where the last outburst had already happened longer time ago often multiple separate components were seen.

In short, the *Planck* early results highlight the importance of low-energy modelling and long-time monitoring in finding the correct properties of the synchrotron spectra and, consequently, the IC-scattering particle populations responsible for the gamma-ray flares.

References
[1] Böttcher M 2010 *Proc. of the Workshop "Fermi meets Jansky: AGN in Radio and Gamma Rays"* ed Savolainen T, Ros E, Porcas R W and Zensus J A (Max-Planck-Institut für Radioastronomie, Bonn, Germany) p 41 (*Preprint* `arXiv:1006.5048`)

[2] Poutanen J and Stern B 2010 *ApJL* **717** L118–L121
[3] Finke J D and Dermer C D 2010 *ApJL* **714** L303–L307
[4] Valtaoja E and Teräsranta H 1996 *A&As* **120** 491
[5] Lähteenmäki A and Valtaoja E 2003 *ApJ* **590** 95–108
[6] León-Tavares J, Valtaoja E, Tornikoski M *et al.* 2011 *A&A* **532** A146
[7] Jorstad S G, Marscher A P, Mattox J R *et al.* 2001 *ApJ* **556** 738–748
[8] Agudo I, Jorstad S G, Marscher A P *et al.* 2011 *ApJL* **726** L13
[9] Kovalev Y Y, Aller H D, Aller M F *et al.* 2009 *ApJL* **696** L17–L21
[10] Pushkarev A B, Kovalev Y Y and Lister M L 2010 *ApJL* **722** L7–L11
[11] Valtaoja E, Tornikoski M, León-Tavares J *et al.* 2010 *Proc. of the Workshop "Fermi meets Jansky: AGN in Radio and Gamma Rays"* ed Savolainen T, Ros E, Porcas R W and Zensus J A (Max-Planck-Institut für Radioastronomie, Bonn, Germany) pp 73–76
[12] Aleksić J, Antonelli L, Antoranz P *et al.* 2011 *ApJL* **730** L8
[13] Tramacere A, Cavazzuti E, Giommi P *et al.* 2010 *AIPC* vol 1223 pp 79–88
[14] Dermer C D, Finke J D, Krug H and Böttcher M 2009 *ApJ* **692** 32–46
[15] Lindfors E J, Türler M, Valtaoja E *et al.* 2006 *A&A* **456** 895–903
[16] Weidinger M and Spanier F 2010 *A&A* **515** A18
[17] Weidinger M, Rüger M and Spanier F 2010 *ASTRA* **6** 1–7
[18] Abdo A A, Ackermann M, Ajello M *et al.* 2009 *ApJ* **699** 817–823
[19] Marscher A P and Gear W K 1985 *ApJ* **298** 114
[20] Valtaoja E, Teräsranta H, Urpo S *et al.* 1992 *A&A* **254** 71
[21] Aatrokoski J, Ade P, Aghanim N *et al.* 2011 *A&A in press* (*Preprint* `arXiv:1101.2047`)
[22] Cotton W D, Wittels J J and Shapiro I I 1980 *ApJL* **238** L123–L128
[23] Valtaoja E, Haarala S, Lehto H and Valtaoja L 1988 *A&A* **203** 1–20
[24] Hughes P A, Aller H D and Aller M F 1991 *ApJ* **374** 57–71
[25] Gear W K, Stevens J A, Hughes D H *et al.* 1994 *MNRAS* **267** 167
[26] Tammi J 2008 *Proc. Int. Cosmic Ray Conf.* vol 2 pp 251–254
[27] Tammi J 2008 *Int. J. of Modern Phys. D* **17** 1811–1817
[28] Virtanen J J P and Vainio R 2005 *ApJ* **621** 313–323
[29] Tammi J and Duffy P 2009 *MNRAS* **393** 1063–1069

International Workshop on Beamed and Unbeamed Gamma-Rays from Galaxies IOP Publishing
Journal of Physics: Conference Series **355** (2012) 012015 doi:10.1088/1742-6596/355/1/012015

Blazar Compton Efficiencies

J. A. Gupta, I. W. A. Browne, M. W. Peel

Jodrell Bank Centre for Astrophysics, Alan Turing Building, School of Physics and Astronomy, The University of Manchester, Oxford Road, Manchester, M13 9PL

E-mail: jgupta@jb.man.ac.uk

Abstract. The *Fermi* gamma-ray space telescope has dramatically increased the number of gamma-ray blazars known and means that for the first time, a large sample of blazars selected by the strength of their inverse Compton emission exists. We have cross-identified the blazars listed in the first *Fermi*-LAT catalog (1FGL) with the CRATES radio catalogue. Using the 8.4 GHz flux density as a proxy for the jet power, we have computed their Compton efficiencies, a measure of the ability of the jet to convert the power in the ultrarelativistic jet electrons into gamma-rays through the inverse Compton process. We have compared the Compton efficiencies of the two blazar subsets, BL Lacs and FSRQs, and find no evidence that they are different. We also do not find an anti-correlation between Compton efficiency and synchrotron peak frequency.

1. Introduction

The blazar sub-set of radio-loud active galactic nuclei represent some of the most energetic objects in the universe and are seen to dominate the extragalactic gamma-ray sky (e.g. Abdo et al. 2010a). However the nature of high energy blazar emission is still unclear. The distinct two-peak shape in log-space of blazar spectral energy distributions reveals that there are two emission mechanisms at work. The low energy SED is attributed to synchrotron emission produced by ultra-relativistic jet electrons. The origin of the high energy emission is more contentious, and both hadronic and leptonic emission models have been proposed (see contributions by Dermer and Reimer in these proceedings). This work does not consider contributions from hadronic emission processes and instead focuses on the leptonic source of high-energy blazar emission, i.e. inverse Compton scattering of low energy photons by the synchrotron electrons.

There are two possible origins for the seed photons required by the inverse Compton process. They can be produced in the jet via the synchrotron emission process (synchrotron self-Compton, SSC) or they can be from AGN regions external to the jet, such as the broad line region or the accretion disk (external Compton, EC). One open issue in blazar research is the importance of external photons; it has been proposed that the ratio of EC to SSC emission can account for the range of spectral energy distributions seen amongst blazars and subsequently give rise to the "blazar sequence", an observed anti-correlation between synchrotron peak frequency and radio luminosity (Fossati et al. 1998).

The theoretical explanation of the blazar sequence uses the abundance of external photons that are available to be upscattered through the inverse Compton process (Ghisellini et al. 1998). Although more recently this has been extended to include black hole mass and accretion rate (Ghisellini et al. 2008), in this work we only consider the role of external photons. The argument is that increasing the number of photons available to be scattered decreases the synchrotron cut-off frequency, and hence the SED synchrotron peak frequency. Therefore the more external photons, the lower the synchrotron peak frequency, as seen between the two types of blazar: flat spectrum radio quasars (FSRQs) whose optical spectra show strong broad emission lines and BL Lac objects where there is no evidence for an external source of photons. This has been interpreted both as strong evidence for the blazar sequence (e.g. Fossati et al. 1998, Ghisellini et al. 2008) and as evidence that BL Lacs and FSRQs are physically different classes of objects (e.g. Anton & Browne 2005).

The *Fermi* gamma-ray space telescope has dramatically increased the number of known gamma-ray blazars. This means that for the first time, we can investigate large samples of blazars selected solely on the strength of their inverse Compton emission.

2. Compton efficiency

We introduce the concept of "Compton efficiency" as a measure of how efficiently a blazar converts power in the form of relativistic jet electrons into high-energy emission via the inverse Compton emission mechanism. We picture a jet where the energy losses are dominated by SSC emission and then increase the number of external photons. This will result in the integrated gamma-ray luminosity increasing and the synchrotron peak frequency decreasing, however the low frequency radio synchrotron emission is unaffected. Therefore, we define Compton efficiency, ϵ, as the ratio of the emission at the high energy (inverse Compton) SED peak to the compact nuclear radio emission, i.e.

$$\epsilon = \log_{10} \frac{(\nu S_\nu)_{\mathrm{IC,\ peak}}}{(\nu S_\nu)_{\mathrm{GHz}}}. \tag{1}$$

In our definition of Compton efficiency, we use the radio emission as a proxy for the jet power because emission at such low frequencies is unaffected by the energy loss processes. We use $(\nu S_\nu)_{\mathrm{IC,\ peak}}$ to measure the strength of the gamma-ray emission, where $\nu_{\mathrm{IC,\ peak}}$ and $S_{\nu_{\mathrm{IC,\ peak}}}$ are the frequency and the flux density respectively at the SED inverse Compton peak. This is so that we are probing the maximum energy output; using a gamma-ray flux in a specific band would probe different parts of the SED from blazar to blazar. In the context of the blazar sequence, ignoring differences in black hole mass and accretion rate, FSRQs should have higher ϵ than BL Lacs because there are more photons available to reduce the jet energy.

Note that our Compton efficiency measure is different to the 'Compton dominance' that has been used by e.g. Ghisellini et al. (2008). Compton dominance measures the ratio of the output at the two SED peaks and is perhaps a better measure of the importance of external photons. However, synchrotron peak measurements are only available for $< 20\%$ of the 1FGL blazars from Abdo et al. (2010a) and we feel that only considering these bright sources would introduce additional selection biases to our results. In addition, we note that Abdo et al. (2010a) present a relationship between then sychrotron peak output, the synchrotron peak frequency and the radio flux density, indicating that the radio emission can be taken as a good proxy for the jet power.

3. Sample

We use the first *Fermi*-LAT (Large Area Telescope) catalogue (1FGL) (Abdo et al. 2010b) and cross-correlate with the Combined Radio All-Sky Targeted Eight GHz Survey (CRATES) (Healey et al. 2007). This gives a sample of 224 FSRQs and 167 BL Lacs which are compact flat spectrum radio sources and have been detected in the 0.1–100 GeV energy range during the first 11 months of science observations with *Fermi*. Further details of the selection process are given in Gupta et al. (2011).

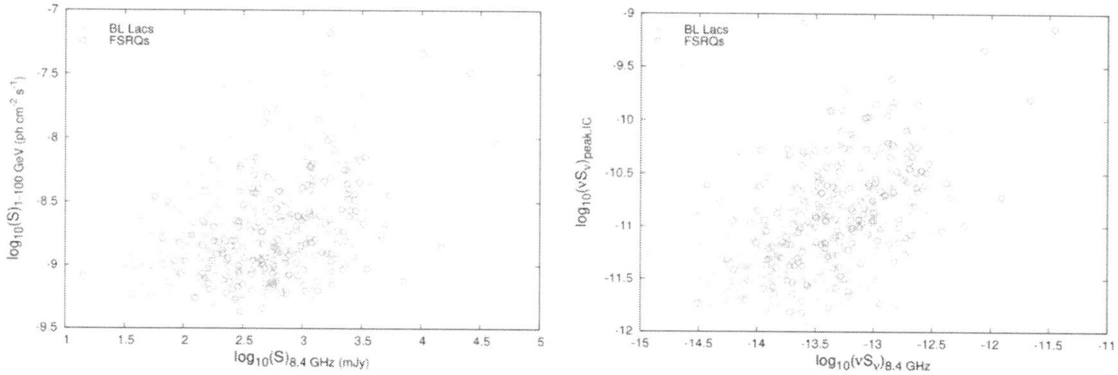

Figure 1. **Left:** CRATES 8.4 GHz flux density against *Fermi* 1–100 GeV flux for BL Lacs (green filled triangles) and FSRQs (blue unfilled circles). There is a weak correlation. **Right:** As left panel but 8.4 GHz radio flux density (multiplied by ν for consistency) against νS_ν at the peak of the inverse Compton SED regime. There is a correlation.

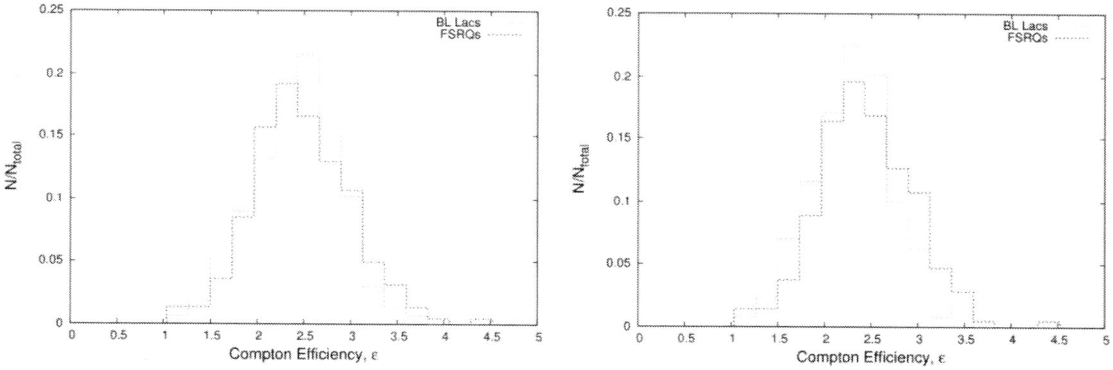

Figure 2. **Left:** Histogram showing the distribution of Compton efficiencies for BL Lacs (green dashed) and FSRQs (blue dotted). **Right:** As left panel but for sources stronger than 100 mJy at 8.4 GHz.

4. Results

The left panel of Figure 1 plots the 1FGL 1–100GeV flux against the CRATES 8.4 GHz flux density for FSRQs and BL Lacs. There is a weak positive correlation (Spearman rank test coefficient $\rho = 0.293$) in agreement with the same correlation reported by elsewhere (e.g. Abdo et al. 2010b, Ghirlanda et al. 2010, Peel et al. 2011). The right panel of Figure 1 plots (νS_ν) at the inverse Compton peak against (νS_ν) at 8.4 GHz. These quantities show a stronger correlation (Spearman rank test coefficient $\rho = 0.534$), indicating that the correlation between radio and gamma-ray emission is real and that the spread in the left panel is partly due to the spread in inverse Compton peak frequencies. In both panels of Figure 1 there is an indication that, for a given gamma-ray measurement, BL Lacs on average have a lower radio measurement compared to the FSRQs. This would suggest that BL Lacs have higher Compton efficiencies than FSRQs.

The left panel of Figure 2 plots the Compton efficiency distributions for BL Lacs and FSRQs. There is no separation between the two types of blazar and the K-S test gives a probability $P = 38\%$ that the two distributions are drawn from the same parent population. The median

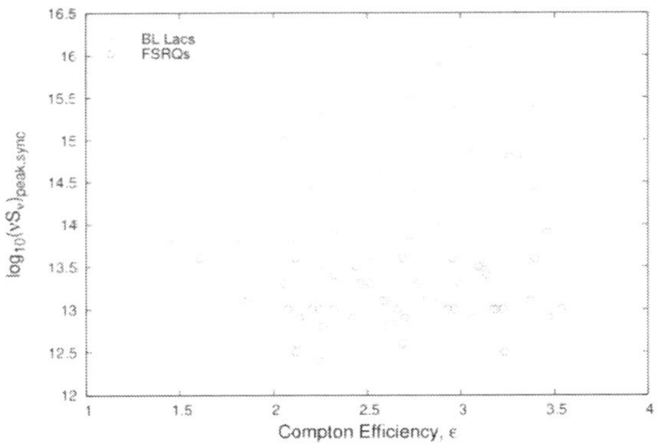

Figure 3. Synchrotron peak frequency against Compton efficiency for BL Lacs (green filled triangles) and FSRQs (blue unfilled circles). A positive correlation is seen for BL Lacs while there is no correlation for FSRQs.

Compton efficiencies are $\epsilon = 2.44$ for the FSRQs and $\epsilon = 2.47$ for the BL Lacs and a Wilcoxon rank sum test on the two distributions does not reject the null hypothesis that the two medians are equal (at the 5% level, $P = 0.51$). These results indicate that BL Lacs and FSRQs are indistinguishable in terms of Compton efficiences, although we do also find evidence that this could be due to selection effects (shown in right panel of Figure 2, see Section 5).

Figure 3 plots Compton efficiency against synchrotron peak frequency for a sub-sample of our objects that are included in the spectral energy distribution fitting by Abdo et al. (2010a). There is no correlation for the FSRQS (Spearman rank test coefficient $\rho = 0.088$) and a positive correlation for the BL Lacs (Spearman rank test coefficient $\rho = 0.532$).

5. Discussion

Our results indicate that BL Lacs and FSRQs are indistinguishable in terms of Compton efficiences. In addition to this, we do not see an anti-correlation between Compton efficiency and synchrotron peak frequency. Before we draw conclusions from our results, we investigate the possibility that our results are due to selection biases. A more detailed discussion is given in Gupta et al. (2011) but the key points are summarised below.

5.1. Selection effects

We have only considered sources that have been identified as blazars in the 1FGL catalogue. There are 373 1FGL sources at $|b| > 10°$ which have no association, so our results could be explained if the association process did not identify high Compton efficiency (i.e. low radio flux density) FSRQs. In addition to this, the sensitivity of the LAT depends on photon index, so it detects BL Lacs to a fainter flux limit than FSRQs (Abdo et al. 2010c).

There is evidence for a bias against FSRQs in Figure 1 where the number of FSRQs compared to BL Lacs decreases with radio flux density. We have therefore repeated our analysis only considering objects with radio flux density $S_{8.4\ \mathrm{GHz}} > 100$ mJy, shown in the right panel of Figure 2. There is now a difference in the Compton efficiency distributions for FSRQs and BL Lacs: there is only a 2.2% probability that the distributions have the same parent population (K-S test) and the Wilcoxon rank sum test rejects the null hypothesis that the medians ($\epsilon = 2.33$ for BL Lacs and $\epsilon = 2.40$ for FSRQs) are the same ($P = 0.012$). This implies that our result that the Compton efficiencies of BL Lacs and FSRQs are indistinguishable could just be a result of incompleteness at low radio flux densities (especially for FSRQs). However, further investigation is currently underway, investigating the effects of other cuts (e.g. in gamma-ray flux).

5.2. Variability
Blazars are characterised by their variability, both on short and long time-scales. The CRATES radio measurements were made before the 1FGL observations so radio variability will increase the spread in Compton efficiencies. However, on these timescales ($\sim$ 10 years) radio flux densities are unlikely to vary by more than a factor of two (e.g. Jackson et al. 2010) which would not be able to account for the dispersion in our Compton efficiencies. This implies that our results cannot be explained by variability, although it would be useful to have simultaneous radio and gamma-ray measurements.

6. Summary and conclusion
The Compton efficiency parameter has been introduced as a way to quantify the amount of jet energy that is converted into high energy emission via the inverse Compton process. We cross-correlate the first *Fermi*-LAT catalogue with the CRATES radio catalogue and find that the Compton efficiencies for BL Lacs and FSRQs are not statistically different. We also find no evidence for an anti-correlation between Compton efficiency and synchrotron peak frequency. However, we do see evidence that our sample is incomplete at low radio flux densities which could influence, but not fully account for, our results.

Our results can be interpreted as evidence against the existence of the blazar sequence. However, a number of assumptions have been made due to the limitations of the data that is currently available. The release of larger and better defined samples from the *Fermi* mission will significantly improve these results. In particular, we await the release of improved SEDs of *Fermi*-detected blazars so that our Compton efficiency measure can be compared to the Compton dominance.

References
Abdo A. A. et al., 2010a, *ApJ*, **716**, 30
Abdo A. A. et al., 2010b, *ApJS*, **188**, 405
Abdo A. A. et al., 2010c, *ApJ*, **715**, 429
Anton S., Browne I. W. A, 2005, *MNRAS*, **356**, 225
Fossati G., Maraschi L., Celotti A., Comastri A., Ghisellini G., 1998, *MNRAS*, **299**, 433
Ghirlanda G., Ghisellini G., Tavecchio F., Foschini L., 2010, *MNRAS*, **407**, 791
Ghisellini G., Celotti A., Fossati G., Maraschi L., Comastri A., 1998, *MNRAS*, **301**, 451
Ghisellini G., Tavecchio F., 2008, *MNRAS*, **387**, 1669
Gupta J. A., Browne I. W. A., Peel M. W., *Preprint* astro-ph/1106.5172
Healey, S. E., Romani, R. W., Taylor, G. B., Sadler, E. M., Ricci, R., Murphy, T., Ulvestad, J. S., Winn, J. N., 2007, *ApJS*, **171**, 61
Jackson N., Browne I. W. A., Battye R. A., Gabuzda D., Taylor A. C., 2010, *MNRAS*, **401**, 1388
Peel M. W. et al., 2011, *MNRAS*, **410**, 2690

Journal of Physics: Conference Series **355** (2012) 012016 doi:10.1088/1742-6596/355/1/012016

On the role of black hole spin and accretion in powering relativistic jets in AGN

Laura Maraschi[1], **Monica Colpi**[2], **Gabriele Ghisellini**[3], **Albino Perego**[2], **Fabrizio Tavecchio**[3]
[1]INAF – Osservatorio Astronomico di Brera, Via Brera 28, I–20100 Milano, Italy;
[2]INFN, Sezione di Milano–Bicocca, I–20126 Milano, Italy;
[3]INAF – Osservatorio Astronomico di Brera, Via Bianchi 46, I–23807 Merate, Italy

E-mail: laura.maraschi@brera.inaf.it

Abstract. Motivated by the remarkable results on the general properties of blazars obtained from *Fermi* observations we discuss the role of spin and accretion in powering relativistic jets in AGNs. We consider studies of the Blandford–Znajek mechanism taking into account advection of magnetic flux from an accretion disk surrounding the Black Hole (BH) which indicate that the mechanism is inhibited for values of the BH spin $a \lesssim 0.5$. However large scale propagation of the jet with highly relativistic bulk acceleration likely requires even larger spin values. The Blandford–Znajek mechanism then predicts a tight connection between jet power and accretion power, accounting for the correlation between γ–ray luminosity and accretion disk luminosity found in the bright *Fermi* blazar sample. The suggested spin threshold would lead to a grand unification of AGN based on only two intrinsic parameters: the BH spin and the accretion rate in Eddington units.

1. Introduction

It is now commonly accepted that the non thermal emission from radio loud Active Galactic Nuclei (AGN) is powered by relativistic jets emanating from the nucleus. (Begelman, Blandford & Rees 1984, Blandford 1990).

However the physical mechanism producing relativistic jets is not yet completely understood (see e.g. Tchekhovskoy, Narayan & McKinney 2010 and references therein). Present investigations focus on the mechanism first studied by Blandford & Znajek (1977) (hereafter BZ mechanism) which still appears to be the most promising candidate. This process, purely electromagnetic, rests on the differential rotation between a spinning Black Hole (BH) and its magnetosphere. In the case of AGN advection of the necessary magnetic flux can be naturally associated with the accretion flow, leading to the generation of a collimated Poynting flux *at the expense of the BH rotational energy.*

Particles in the jet can be accelerated up to bulk Lorentz factors Γ of 10–100 at $r \simeq$100–1000 R_S with collimation consistent with jet phenomenology (McKinney 2006 and references therein). Thus in general the mechanism appears successful in accounting for the properties of relativistic jets in AGN. However the efficiency of the process, in particular its dependency on the spin parameter $a = Jc/GM^2$ (where J is the BH angular momentum and M the BH mass), is still under discussion. Spin dependencies in the range a^2–a^6 have been obtained (Mc Kinney

2005,Tchekhovskoy et al. 2010), while a global efficiency of $\simeq 10\%$ seems attainable for maximal BH spin (McKinney 2006).

On the observational side the results provided by the *Fermi* satellite, continuously surveying the whole sky in gamma-rays (100 MeV - 10 GeV) with the LAT instrument have opened a new era for the study of relativistic jets. AGN represent the dominant population in the gamma-ray sky, the vast majority being classified as "blazars", radio loud AGN whose relativistic jets closely aligne with the terrestrial line of sight. Most of their radiative luminosity is emitted in the gamma-ray range. The results of the first three months of observations, the "bright AGN survey" performed by LAT (LBAS, Abdo et al. 2009) are still of particular importance since this sample is completely identified and complemented by multifrequency data allowing to study in a systematic way the general properties of the population (Abdo et al. 2010, Ghisellini et al. 2010; hereafter G10).

In the following we will argue that the correlation between gamma-ray luminosity and accretion disk luminosity found in the three month *Fermi* blazar sample (G10) supports the hypothesis that relativistic jets in AGN are generated only for high spin values. In §2 we summarize existing studies of the dependency of the strength of the BZ process on the BH spin. In §3 we discuss the correlation mentioned above with minimal assumptions on radiation models. We conclude in §4 outlining a scheme for a "grand" unification" of AGN.

2. Arguments for an effective spin threshold in the Blandford–Znajek mechanism

The analytic formula for the power generated via the BZ mechanism as derived by e.g. MacDonald and Thorne (1982) is given by:

$$L_{\rm BZ} \simeq k\,\frac{\Omega_{\rm F}(\Omega_{\rm H}-\Omega_{\rm F})}{\Omega_{\rm H}^2}a^2 \times B_{\rm H}^2 r_{\rm H}^2 c \qquad (1)$$

where $\Omega_{\rm F}$ and $\Omega_{\rm H}$ are the angular velocities of the magnetic field lines and of the BH respectively, $B_{\rm H}$ is the magnetic field at the horizon, $r_{\rm H}$ is the horizon radius.

This formula can be read as the product of two terms: one, with physical dimension of power, $(B_{\rm H}^2 r_{\rm H}^2 c)$ corresponds to the magnetic energy flux at the horizon, the other, dimensionless, includes the dependency on the BH spin a and on the different angular velocities of the spinning BH and the surrounding magnetosphere.

Eq. (1) shows that, in order to produce a positive power, $\Omega_{\rm H}$ must be larger than $\Omega_{\rm F}$, corresponding to the physical intuition that, in order to brake the spinning BH, the rotation of external field lines should be slower than the hole itself. In the opposite case the process would correspond to absorption rather than emission of power. BZ suggested that the field angular velocity will adjust to $\Omega_{\rm F} \simeq \Omega_{\rm H}/2$ as found in their approximate analytic solution. For this value of $\Omega_{\rm F}$ the extracted power is maximized. This result is confirmed by Komissarov (2001, 2004) via direct numerical simulations and with an improved description of the electrodynamics around a spinning black hole in the force-free approximation.

The above models however do not include a description of how the field is advected to the BH from an *accretion flow*, the only possible source of magnetic flux in the AGN case. In the latter case, one could assume, approximately, that the poloidal component of the field, frozen in the disk, rotates with the same angular velocity as the rotating gas at the innermost stable corotating orbit (ISCO), i.e., $\Omega_{\rm F} \sim \Omega_{\rm ISCO}(a)$. For low values of the BH spin, $\Omega_{\rm ISCO}(a)$ exceeds $\Omega_{\rm H}(a)$: the crossing occurs at $a_{\rm cr} = 0.35$. For $a < a_{\rm cr}$ the analytic expression for the BZ power (Eq. 1) suggests that the process would *not* take place as long as $\Omega_{\rm F} > \Omega_{\rm H}$. This is physically plausible: in fact the interaction of the *slowly* spinning BH with the *faster* rotating infalling magnetized plasma, should result in a torque of opposite sign with respect to that acting in the case of the BZ process, as previously noted and discussed by Li (2000a,b) and Li & Paczyinski (2000).

Computing the analytic expression of the BZ power for $\Omega_F = \Omega_{ISCO}$ for different values of a one can verify that the difference with respect to the BZ hypothesis is minor for intermediate values of a, but the power becomes becomes *negative* for values of $a \lesssim 0.4$. In other words, for $a \lesssim 0.4$ the conditions necessary for the BZ process to work may not be met.

This rather simplistic approach, finds some support in appropriate GRMHD simulations. In fact Mc Kinney and Gammie (2004) studied the self consistent evolution of a rotating BH surrounded by a weakly magnetized torus ($H/R = 0.1$), a situation close to that described above. We mention two aspects of the results discussed in their paper: one is the finding that Ω_F tends to Ω_{ISCO} at the equator and remains constant troughout the plunging region; the other one is that the ratio of "electromagnetic luminosity" to nominal "accretion luminosity", i.e. the efficiency of the BZ process is very small below $a = 0.6$ and becomes negative for $a < 0.5$, i.e. the BZ effect is not operating. Thus an "effective" spin threshold is found in the above paper at $a > 0.5$. This threshold does not seem to be present in the simulations of Tchekhovskoy et al. (2010). More general estimates of the power output from a rapidly rotating black hole with self–consistently determined geometry and magnitude of the magnetic field for a realistic accretion disk are summarized in Mc Kinney (2005). Due to the different structure of the magnetosphere which depends on the BH spin, a higher fraction of the power is emitted in the polar regions (jet) for high spin. In any case below $a = 0.5$ the jet efficiency is less than 10^{-4}.

For a rapidly rotating BH a larger magnetic flux is generated close to the nearly force free poles. Due to the low density and high magnetic field these are the only regions that can give rise to highly relativistic jets, generating particle flows with bulk Lorentz factors reaching 10 - 100 on large scales (10–1000 r_g) and collimation properties consistent with what inferred from observations (Mc Kinney 2006). Note however that the latter results are valid only if the pressure of the surrounding medium is small compared to the magnetic pressure in the jet. If, for the same value of the accretion rate the BH has a relatively low spin, $0.5 \lesssim a \lesssim a_{\rm lim}$, the magnetic pressure in the forming jet could be insufficient to satisfy the condition for large scale bulk acceleration. The production of highly relativistic jets may therefore require a higher "effective" spin threshold $a_{\rm lim} > 0.5$.

The jet power efficiency in terms of accretion power for $a > 0.5$ is estimated as $\eta_{\rm jet} = 6.8 \times 10^{-2}[\Omega_{\rm H} \,/\, \Omega_{\rm H}(a = 1)]^5$ (Mc Kinney 2005). Since $\Omega_{\rm H}(a)$ decreases by a factor 2 between $a = 1$ and $a = 0.8$, with a predicted decrease of a factor 30 in jet power for the same accretion power. A further drop of a factor 30 in efficiency is predicted between $a = 0.8$ and $a = 0.5$. Given the large decrease in efficiency we provisionally adopt an "effective" spin threshold for relativistic jet production at $a_{\rm lim} \simeq 0.8$. The threshold value may also depend on the accretion rate. For larger accretion rates the ambient pressure may be higher and a higher field (spin) may be required to launch large scale relativistic jets.

3. The power of relativistic jets and accretion

The physical scale of the power produced in the form of collimated Poynting flux is given by the dimensional factor in the BZ formula, which corresponds to the magnetic energy flux at the horizon. Assuming that beyond $R_{\rm ISCO}$, i.e. in the nearly free fall or "plunging" region, the flow is quasi–spherical and the magnetic and kinetic energy densities are in near equipartition, the value of the magnetic field at the horizon can be directly linked to the accretion rate (Krolik 1999, Maraschi 2001, Levinson 2010). The resulting equation shows that the physical scale of the process is *proportional* to the accretion power.

$$B_{\rm H}^2 r_{\rm H}^2 c \sim 8\pi\, \rho c^2 r_{\rm H}^2 c \simeq 2\, \dot{M} c^2 \tag{2}$$

Eq. (2) does not include the dependence of the jet power on BH spin. If relativistic jets are generated only for a limited range of spin ($a > 0.8$), a direct correlation between jet power and accretion power is predicted.

Relativistic jets in AGN emit most their radiative power in the γ–ray band as indicated already by the results obtained with the *Gamma Ray Observatory* (e.g., von Montigny et al. 1995). Early estimates of jet power and accretion power in few brightest quasars and BL Lac objects led to the suggestion that the two powers were comparable and a connection was present, though with large uncertainties (Maraschi 2001, Ghisellini & Celotti 2002, Maraschi and Tavecchio 2003, Celotti & Ghisellini 2008).

The ongoing all sky survey of the *Fermi* satellite with the LAT instrument has opened a new era in γ–ray astronomy providing rich and unbiased samples of γ–ray selected blazars. The "Bright AGN sample" based on the 3 months survey by the LAT instrument on board *Fermi* (LBAS, Abdo et al. 2009), completely identified and complemented by a rich set of multifrequency data has been fully modelled by G10 who found a significant correlation between jet power and accretion power for the FSRQ subsample in agreement with the expectation from Eq. 2. The correlation was further extended by analogous modelling of a group of 28 FSRQ with high ($z > 2$) redshift (Ghisellini et al. 2011).

Here we consider observed quantities, i.e. the observed γ–ray luminosities vs. the accretion disk luminosities or upper limits (Fig. 1). This choice is intended to show that a correlation exists *prior* to modelling. Few points should be recalled:

(i) The accretion disk luminosity is derived from the measured luminosity of a blue bump which is present in the SED of FSRQs. If a blue bump is not apparent, as is typical for BL Lac objects, an upper limit is reported.

(ii) The observed γ–ray emission is beamed, therefore depends on the angle to the line of sight. However it can be shown (see Celotti & Ghisellini 2008) that the total power spent by the jet to produce the observed γ–ray emission is $P_\gamma \sim L_\gamma/\Gamma^2$, where Γ is the bulk Lorentz factor of the jet. Since L_γ often dominates the bolometric output, especially for FSRQs, it represents a good proxy for the radiative power of the jet $P_{\rm rad} \sim L_{\rm bol}/\Gamma^2$, which yields in turn a robust lower limit to the jet power $P_{\rm jet}$.

(iii) Excluding upper limits, L_γ and $L_{\rm d}$ correlate (see Fig. 1), with a random probability of $P_{\rm random} = 5.6 \times 10^{-3}$ after subtracting off the common redshift dependence (see Padovani 1992 for the partial correlation analysis used here). The analogous correlation derived in G10, where each source was modelled yielding individual values for the bulk Lorentz factor, has higher significance, however the existence of a correlation between observed quantities shows that it is robust, and confirms, a posteriori, the relatively narrow range of Γ derived earlier for these sources: $10 \leq \Gamma \leq 17$ with an average of $\langle\Gamma\rangle = 13$. From Fig. 1 one derives $L_\gamma \simeq 30\, L_{\rm d}$ and this *implies* $P_\gamma \sim L_\gamma/\Gamma^2 \sim 0.18\, L_{\rm d}$.

The bottom line is that also from this simple, model independent analysis one derives that for FSRQs jet power and accretion power are correlated. This is expected if the jet is produced through the BZ mechanism because the magnetic field at the horizon which determines the jet power is naturally related to the accretion energy density as discussed above (eq. 2).

Moreover the relation $P_\gamma \simeq 0.18 L_{\rm d}$ allows to estimate the efficiency of the jet production mechanism. Assuming that the total power carried by the jet is $P_{\rm jet} = 10\, P_\gamma$ one derives $P_{\rm jet} = 1.8\, L_{\rm d} = 1.8\eta_{\rm acc}, \dot{M}c^2$. The efficiency of the accretion process $\eta_{\rm acc}$ is estimated to be 6% for a "standard" accretion disk around a non rotating BH therefore for the FSRQs in this sample we derive $P_{\rm jet} = 0.108\, \dot{M}c^2$ i.e. a jet production efficiency $\eta_{\rm jet} \simeq 0.1$.

Comparing with the results of GRMHD simulations we find that a 10% efficiency is at the top of the expected values. Moreover all the simulations predict that the efficiency is strongly dependent on the value of the BH spin (e.g. Martínez-Sansigre & Rawlings 2011 for a direct comparison of the results of different simulations). Hence the observed correlation implies that i) the jet production efficiency is high, i.e. $\eta_{\rm jet} \simeq 10^{-1}$, and ii) all these sources should have close to maximal spin.

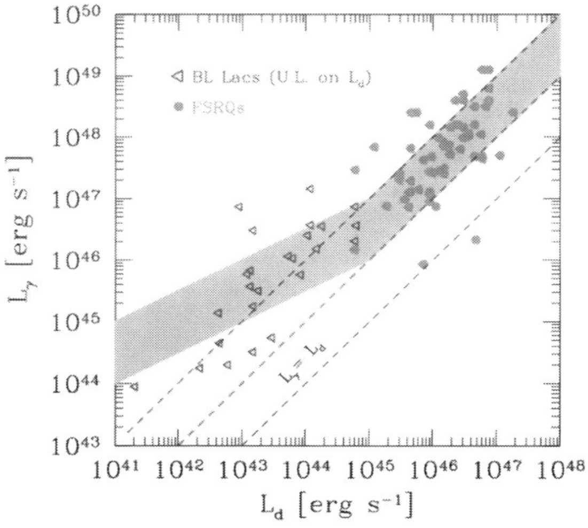

Figure 1. The γ–ray luminosity L_γ as a function of the accretion disk luminosity $L_{\rm d}$ for the blazars discussed in G10. Triangles correspond to upper limits of $L_{\rm d}$. The grey stripe, which is only indicative, corresponds to $L_{\rm d} \propto \dot{M}$ above a critical $\dot{M}/\dot{M}_{\rm Edd}$ value, and $L_{\rm d} \propto \dot{M}^2$ below, while $L_\gamma \propto \dot{M}$ for any value of $\dot{M}$.

Clearly these estimates suffer from uncertainties. If the average jet Lorentz factors in the emission regions were higher the jet powers could be lower. However we have assumed for the accretion disk the "standard" efficiency, while in the case of a spinning BH the accretion efficiency could be larger, implying even larger jet production efficiency.

A second important point concerns the comparison of the properties of FSRQ and BL Lac objects: FSRQs populate the luminosity range $10^{47} - 10^{45}$ erg/s and disappear below this limit, while BL Lac objects populate the lower luminosity region. The luminosity separation was interpreted as due to a population of AGN with similar BH mass ($10^9 M_\odot$) but decreasing accretion rates (see Ghisellini et al. 2009). This is consistent with physical models of AGN evolution (Cavaliere et al. 1985). Accretion disk models predict that for $\dot{m} \equiv \dot{M}/\dot{M}_{\rm Edd}$ below 10^{-2} , due to the reduced opacity, the radiative efficiency of the disk diminishes (e.g. Narayan, Garcia & McClintock 1997) causing the disappearence of the optically thick emission (Blue Bump) and of the Broad Line Region, turning the jetted AGN into a BL Lac object. However, according to Eq. 2, the BZ power (i.e. the jet power) is still proportional to the accretion rate. The jet radiative efficiency should not change substantially since particle acceleration and radiation processes remain similar. As a result *in this regime the jet power and L_γ still trace the accretion rate*, while the accretion disk luminosity drops faster, causing a flattening of the expected L_γ–$L_{\rm disk}$ relation. The predicted behaviour is shown as a grey stripe in Fig. 1.

4. Conclusions: a grand unification scheme for AGN ?

Based on the results of some GRMHD simulations we argued for the adoption of an "effective" spin threshold to launch relativistic jets in an accreting SMBH. The threshold value should be above $a = 0.5$ and possibly as high as $a \sim 0.8$. The correlation between γ-ray luminosity and accretion disk luminosity found in the *Fermi* Lat Bright AGN Sample supports this hypothesis. The relation between the power produced in γ-rays and the luminosity of the accretion disk is found to be $P_\gamma \sim 0.2 L_{\rm d}$ implying $P_{\rm jet} \sim 2 L_{\rm d}$ i.e. an efficiency of the jet production mechanism

Journal of Physics: Conference Series **355** (2012) 012016 doi:10.1088/1742-6596/355/1/012016

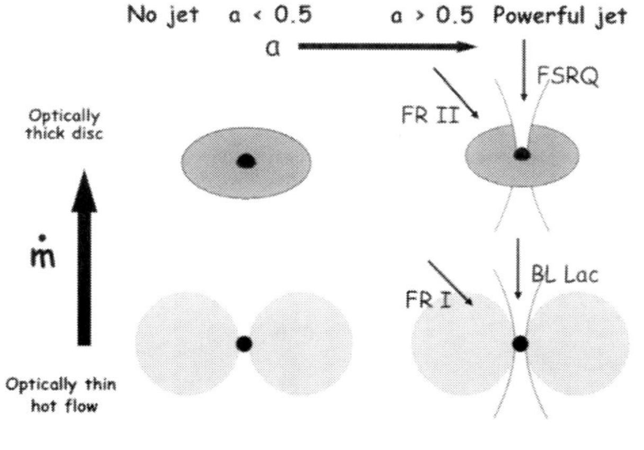

Figure 2. Grand Unification Scheme for AGN: the four quadrants qualitatively depict the properties of AGN for different accretion rates and BH spin intervals. The vertical axis represents the value of $\dot{m}$: the pink circles indicate that the accretion flow occurs in the ADAF mode (low $\dot{m}$), the blue ellipses indicate a standard accretion disk (high $\dot{m}$). The horizontal axis represents the value of the BH spin a: for a larger than 0.5-0.8 a relativistc jet is generated. The associated radio source is likely FRII for high values of $\dot{m}$ and FR I for low values of $\dot{m}$. If viewed at small angle the AGN will be classified as FSRQ or BL Lac respectively.

close to the radiative efficiency of the accretion disk. Given the efficiencies for jet production estimated by Mc Kinney (2005) our result requires that all the sources in the sample have close to maximal spin. Deeper γ–ray samples, which will be obtained in few years from now, may probe this hypothesis in more depth but will require extensive work for identification of the optical counterparts and study of their properties.

The link between accretion power and jet power together with the limited range of spin for which the process can actually take place naturally suggests a Grand Unification Scheme for AGNs in terms of the main *intrinsic* astrophysical properties, i.e. mass, spin and accretion rate. We will not consider the mass explicitly but will use mass normalized quantities, i.e. the adimensional angular momentum a and the accretion rate in Eddington units $\dot{m}$.

The hypothesis that the BH spin was the physical parameter underlying the radio-loud radio quiet dicotomy of AGN has been repeatedly discussed in the past without definitive answers (e.g Wilson and Colbert 1995, Blandford 1990, Meier 1999, 2002, Sikora et al. 2007). The existence of an effective spin threshold for the BZ process at $a_{\rm lim} \simeq 0.8$ introduces a definite separation between radio loud and radio-quiet AGN with the following properties:

- In luminous radio loud AGN ($a > a_{\rm lim}$), accreting at $\dot{m} > 10^{-2}$ (where $\dot{m}$ is the accretion rate measured in Eddington units, $\dot{m} \equiv \dot{M}/\dot{M}_{\rm Edd}$, $\dot{M}_{\rm Edd} \equiv L_{\rm Edd}/(\eta c^2)$ and $\eta = 0.1$) highly relativistic jets are able to carry significant non thermal powers to large or intermediate distances from the nucleus; The luminosity of the accretion disk and the power carried by the jet are comparable. The radio source morphology should be FR II. Viewed along the jet axis these objects would be classified as Flat Spectrum Radio Quasars.
- Luminous radio quiet AGNs do not have relativistic jets; they may emit weakly in the radio band (compared to their optical luminosity) through aborted jets or non relativistic outflows (e.g. the Blandford & Payne mechanism) or other mechanisms. The (weak) radio power may still correlate with the luminosity of the accretion due to the common dependence on

the accretion rate, but their ratio should be $< 10^{-2}$.

- For accretion rates $\dot{m} < 10^{-2}$ the appearence of the AGN nucleus in the optical-UV-X-ray changes, due to the transition to a different, radiatively inefficient, accretion regime (RIAF). If the central BH has high spin, the jet power (still proportional to $\dot{m}$) can be larger than the observed thermal luminosity. These objects may therefore be very radio loud, but their absolute powers both thermal and non thermal will be less than those of the first two classes for the same mass of the SMBH. If viewed along the jet axis these objects would likely be classified as BL Lacs. The radio-quiet counterparts (low spin) could correspond to the LINERS AGN category.

A more detailed discussion of these ideas will be given elsewhere.

[1] Abdo A. A., et al., 2009, ApJ, 700, 597
[2] Abdo A. A., et al., 2010, ApJ, 716, 30
[3] Begelman M. C., Blandford R. D., Rees M. J., 1984, Rev. Mod. Phys., 56, 255
[4] Blandford R. D., 1990, Saas-Fee Advanced Course 20
[5] Blandford R.D. & Znajek R.L., 1977, MNRAS, 179, 433
[6] Blandford R. D., Payne D. G., 1982, MNRAS, 199, 883
[7] Cavaliere A., et al. 1985, ApJ, 296, 415
[8] Celotti A., Ghisellini G., 2008, MNRAS, 385, 283
[9] Ghisellini G., Celotti A., 2002, ASPC, 258, 273
[10] Ghisellini G., Maraschi L. & Tavecchio F., 2009, MNRAS, 396, L105
[11] Ghisellini G., et al. , 2010, MNRAS, 402, 497 (G10)
[12] Ghisellini G., et al., 2011, MNRAS, 411, 901
[13] Komissarov S. S., 2001, MNRAS, 326, L41
[14] Komissarov S. S., 2004, MNRAS, 350, 1431
[15] Krolik J. H., 1999, ApJ, 515, L73
[16] Levinson A., 2010, IJMPD, 19, 649
[17] Li L.-X., Paczyński B., 2000, ApJ, 534, L197
[18] Li L.-X., 2000a, PhRvD, 61, 084016
[19] Li L.-X., 2000b, ApJ, 533, L115
[20] MacDonald D., Thorne K. S., 1982, MNRAS, 198, 345
[21] Maraschi L., 2001, AIPC, 586, 409
[22] Maraschi L., Tavecchio F., 2003, ApJ, 593, 667
[23] Martínez-Sansigre A., Rawlings S., 2011, MNRAS, 414, 1937
[24] McKinney J. C., Gammie C. F., 2004, ApJ, 611, 977
[25] McKinney J. C., 2005, ApJ, 630, L5
[26] McKinney J. C., 2006, MNRAS, 368, 1561
[27] Meier D. L., 1999, ApJ, 522, 753
[28] Meier D. L., 2002, NewAR, 46, 247
[29] Narayan R., Garcia M. R., McClintock J. E., 1997, ApJ, 478, L79
[30] Padovani P., 1992, A&A, 256, 399
[31] Sikora M., Stawarz Ł., Lasota J.-P., 2007, ApJ, 658, 815
[32] Tchekhovskoy A., Narayan R., McKinney J. C., 2010, ApJ, 711, 50
[33] Wilson A. S., Colbert E. J. M., 1995, ApJ, 438, 62

Journal of Physics: Conference Series **355** (2012) 012017 doi:10.1088/1742-6596/355/1/012017

Multi-wavelength Observations of the HBL 1ES 1011+496 in Spring 2008

Riho Reinthal[1], Stefan Rügamer[2], Elina J. Lindfors[1], Daniel Mazin[3], Antonio Stamerra[4,*] for the MAGIC collaboration, Francesco Longo[5], Fabrizio Lucarelli[6], Carlotta Pittori[6] for the AGILE team and Anne Lähteenmäki[7]

[1] *Tuorla Observatory, University of Turku, FI-21500 Piikkiö, Finland*
[2] *Universität Würzburg, D-97074 Würzburg, Germany*
[3] *Institut de Fisica d'Altes Energies (IFAE), E-08193 Bellaterra (Barcelona), Spain*
[4] *Deutsches Elektronen-Synchrotron (DESY), Zeuthen, Germany*
[5] *Università di Trieste and INFN Trieste, I-34127 Trieste, Italy*
[6] *Agenzia Spaziale Italiana (ASI) Science Data Center, I-00044, Frascati, Italy*
[7] *Aalto University Metsähovi Radio Observatory, FI-02540 Kylmälä, Finland*
[*] *For the Swift team*

Väisäläntie 20, FI-21500 Piikkiö, Finland

E-mail: `rirein@utu.fi`

Abstract. In the spring of 2008 MAGIC organised multi-wavelength (MWL) observations of the blazar 1ES 1011+496. 1ES 1011+496 is a high-frequency peaked BL Lac object (HBL) discovered at VHE γ-rays by MAGIC in spring 2007 during an optical outburst reported by the Tuorla Blazar Monitoring Programme. MAGIC re-observed the source during the 2008 MWL campaign which also included the Metsähovi, KVA, Swift and AGILE telescopes. This was the first MWL campaign on this source that also included VHE coverage. MAGIC observed 1ES 1011+496 from March 4th to May 24th 2008 for a total of 27.9 hours, of which 20 h remained after quality cuts. The observations resulted in a detection of the source at $\sim$7 σ significance level with a mean flux and spectral index similar to those during the discovery.

Here we will present the results of the MAGIC observations of the source in combination with contemporaneous observations at other wavelengths (radio, optical, X-rays, high energy γ-rays) and discuss their implications on the modelling of the spectral energy distribution.

Keywords: AGNs, BL Lac objects, 1ES 1011+964, multi-wavelength campaigns, VHE γ-rays, MAGIC

1. Introduction

Blazars are an extreme class of Active Galactic Nuclei (AGN) in which the jet of relativistic particles – particles travelling at nearly the speed of light – originating from close to the central black hole is pointed close to our line of sight. This causes an apparent increase in the jet luminosity and a decrease in the variability time scales of the AGN core across all wavelengths due to relativistic beaming effects. Blazar emissions are dominated by non-thermal radiation and their spectral energy distributions (SED) are characterised by two broad peaks of which the lower energy one is believed to originate from synchrotron emission of charged particles in the jet. The higher energy peak is most commonly explained by inverse Compton scattering of either the synchrotron (synchrotron self Compton - SSC, see e.g. [1, 2]) or external (external Compton - EC, [3, 4]) seed photons by the electrons and positrons in the jet but hadronic models [5, 6] are also considered. Blazars are divided into flat spectrum radio quasars and BL Lacertae objects (see e.g. [7]). The latter are further subdivided into low-, intermediate- and high-frequency peaked BL Lac (LBL, IBL and HBL, respectively) objects according to the frequency of the first peak which in the case of HBLs is located in the UV to soft X-ray regime. The second peak in HBLs is located in the GeV to TeV range which makes them good candidates to detect very high energy (VHE, >100 GeV) gamma-rays from.

Blazars show variability in flux at all wavelengths as well as spectral shape on time scales ranging from months to a few minutes. Therefore, in order to shed light on the VHE emission mechanisms of blazars, simultaneous observations of these sources across multiple wavelengths are required. It is particularly important to study the connection between the variability between different wavebands. With the increased sensitivity of the latest generation of Imaging Atmospheric Cherenkov Telescopes (IACTs) such campaigns have become feasible for a larger number of sources. Three campaigns of this type were organised by MAGIC in 2008: one on 1ES 1011+496 described in this contribution and the other two on Mrk 180 and 1ES 2344+514.

1ES 1011+496 is an HBL located at a medium redshift of z=0.212. It was discovered at VHE by MAGIC in 2007 following an optical high state reported by the Tuorla Blazar Monitoring Programme[1] [8]. Previous observations of this source at VHE showed only a hint of signal (see e.g. [9]) and the results presented here constitute the first follow-up detection of the source. The object shows frequent fluctuations in its optical brightness with the core flux increasing by up to a factor of ~2-3 over the lowest states during flares. It has also been observed in the X-rays by Einstein [10] and the Swift/XRT [11] telescopes always showing a steep spectrum. There was no definite detection by EGRET [12] but it has been clearly detected by Fermi and was included already in the Fermi 3-month catalogue. The Fermi observations showed a very hard γ-ray spectrum [11]. The MWL campaign was the first of its kind on this source.

2. Observation Campaign and Results

MAGIC (Major Atmospheric Gamma-ray Imaging Cherenkov) is a system of two 17 m IACTs located on the Canary island of La Palma, Spain at ~2200 m asl. At the time of the 2008 campaign the second telescope was still under construction and observations were carried out using only MAGIC-I which has been in operation since 2004. Thanks largely to its 236 m^2 reflective area MAGIC-I achieved an energy threshold of ~60 GeV (~30 GeV with a special trigger) – the lowest of any IACT. It reached a sensitivity of ~1.6% of Crab nebula flux (5 σ in 50 h) >200 GeV with an energy resolution of ~20-30% and an angular resolution of ~0.1° [13].

The MWL observations were centered around common observation windows of the AGILE [14] satellite and the MAGIC-I telescope in spring 2008. Optical coverage was provided by the KVA [15] telescope which is operated concurrently with the MAGIC telescopes. Observations in the radio were performed by the Metsähovi radio telescope [16] and in the X-rays by the Swift [17] satellite.

MAGIC was able to observe the object on 25 nights between March 4th and May 24th 2008. The observation period was plagued by poor weather conditions at La Palma with frequent clouds, rainfall, strong wind and calima towards the end of the observation window. Nevertheless

[1] http://users.utu.fi/kani/1m

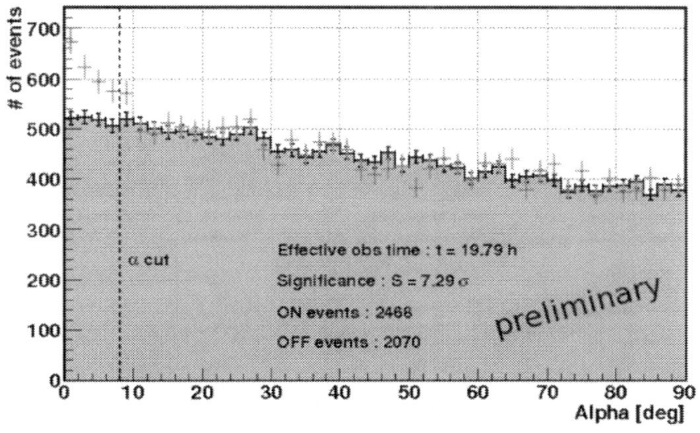

Figure 1. Distribution of α angle for ON-region and OFF-region data. ON-data is marked by the red crosses while the filled region is the OFF-data. Signal region is marked by the dashed line.

a total of 27.9 hours of data was collected of which 8 hours had to be removed due to adverse weather conditions. The remaining 20 hours were not or only marginally affected by bad weather and survived the quality cuts. The observations were done in wobble mode between zenith angles of 20° and 37° and the data was analysed using the MAGIC standard analysis (for a comprehensive overview see e.g. [13]). The cut on α (¡ 8 deg) to determine the significance of the signal and the number of excess events was optimised on Crab data from the same period. The MAGIC observations resulted in the detection of 398 events over 2070 background events in the signal region corresponding to a significance of the excess of 7.29 standard deviations using equation (17) in [18]. This constitutes the first confirmation of the source as a VHE emitter (see Figure 1). The energy threshold of this analysis was $\sim$ 150 GeV which allowed for signal extraction down to about 100 GeV (see figure 2).

3. Discussion

The MAGIC observations showed a mean flux level similar to that measured during the discovery in 2007 [8]. Also the derived spectrum was compatible within the error bars although weak indications for a spectral hardening in 2008 were seen (see Figure 2).

The VHE γ-ray spectrum can be described by the simple power law (with the differential

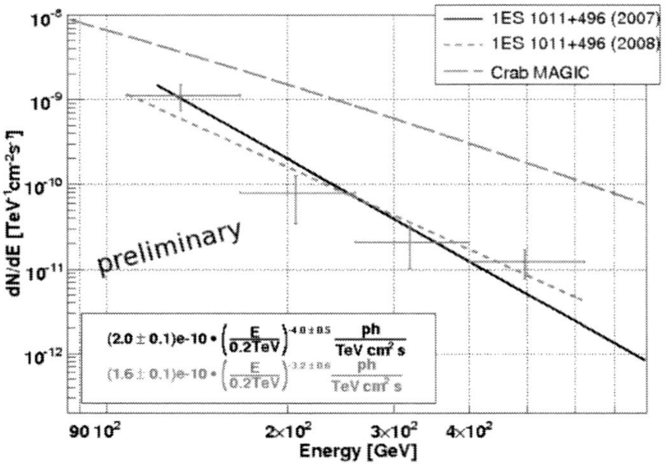

Figure 2. The measured spectrum. The dotted line is the power law fit to the spectrum while the solid line is the spectrum measured during the discovery. The crosses are the data points from 2008. For comparison the Crab Nebula spectrum is shown (the grey dashed line).

Journal of Physics: Conference Series 355 (2012) 012017 doi:10.1088/1742-6596/355/1/012017

flux given in units of $\mathrm{TeV^{-1}m^{-2}s^{-1}}$):

$$\frac{dN}{dE} = (1.6 \pm 0.1) \times 10^{-10} (\frac{\mathrm{E}}{200\,\mathrm{GeV}})^{-3.20\pm0.58}$$

No significant variability of the VHE flux was detected during the campaign (see Figure 3). A constant fit to the data yielded a χ^2/d.o.f. of 13.9/18, corresponding to a probability of 73%. Nevertheless in one night (MJD 54561.9) a flux has been found which is >4 sigma off the fit line.

Of the other telescopes Metsähovi and AGILE did not detect the source within the given time windows. In the optical R band the source was at a relatively low state for most of the time with a major flare occurring at the end of the observation period. From the end of April to the beginning of May the source was also observed in the optical V and B band. The fluxes in these bands followed the same trends as in the R-band. In the X-rays a flare could be clearly detected as can be seen in the light curve in Figure 3. The observation sampling prevented to define a baseline flux level, therefore the rise and fall times of the flare could not be evaluated. The Swift UVOT data are still being analysed.

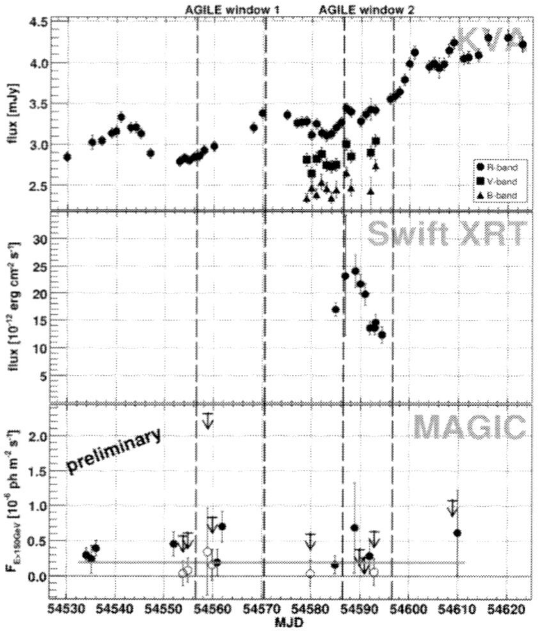

Figure 3. The MWL light curve of 1ES 1011+496 with daily binning. The optical data (including the B and V band) are from the KVA telescope and are not host-galaxy corrected. In the MAGIC data, for points consistent with zero flux 99.7% upper limits are shown and the grey horizontal line denotes a constant fit to the data points. The full circles are the significant data points while the open ones represent fluxes consistent with zero but with significance >0.

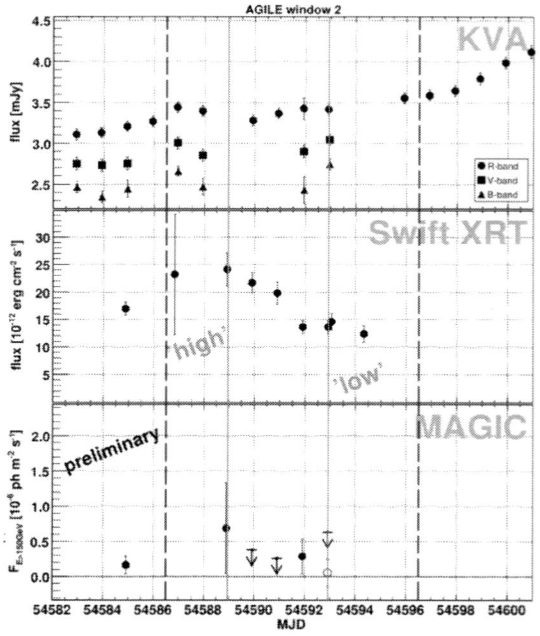

Figure 4. A zoom into the MWL light curve around the second AGILE window. The same properties apply to the data points as in Figure 3. A 'high' and 'low' state has been defined according to the X-ray flux level which are used to compose simultaneous SEDs (see Figure 5).

B[G]	δ	R[cm]	K [cm^{-3}]	e_1	e_2	γ_{min}	γ_{break}	γ_{max}
0.19	25	1.E+16	6.6E+03	2.0	4.1	1	4.E+04	6.E+05

Table 1. Model parameters resulting from the modelling. See text for details.

There does not appear to be any obvious connection between the optical and X-ray, X-ray and VHE or optical and VHE light curves (see Figure 4). However the statistics of the individual data points of the VHE light curve are too low to draw definitive conclusions.

Two separate data sets, 'high' and 'low', defined according to the X-ray flux state of the source as can be seen in Figure 4, were used to construct a simultaneous SED (see Figure 5). Due to the small second observation window, no AGILE upper limit could be extracted for this period. The data has been modelled using a one-zone SSC [19] model. The small difference between the SEDs of the low and high X-ray flux states does not make a dedicated modelling of both seem meaningful. Therefore, only one model fit was applied to the data for the time being.

The model assumes an emission region characterised by its radius R, magnetic field B and Doppler factor δ containing an electron population. The electrons with density K (defined at $\gamma = 1$) with which they are injected into the acceleration region follow a broken power law distribution with index e_1 for $\gamma_{min} < \gamma < \gamma_{break}$ and e_2 for $\gamma_{break} < \gamma < \gamma_{max}$. The spectral indices and power law slopes are derived then self-consistently from cooling and acceleration processes. The model parameters are given in table 1 and the results represent standard values for HBLs. The model fits the data rather well indicating that SSC mechanisms are the likely explanation for the observed emission.

Note that even though the Fermi bow-tie was not measured simultaneously to the other data it still fits relatively well into the applied model scenario. Taking into account that also the VHE spectrum seems not to have changed considerably from the time of the discovery to these observations, 1ES 1011+496 seems to show a rather constant emission at gamma-ray energies. At the lower energies, however, the average optical flux of ~2.8 mJy from the core was ≈ 40% lower than the one measured during the discovery and equal to the one measured during the 2006 MAGIC observations. The X-ray flux was measured at almost 10x higher level than the one derived from archival observations in the VHE discovery paper. The low (non-simultaneous) X-ray flux constraining the SED modelling led [8] to the conclusion that in this

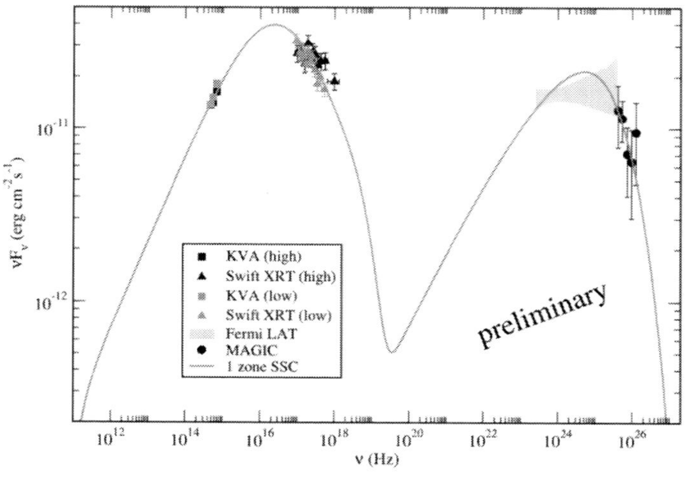

Figure 5. MWL SED of 1ES 1011+496. The 'high' and 'low' data correspond to the observation times defined in Fig 4. The KVA data are host-galaxy corrected following [20]. The Fermi bow-tie deduced from the LAT 1-year Catalog [21] is added for illustrative purposes. The MAGIC data have been corrected for EBL absorption using [22].

source, the inverse-Compton component would dominate over the synchrotron component. On the contrary, the simultaneous SEDs from these observations indicate that this interpretation may not be true, corroborating that this source is synchrotron dominated like most HBLs.

4. Conclusion

We organised the first MWL campaign including VHE coverage on the HBL 1ES 1011+496. The campaign resulted in the first confirmation of the source at VHE at a flux level similar to that measured during the discovery and a hint of hardening of the VHE spectrum compared to the discovery. The VHE flux was consistent with being constant, though significant variability in both the optical and X-ray wavelengths was detected. No apparent correlation between the flux levels in different wavebands was found. The data was modelled with a one-zone SSC model utilising typical parameters for BL Lac objects. The model provides a good fit to the available spectral data indicating that this scenario is sufficient to describe the emission mechanisms of the source. However data analysis and more refined modelling is still ongoing.

Acknowledgements: The work of RR and EL has been supported by grant 127740 of the Academy of Finland. We would also like to thank the Instituto de Astrofísica de Canarias for the excellent working conditions at the Observatorio del Roque de los Muchachos on La Palma. The support of the German BMBF and MPG, the Italian INFN, the Swiss National Fund SNF, and the Spanish MICINN is gratefully acknowledged. This work was also supported by the Marie Curie program, by the CPAN CSD2007-00042 and MultiDark CSD2009-00064 projects of the Spanish Consolider-Ingenio 2010 programme, by grant DO02-353 of the Bulgarian NSF, by the YIP of the Helmholtz Gemeinschaft, by the DFG Cluster of Excellence "Origin and Structure of the Universe", by the DFG Collaborative Research Centers SFB823/C4 and SFB876/C3, and by the Polish MNiSzW grant 745/N-HESS-MAGIC/2010/0. The AGILE Mission is funded by the Italian Space Agency (ASI) with scientific and programmatic participation by the Italian Institute of Astrophysics (INAF) and the Italian Institute of Nuclear Physics (INFN).

We gratefully acknowledge N. Gehrels for approving this set of ToOs and the entire *Swift* team, the duty scientists and science planners for the dedicated support, making these observations possible.

References

[1] Maraschi, L., Ghisellini, G., Celotti, A., ApJ, 1992, **397**: L5-L9
[2] Costamante, L., & Ghisellini, G. A&A, 2002, **384**: 56-71
[3] Dermer, C. D., & Schlickeiser, R. ApJ, 1993, **416**: 458-484
[4] Ghisellini, G., Tavecchio, F., Chiaberge, M., A&A, 2005, **432**: 401-410
[5] Mannheim, K., A&A, 1993, **269**: 67-76
[6] Mücke, A., *et al*, Astro-ph, **18**: 593-613
[7] Ghisellini, G. *et al* , MNRAS, 1998, **301**: 451-468
[8] Albert, J., *et al* (the MAGIC collaboration), ApJ, 2007, **667**: L21-L24
[9] Albert, J., *et al* (the MAGIC collaboration), ApJ, 2008, **681**: 944-953
[10] Elvis, M., *et al*, ApJS, 1992, **80**: 257-303
[11] Abdo, A. A., *et al* (the Fermi-LAT collaboration), ApJ, 2010, **716**: 30-70
[12] Sowards-Emmerd, D., Roman, R. W., & Michelson, P. F., ApJ, 2003, **590**: 109-122
[13] Albert, J., *et al* (the MAGIC collaboration), ApJ, 2008, **674**: 1037-1055
[14] Tavani, M., *et al*, NIMPA, 2008, **588**: 52-62
[15] Aleksić, J., *et al* (the MAGIC collaboration), A&A, 2011, **530**: A4
[16] Teräsranta, H., *et al*, A&AS, 1998, **132**: 305-331
[17] Gehrels, N., *et al*, ApJ, 2004, **611**: 1005-1020
[18] Li, T.-P. and Ma, Y.-Q., 1983, ApJ, **272**: 317-324
[19] Maraschi, L. & Tavecchio, F., ApJ, 2003, **593**: 667-675
[20] Nilsson, K., *et al*, A&A, 2007, **475**: 199-207
[21] Abdo, A. A., *et al* (the Fermi-LAT collaboration), ApJ, 2009, **707**: 1310-1333
[22] Kneiske, T. M. & Dole, H., A&A, 2010, **515**: A19

International Workshop on Beamed and Unbeamed Gamma-Rays from Galaxies IOP Publishing
Journal of Physics: Conference Series **355** (2012) 012018 doi:10.1088/1742-6596/355/1/012018

Constraints given by the MAGIC discovery of the Flat Spectrum Radio Quasar PKS1222+21 in VHE Gamma rays

J. Becerra-González[1,2], A. Stamerra[3], K. Saito[4], D. Mazin[5], F. Tavecchio[6], L. Maraschi[6], T. Terzic[7]E. Prandini[8] and J. Sitarek[9] on behalf of the MAGIC collaboration.

[1] Instituto de Astrofísica de Canarias, [2] Universidad de La Laguna, [3] Universita di Siena and INFN Pisa, [4] Max-Planck-Institut fur Physik, [5] Institut de Fisica d'Altes Energies, Barcelona, [6] INAF National Institute for Astrophysics, [7] University of Rijeka, [8] Universita di Padova and INFN, [9] University of Lodz

E-mail: jbecerra@iac.es

Abstract. The MAGIC telescopes discovered very high energy (VHE, E>100 GeV) gamma-ray emission coming from the distant Flat Spectrum Radio Quasar (FSRQ) PKS 1222+21 (4C +21.35, z=0.432). It is the second most distant VHE gamma-ray source, with well measured redshift, detected until now. The observation was performed on 2010 June 17 (MJD 55364.9) using the two 17 m diameter imaging Cherenkov telescopes on La Palma (Canary Islands, Spain). The MAGIC detection coincides with high energy MeV/GeV gamma-ray activity measured by the Large Area Telescope (LAT) on board the Fermi satellite. The averaged integral flux above 100 GeV is equivalent to 1 Crab Nebula flux. The VHE flux measured by MAGIC varies significantly within the 30 minutes of exposure implying a flux doubling time of about 10 minutes. The VHE and MeV/GeV spectra, corrected for the absorption by the extragalactic background light, can be described by a single power law with photon index 2.72 ± 0.34 between 3 GeV and 400 GeV, consistent with gamma-ray emission belonging to a single component in the jet. The absence of a spectral cut-off constrains the gamma-ray emission region to lie outside the broad line region, which would otherwise absorb the VHE gamma-rays. Together with the detected fast variability, this challenges present emission models from jets in FSRQs.

1. Introduction

Flat Spectrum Radio Quasars (FSRQs) are blazars which present broad emission lines often accompanied by a 'big blue bump" in the optical-UV region, associated with the thermal emission from the accretion disk. Only two FSRQs had been detected in Very High Energy (VHE) γ-rays, 3c279 [1] and PKS 1510-089 [2]. The detection of the FSRQ PKS 1222+21 (4C +21.35, $z = 0.432$,[3]) by MAGIC makes it the second most distant object with known redshift (after 3C 279, the most distant VHE source detected until now, z=0.536 [1, 4]) ever detected at VHE[1].

In this proceeding, we present the MAGIC discovery of this source in the VHE [5], which was detected during a gamma-ray flare announced by the Fermi/LAT collaboration, and its physics implications.

[1] The redshift measurement ($z = 0.444$) of the VHE BL Lac 3C 66A has large uncertainties [6].

2. MAGIC observations

MAGIC (Major Atmospheric Gamma-ray Imaging Cherenkov) consists of two 17 m diameter Imaging Atmospheric Cherenkov Telescopes (IACT) located at the Roque de los Muchachos, Canary Island of La Palma (28°46′ N, 17°53′ W), at the height of 2200 m a.s.l. MAGIC observed PKS 1222+21 from 2010 May 3, but the source was not detected until June 17, 2010 (MJD 55364). The observation carried out on June 17 lasted 30 minutes and was performed as part of a Target of Opportunity program triggered by an increased flux in the *Fermi* energy band [7]. During this detection by MAGIC the source was close to the brightest flare ever observed by the *Fermi* Large Area Telescope (LAT) [8]. The data were taken at zenith angles between 26° and 35°. Stereoscopic events, triggered by both telescopes, were analyzed in the MARS analysis framework [9]. Details on the analysis can be found in [10] whereas the performance of the MAGIC telescopes stereo system will be discussed in detail in a forthcoming paper [11].

The signal evaluation was performed using the θ^2 distribution (squared angular distance between the true and reconstructed source position), see the Figure 1. We got an excess of 190 γ-like events ($6\,\gamma$/min.) above a background of 86 events, which corresponds to a statistical significance of $10.2\,\sigma$ using eq. 17 in [12]. The energy threshold of this analysis is ≈ 70 GeV.

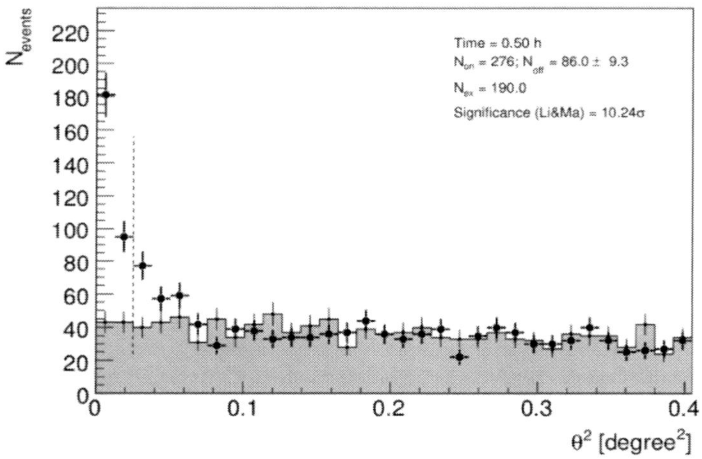

Figure 1. Distribution of the squared angular distance (θ^2) for events in the direction of PKS 1222+21 (black points) and normalized off-source events (gray histogram). The signal is extracted in the θ^2-region denoted by the vertical dashed line.

3. The Very High Energy Spectrum

The differential energy spectrum of PKS 1222+21 measured by MAGIC extends from 70 GeV to 400 GeV (Figure 2), and it is well-described by a simple power law of the form:

$$\frac{dN}{dE} = N_{200}\left(\frac{E}{200\,\mathrm{GeV}}\right)^{-\Gamma} \tag{1}$$

with a photon index $\Gamma = 3.75 \pm 0.27_{\rm stat} \pm 0.2_{\rm syst}$ and a normalization constant at 200 GeV of $N_{200} = (7.8 \pm 1.2_{\rm stat} \pm 3.5_{\rm syst}) \times 10^{-10}\mathrm{cm^{-2}s^{-1}TeV^{-1}}$, yielding an integral flux $(4.6 \pm 0.5) \times 10^{-10}\,\mathrm{cm^{-2}s^{-1}}$ (≈ 1 Crab Nebula flux) at E> 100 GeV. For energies higher than

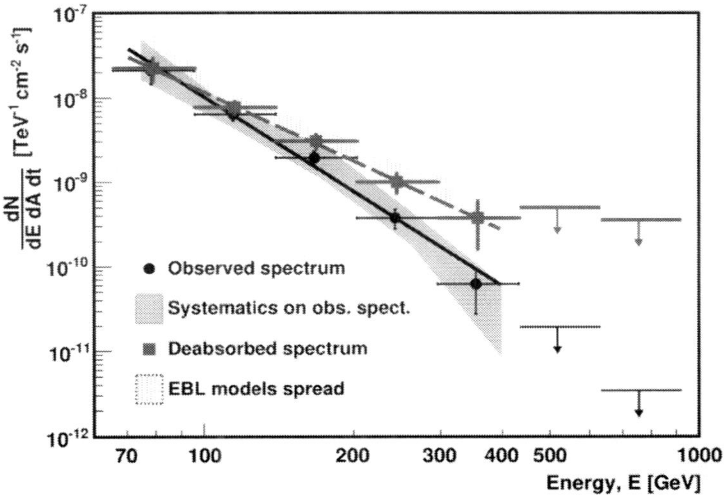

Figure 2. Differential energy spectrum of PKS 1222+21 as measured by MAGIC on 2010 June 17. Differential fluxes are shown as black points, upper limits (95% C.L.) as black arrows. The black line is the best fit to a power law. The gray shaded area represents the systematic uncertainties of the analysis. The absorption corrected spectrum and upper limits using the EBL model of [13] are shown by the blue squares and arrows; the dashed blue line is the best fit power law. The blue-striped area illustrates the uncertainties due to differences in the EBL models of [14, 15, 16] and [1].

400 GeV, no significant excess was found but two upper limits corresponding to 95% confidence level (C.L.) have been derived. The method used for the spectral reconstruction is the so-called "Tikhonov" unfolding algorithm [17], which takes into account the finite energy resolution of the instrument and the biases in the energy reconstruction. The systematic uncertainty of the analysis (studied by using different cuts and different unfolding algorithms) is shown by the gray area.

To derive the intrinsic spectrum emitted at the source the VHE spectral data is de-absorbed using the EBL model described in [13] (Fig. 2 blue points). It is well fitted by a power law with an intrinsic photon index of $\Gamma_{\rm intr} = 2.72 \pm 0.34$ between 70 GeV and 400 GeV. Uncertainties caused by the differences between the EBL models are investigate utilizing different EBL models [13, 14, 15, 16, 1]. The corresponding spread (Fig. 2 blue shaded are) is smaller than the systematic uncertainties of the MAGIC data analysis.

The possibility of a cut-off in the energy spectrum was studied using the χ^2 difference method. Fitting the spectrum with a broken power law with different photon indexes and values for the cut-off, we cannot exclude the presence of a cut-off above 130 GeV for a photon index 2.4 (the lowest possible value compatible with fit uncertainties and with the *Fermi*/LAT data) or above 180 GeV for a photon index 2.7, with the available statistics.

4. The Spectral Energy Distribution

The high-energy Spectral Energy Distribution (SED) is shown in Figure 3 where the MeV/GeV energy range spectrum from *Fermi*/LAT and the GeV/TeV spectrum measured by MAGIC are represented. The source showed a significant flare in the *Fermi* band lasting ~3 days, with a peak flux on 2010 June 18 (MJD 55365) [8]. During the 1/2 hr MAGIC observation there is

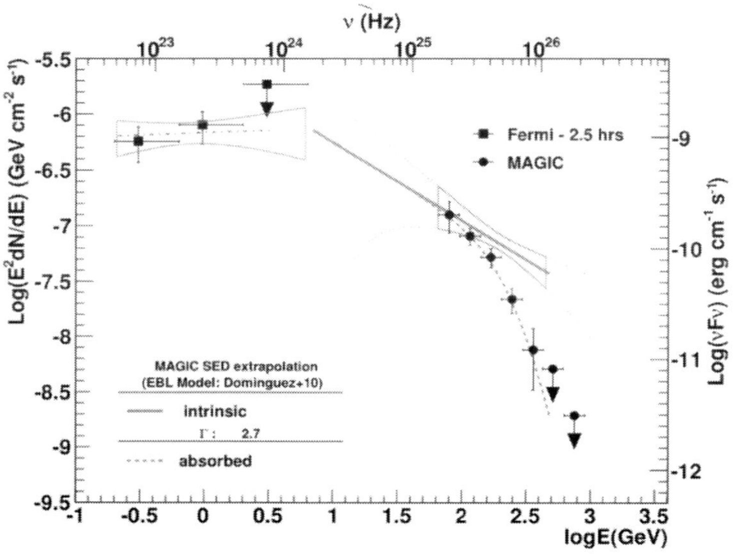

Figure 3. High energy SED of PKS 1222+21 during the flare of 2010 June 17 (MJD 55364.9), showing *Fermi*/LAT (squares) and MAGIC (circles) differential fluxes. A red bow tie in the MeV/GeV range represents the uncertainty of the likelihood fit to the *Fermi*/LAT data. The unfolded and deabsorbed spectral fit of the MAGIC data is also shown as a red bow tie, extrapolated to lower and higher energies (dotted lines). A thick solid line (photon index $\Gamma = 2.7$) indicates a possible extrapolation of the MAGIC deabsorbed data to lower energies. The thick dashed line represents the EBL absorbed spectrum obtained from the extrapolated intrinsic spectrum using the model of [13].

a gap in the LAT exposure, so we analyzed the closest data available, a period of 2.5 hr (MJD 55364.867 to 55364.973) centered around the MAGIC observation. Given the short observation time (chosen in order to be as much contemporaneous as possible with the MAGIC data) there is no detection above 2 GeV in the *Fermi*/LAT data, but an upper limit at the 95% C.L. in the energy range 2 – 6.3 GeV has been calculated together with the spectral points up to 2 GeV (details on the analysis can be found in [8]) and combined with the MAGIC data in the SED shown in Fig. 3.

Extrapolating the intrinsic MAGIC spectrum to lower energies we can see that there is a potentially smooth connection between the *Fermi*/LAT and MAGIC extrapolated data and the photon index steepens from 1.9 in the *Fermi*/LAT range to 2.7 in the MAGIC range. These results agree with the results from the analysis of Fermi-LAT data of larger temporal intervals during this flare and during the whole active period, in which the source spectrum is well described by a broken power law with an energy break between 1 and 3 GeV [8]. Furthermore it is found that the high energy tail ($E > 2$ GeV) of the *Fermi*/LAT spectrum of PKS 1222+21 extends up to 50 GeV, with a photon index in the range 2.4-2.8.

5. Source Variability

Due to the strength of the signal, even if the observation time is as short as 30 minutes, a variability study is possible. In Fig. 4, the MAGIC VHE light curve binned in 6 minutes long intervals is shown. It reveals a flux variation within the 30 minutes of observation time. The hypothesis of a constant flux is rejected ($\chi^2/NDF = 28.3/4$) with high confidence (probability $< 1.1 \times 10^{-5}$). The flux of the background events surviving the γ/hadron selection cuts

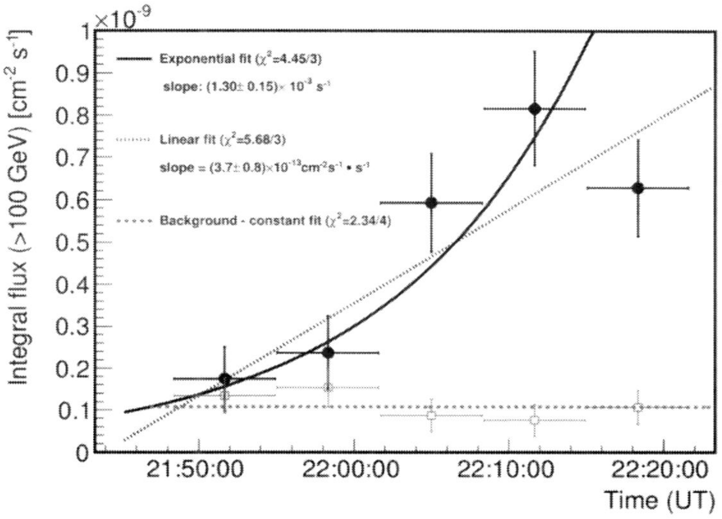

Figure 4. PKS 1222+21 light curve above 100 GeV, in 6 minutes bins (black filled circles). The observation was carried out on MJD 55364. The black solid line is a fit with an exponential function and the black dotted line a fit with a linear function. The gray open squares denote the fluxes from the background events and the gray dashed line is a fit with a constant function to these points.

is compatible with being constant and hence we can exclude a variation of the instrument performance during the observation.

To quantify the variability time scale, we performed an exponential fit (solid black line in Fig. 4). A linear fit is also acceptable but does not allow us to define a time scale unambiguously. For the exponential fit the doubling time of the flare is estimated to $8.6^{+1.1}_{-0.9}$ minutes. The derived timescale corresponds to the fastest time variation ever observed in an FSRQ in the VHE range and in any other energy range [21], and is amongst the shortest measured from any TeV emitting source [22, 23].

6. Conclusions

As discussed, the almost simultaneous VHE and GeV spectra are consistent with a single power law with index $\sim 2.7 \pm 0.3$ between 3 GeV to 400 GeV, without a strong intrinsic cut-off. This evidence suggests that the 100 MeV - 400 GeV emission belongs to a unique component, peaking at $\approx 2-3$ GeV, produced in a single region of the jet. Considering the inverse Compton scattering on external photons and relativistic electrons in the jet as the emission process, as it is usually assumed, we have two possible scenarios. The emitting jet region could be inside of the Broad Line Region (BLR), where the external photons field would be the UV photons from the BLR, or we can assume this emitting region to be outside of the BLR, where the photon field would be composed of the IR photons coming from the torus. There are two important effects that need to be taken into account: the decreased efficiency of the IC scattering occuring in the Klein-Nishina (KN) regime and the absorption of γ-rays through pair production.

The energy above which the KN effects and γ-ray absorption become important in the case of UV external photons from the BLR is $\sim$ tens of GeV, and thus we would expect a cut-off at these energies, if the emitting jet region is inside the BLR. But if we consider as target photons the ones coming from the IR torus this effect is expected to appear at much higher energies above ≈ 1 TeV, because this effect depends on the frequency of the external target photons. Since there is no evidences of a cut-off at low energies we can conclude that the emission should come from outside of the BLR, as has been proposed by the "far dissipation" scenarios (e.g. [20]).

Besides, the other important result of the MAGIC observation is the fast variability, $t_{\rm var}\sim 10$ minutes which indicates an extremely compact emission region. Fast variability suggests that the emission takes place close to the black hole and thus inside of the BLR.

Some explanations have been already proposed, in order to explain this kind of observational facts, invoking the presence of very compact emission regions embedded within the large scale jet [24], turbulent cell model [25] or the possibility of a very strong jet recollimation (e.g. [18]).

In conclusion, the MAGIC observations of VHE emission from the FSRQ PKS 1222+21 set severe constraints on the emission models of blazar jets.

7. Acknowledgments

We would like to thank the Instituto de Astrofísica de Canarias for the excellent working conditions at the Observatorio del Roque de los Muchachos in La Palma. The support of the German BMBF and MPG, the Italian INFN, the Swiss National Fund SNF, and the Spanish MICINN is gratefully acknowledged. This work was also supported by the Marie Curie program, by the CPAN CSD2007-00042 and MultiDark CSD2009-00064 projects of the Spanish Consolider-Ingenio 2010 programme, by grant DO02-353 of the Bulgarian NSF, by grant 127740 of the Academy of Finland, by the YIP of the Helmholtz Gemeinschaft, by the DFG Cluster of Excellence "Origin and Structure of the Universe", and by the Polish MNiSzW grant 745/N-HESS-MAGIC/2010/0.

[1] J. Albert et al., Science,2008, 320, 1752
[2] S. Wagner & B. Behera, 10th *HEAD Meeting*, BAAS, 42, 2, 07.05
[3] D. E. Osterbrock & R. W. Pogge, ApJ,1987,323, 108
[4] J. Aleksic et al., A&A, 2011, 530, A4
[5] J. Aleksić et al., ApJL, 2011, 730, L8
[6] D. A. Bramel,*ApJ* (2005), 629, 108
[7] D. Donato, The Astronomer's Telegram, #2584
[8] Y.T. Tanaka, ApJ, 2011, 733 (1), 19
[9] A. Moralejo et al., Proc. of 31st ICRC, 2009, [arXiv:0907.0943]
[10] J. Aleksic et al., ApJL, 2010, 726, 58, [arxiv:1010.0550]
[11] J. Aleksic et al., *in prep.*
[12] T.-P. Li & Y.-Q. Ma, ApJ, 1983, 272, 317
[13] A. Dominguez et al., MNRAS, 2011, 410, 2556
[14] T. M. Kneiske & H. Dole, A&A, 2010, 515, 19
[15] R. Gilmore et al., MNRAS, 2009, 399, 1694
[16] A. Franceschini et al., A&A, 2008, 487, 837
[17] J. Albert et al., Nucl. Instr. Meth., 2007, A 583, 494
[18] K. Nalewajko & M. Sikora, MNRAS, 2009, 392, 1205
[19] G. Ghisellini & F. Tavecchio, MNRAS, 2008, 386, 28
[20] M. Sikora et a., ApJ, 2008, 675, 71
[21] L. Foschini et al., A&A, 2011, 530, A77
[22] Aharonian et al., ApJ, 2007,664, L71
[23] J. Albert et al,. ApJ, 2007, 663, 125
[24] F. Tavecchio et al., Submitted to A&A, 2011, [arXiv:1104.0048v1]
[25] A. P. Marscher and S. G.Jorstad, 2010, [arXiv:1005.5551]

Journal of Physics: Conference Series **355** (2012) 012019 doi:10.1088/1742-6596/355/1/012019

Evidence for Helical Magnetic fields in Kiloparsec-Scale AGN Jets and the Action of a Cosmic Battery

D.C. Gabuzda[1], D.M. Christodoulou[2], I. Contopoulos[3], D. Kazanas[4]

[1]Physics Department, University College Cork, Cork, Ireland
[2] Department of Mathematical Sciences, University of Massachusetts Lowell, Lowell 01854
[3] Research Center for Astronomy, Academy of Athens, Athens 11527, Greece
[4] NASA/GSFC, Code 663, Greenbelt, MD 20771

E-mail: d.gabuzda@ucc.ie

Abstract. A search for transverse kiloparsec-scale gradients in Faraday rotation-measure (RM) maps of extragalactic radio sources in the literature has yielded 6 AGNs displaying continuous, monotonic RM gradients across their jets, oriented roughly orthogonal to the local jet direction. The most natural interpretation of such transverse RM gradients is that they are caused by the systematic change in the line-of-sight components of helical magnetic fields associated with these jets. All the identified transverse RM gradients increase in the counterclockwise (CCW) direction on the sky relative to the centers of these AGNs. Taken together with the results of Contopoulos et al. [7], who found evidence for a predominance of clockwise (CW) transverse RM gradients across parsec-scale (VLBI) jets, this provides new evidence for preferred orientations of RM gradients due to helical jet magnetic fields, with a reversal from CW in the inner jets to CCW farther from the centers of activity. This can be explained by the "Poynting–Robertson cosmic-battery" mechanism, which can generate helical magnetic fields with a characteristic "twist," which are expelled with the jet outflows. If the Poynting–Robertson battery mechanism is not operating, an alternative mechanism must be identified, which is able to explain the predominance of CW/CCW RM gradients on parsec/kiloparsec scales.

1. Introduction

It is widely believed that the physical mechanism that drives the jets of extragalactic radio sources is magnetohydrodynamical: the rotating active galactic nucleus (AGN) is threaded by a magnetic field organized on scales out to tens of kiloparsecs; the field is wound up by the rotation of the plasma (i.e., it becomes helical) and drives a collimated outflow (jet) above and below the symmetry plane of the nucleus. The asymptotic velocity attained by the jet material is of the same order as the rotational velocity at its base. It is now well understood that the galactic nucleus consists of two components, a rotating supermassive ($\sim 10^8 M_\odot$) black hole with its event horizon extending out to ~1 AU, and a surrounding rotating accretion disk extending out to ~1 pc. Therefore, extragalactic jets are also expected to consist of two components: an inner, relativistic, axial jet, and an outer, nonrelativistic, extended disk wind [17, 35]. The theoretical details of the jet launching mechanism have been well studied in the literature (e.g., [4, 10]; see also [37], and references therein), where it has been shown that the field is effectively wound up only beyond the Alfvén distance, which is ~10 times the radial extent of the outflow at its base. Thus, the theoretical models suggest that the magnetic field of the inner jet and the outer extended wind will develop significant toroidal components on distances beyond ~10 AU and ~10 pc, respectively, from the AGN centers.

In the past few years, a mechanism has been proposed that seems capable of generating strong, ordered, large-scale magnetic fields in AGN accetion disks and of providing the physical background needed to understand some of the most prominent features observed in magnetized jets. We call this the *Poynting–Robertson* (PR) *Cosmic Battery* because it relies on the PR drag on plasma electrons to generate large–scale azimuthal currents in the inner disks of AGN [6, 8, 9]. In the vicinity of a $10^8 M_\odot$ black hole, the PR battery can generate an equipartition field on timescales shorter than 100 million years [8]. But the most important feature of this mechanism is that the azimuthal electric current in the accretion disk always flows in the direction of rotation; thus, it supports and maintains a large-scale poloidal magnetic field that is inextricably tied to the rotation of the disk.

The PR battery solves the problem of large-scale field amplification via the continual (secular) generation of poloidal magnetic-field loops around the inner edge of the accretion disk surrounding the central supermassive black hole [9]. The outer footpoint of each loop resides well inside the accretion disk and diffuses outward on the disk's local diffusion timescale, provided that this is at least a factor of ~2 shorter than the local accretion timescale. The other footpoint is dragged inward by the accretion flow and eventually ends up near the black hole. This behavior was predicted in [8] and was demonstrated by the simulations of [6]. The inner and the outer section of the loop are both helically wound in the direction of rotation, but because the poloidal component of the inner (outer) loop is parallel (antiparallel) to the angular-velocity vector, the section in the inner jet (the "inner" field) develops a left–handed twist about the jet axis whereas the section in the outer wind (the "return" field) develops a right–handed twist about the jet axis, irrespective of the direction of the angular velocity [7].

The most important diagnostic tool of the above magnetic-field configuration is the predicted observation of characteristically oriented transverse Faraday RM gradients across the jets and lobes of at least some radio sources that are not too disturbed due to interactions with their environment or other factors. When an RM gradient is dominated by the contribution of the inner field, it should increase in the clockwise (CW) direction on the sky with respect to the AGN center. The opposite holds for a Faraday RM gradient that is dominated by the return field. These results are independent of the location of the observer [7].

With the above theoretical considerations in mind, we set out to search for transverse RM gradients on even larger scales in radio sources with jets extending over many kiloparsecs. The purpose of this study was essentially twofold: (i) to establish whether or not transverse RM gradients reasonably interpreted as reflecting the presence of helical jet magnetic fields are present on kiloparsec scales; and if present, (ii) to establish whether such RM gradients show any evidence for a preferred orientation. We have found clear, systematic, transverse RM gradients associated with kiloparsec-scale jets in 6 extragalactic radio sources, with tentative gradients identified in 4 more sources.

2. Transverse RM Gradients Across Kiloparsec-Scale Jets

Our extensive search for Faraday RM maps of extragalactic radio sources on kiloparsec scales published over the past 12 years produced 86 such maps [1–3, 5, 14–16, 20–24, 26, 30–32, 38, 40]. We were unable to reliably identify large-scale, systematic transverse RM gradients in most of these maps, because the jets appeared to be strongly influenced by their surroundings, the emission was patchy or limited to just a few pixels across the images, or the results were shown in gray scale that did not make the structures in the RM images sufficiently clear. We have focused specifically on objects in which we believe we have been able to identify gradients *transverse* to the jets; objects displaying RM gradients that were predominantly longitudinal were discarded from consideration.

In the end, we were able to identify 6 firm and 4 tentative objects with extended, monotonic, nearly transverse (to within $\sim 20°$ in most cases) RM gradients based on data at three or more wavelengths, listed below. We have made no attempt to measure the *magnitudes* of the RM gradients we have identified, and focus solely on the *presence* of clear transverse RM gradients and their *directions*—either CW or CCW on the sky—relative to the bases of the jets.

Firm transverse RM gradients:

0156−252 [CCW].—There is a clear RM gradient across the Eastern jet (Fig. 3 of [1]).

5C4.114 [CCW].—There is a clear, systematic, extended RM gradient across the entire Northern lobe of the radio structure. The RM distribution is extremely well resolved (Fig. 12 of [5]).

5C4.152 [CCW].—There are clear RM gradients present in both lobes (Fig. 15 of [5]). The gradient across the Southern lobe is fairly orthogonal to the jet direction, while the gradient across the Northern lobe appears to be offset from orthogonality as the jet bends toward the East just before the hotspot.

Cen A [CCW].—After subtraction of the overall mean RM (which is presumably foreground), the residual RM map shows a tendency for positive residuals along the Eastern side and negative residuals along the Western side of the entire Northern lobe (Fig. 7 of [15]). The Southern lobe is strongly bent; although local RM gradients can be identified, no systematic trends are obvious.

A2142A [CCW].—The RM distribution shows a clear tendency for RM values $\simeq -600$ rad m^{-2} along the Southern side of the jet and RM values $\simeq -200$ to -300 rad m^{-2} along the Northern side (Fig. 13 of [22]).

3C465 [CCW].—The Southern jet displays a systematic, extended transverse RM gradient that extends along most of this jet (Fig. 6 of [14]).

Tentative transverse RM gradients:

3C31 [CCW].—There is some evidence for an RM gradient across the Southern lobe (region SP3 in Fig. 2c of [30]); but an appreciable smaller-scale patchy component to the distribution makes it unclear whether a systematic component is present in the RM distribution.

5C4.74 [CCW].—An RM gradient is visible over an extended region in the Northwestern lobe, but it is hard to be sure of the monotonicity of this gradient due to the range of the color scale used (Fig. 11 of [5]). No RM gradient is seen in the Southeastern radio lobe.

5C4.127 [CCW, CW].—There are indications of RM gradients in the Western jet, however these are not very extended along the jet, and it is somewhat unclear whether a systematic component is present in the RM distribution (Fig. 13 of [5]).

A2065A [CW, CCW].—There are some regions with fairly clear transverse RM gradients at the bases of the lobes, but the overall RM distribution is fairly patchy, as is also noted by the authors of [22] (Fig. 14 of [22]).

3. Discussion

Reliability of the Detected Transverse RM Gradients.

In their critique of various parsec-scale Faraday rotation analyses, including that of [7], Taylor & Zavala [39] proposed four criteria for the reliable detection of transverse Faraday rotation gradients: (1) at least three "resolution elements" across the jet, (2) a change in the RM of at least three times the typical error σ, (3) an optically thin synchrotron spectrum at the location of the gradient, and (4) a monotonically smooth (within the errors) change in the RM from side to side. We have explicitly satisfied criteria (3) and (4) in our search, by retaining in our list only monotonic, smooth gradients in optically thin jet regions. Criterion (2) is also satisfied by the images selected here, sometimes with a very large margin; in some cases, the RM difference across the jet corresponds to tens of σ.

Criterion (1) reflects the desire to ensure that it is possible to distinguish properties between regions located on opposite sides of the jets, which are often quite narrow structures. However, the minimum of 3 "resolution elements" is proposed without any justification, and indeed without even defining what is meant by a "resolution element." In fact, quite a few cases are known in which imaged structures can be clearly resolved even when they are separated by distances of about one (or less than one) beamwidth; one such example is the detection of jet components less than 1 mas from the core in many sources in the MOJAVE VLBA monitoring project at 15 GHz, for which the beamwidths are typically about 1 mas (e.g. [27]). In addition, clear transverse polarization structure is often visible in the jets of this sample, even in parts of the jets with transverse sizes of 1–2 beamwidths [33].

Thus, we consider this to be an open issue, and further work is needed to address this question; see also the contribution to these proceedings by Murphy and Gabuzda. Most of the observed RM gradients considered here span 2–3 beamwidths; and those in 3C465, Cen A, and 3C31 are extremely well resolved, and extend across more than 10 beamwidths.

Importance of the Presence of Transverse RM Gradients on Kiloparsec Scales

The most fundamental implication of the above results is that transverse RM gradients—predicted to exist if the jets carry toroidal or helical magnetic fields and detected earlier in some AGN on parsec scales—are also present on kiloparsec scales. This identification of systematic transverse RM gradients on kiloparsec scales not only provides direct evidence that the jets of at least some of these objects carry helical magnetic fields; by inference, it also provides corroborative support to previous reports of parsec-scale helical magnetic fields.

In fact, it is easy to understand why it should be comparatively difficult to detect transverse RM gradients due to the presence of helical magnetic fields in kiloparsec-scale jets. There is an appreciable turbulent, inhomogeneous component to the thermal ambient media surrounding the jets on these scales, which superposes a more or less random pattern over the systematic pattern due to the helical fields. This random component in the RM distribution may dominate in the majority of cases. Thus, it is perfectly natural that most of the observed RM distributions appear random and patchy, but the overall pattern due to the helical fields sometimes comes through. This suggests that, on average, it may be easier to detect the systematic RM component due to helical jet magnetic fields on parsec scales, where the ordered inner field is more dominant, a result that seems to be borne out from the observations.

Predominance of CCW RM Gradients on Large Scales

Our results show a predominance of CCW RM gradients: namely, all 6 firm gradients are CCW. Based on a simple unweighted binomial probability function, the probability for 6/6 firm gradients to be CCW by chance is about 1.6%. This is not quite a 3σ result, and so must be considered tentative. However, we believe it to be significant, in light of the previous results collected and analyzed in [7] for parsec scales, as well as recent RM results for the jets of AGNs on scales of tens of parsec. In partiular, 3C120 [12], Mrk501 [13], 1749+701 [25], and 1633+382 [36] all display transverse CCW RM gradients on scales of tens of parsec in Very Long Baseline Array (VLBA) observations at four wavelengths between 18 and 22 cm.

This preponderance of CCW transverse RM gradients and the presence of helical magnetic fields on large scales can be explained naturally in the framework of the PR cosmic battery: the return helical field dominates the total observed Faraday rotation on relatively large angular scales on the sky, where the detected radio emission extends to fairly large distances from the jet axes. It is much less likely for such observations to be dominated by Faraday rotation due to the inner helical field, which is embedded in the outer layers of the jet itself. In contrast, as was shown by the analysis of [7], this "inner" region of helical field is more likely to dominate the RM measurements on parsec scales.

Königl [28] has argued that the CW RM gradients on parsec scales reported in [7] could be the result of ordered, large-scale, magnetic fields that were produced in the outer weakly-ionized accretion disks by Hall currents, and that were launched from the disks in centrifugally-driven wind outflows. This model could in principle also give rise to a predominance of CCW RM gradients on larger scales, if there is also a "return field". However, Hall currents are usually believed to be unimportant to AGN physics, because, unlike protostellar disks [29], the accretion disks of AGN have relatively higher ionization fractions near the central compact objects, where all the prominent magnetohydrodynamical phenomena are thought to take place (e.g., [18–19]). If, on the other hand, the magnetic field is brought in to the outer, cold, weakly-ionized part of the disk from outside [28], then different polarities are likely to undergo fast reconnection long before a strong toroidal field component can develop due to the much slower differential rotation of the outer disk. Therefore, we consider this model physically less plausible in AGNs than the PR battery model.

4. Summary

We have found 6 extragalactic radio jets across which monotonic RM gradients can firmly be identified on kiloparsec scales, providing direct evidence for the presence of helical jet magnetic fields. All of these RM gradients are monotonic and smooth, occur in optically thin jet regions, and correspond to RM differences of more than 3σ across the jets. Although some span only 2–3 beamwidths across the jets, others extend over dozens of beamwidths. We have also identified four additional cases, in which transverse gradients may be present, but are considered tentative because they could be due to random patchiness of the Faraday-rotating medium.

We find evidence for a preponderance of kiloparsec-scale gradients in which the RMs increase CCW on the sky. Although this result does not have an extremely high statistical significance, it is supported by (i) the tendency for parsec-scale transverse RM gradients to be CW [7] and (ii) the detection of transverse CCW RM gradients in several 18–22 cm VLBA RM images probing scales of tens of parsec. This result seems counterintuitive, since in the simplest models, one would expect transverse RM gradients associated with helical jet magnetic fields not to display any preferred orientation on the sky. The observed predominance of CCW transverse RM gradients essentially demands that the direction of rotation of the accretion disk and the direction of the initial poloidal seed field that is "wound up" by the rotation be coupled. Such a coupling can be provided by the PR cosmic battery.

The poloidal magnetic-field loops generated by the PR cosmic battery around the inner edge of the underlying accretion disk extend out to kiloparsec-scale distances and effectively establish an inner field embedded in the relativistic jet and a return field embedded in the outer extended disk wind. Faraday rotation observations of radio jets on relatively large scales can resolve the return-field sections of such poloidal loops that are not too disturbed by transverse motions, backflows, or turbulence. The associated Faraday RM gradients should increase in the CCW direction on the sky, independent of the orientation of the observer [7].

In the same physical picture, the Faraday RM gradients due to the inner sections of the poloidal loops should increase CW on the sky about the central sources, but such gradients are located much closer to the jet axes and cannot be observed in arcsecond-resolution images because the RMs on such large scales are dominated by the Faraday rotation in the outer extended disk wind. In turn, the "inner" CW RM gradients dominate on the parsec scales observed with VLBI, giving rise to the preponderance of CW transverse RM gradients found in the collected results analyzed in [7].

If the PR battery mechanism is not operating for some reason, another mechanism that can provide the required coupling between the direction of disk rotation and direction of the poloidal seed field must be found. The Hall-current mechanism of [28] may provide one alternative, although it remains unclear whether this mechanism is sufficiently efficient in the jet+disk systems of AGNs.

The results we have presented here must still be considered somewhat tentative; however, if confirmed by further studies, they will be of fundamental importance for our understanding of AGN jets. We are currently investigating ways to better quantify detected transverse RM gradients and to identify them in a more objective, automated fashion.

5. Acknowledgements

We acknowledge insights and assistance with data provided by Drs. Philip Best, George Contopoulos, Federica Govoni, Christian Kaiser, and Preeti Kharb.

[1] Athreya, R. M., Kapahi, V. K., McCarthy, P. J., & van Breugel, W. 1998, A&A, 329, 809
[2] Best, P. N., Carilli, C. L., Garrington, S. T., Longair, M. S., & Röttgering, H. J. A. 1998, MNRAS, 299, 357
[3] Best, P. N., Eales, S. A., Longair, M. S., Rawlings, S., & Röttgering, H. J. A. 1999, MNRAS, 303, 616
[4] Blandford R. D., & Payne, D. G. 1982, MNRAS, 199, 883
[5] Bonafede, A., Feretti, L., Murgia, M., Govoni, F., Giovannini, G., Dallacasa, D., Dolag, K. ,& Taylor, G. B. 2010, A&A, 513, 30
[6] Christodoulou, D. M., Contopoulos, I., & Kazanas, D. 2008, ApJ, 674, 388
[7] Contopoulos, I., Christodoulou, D. M., Kazanas, D., & Gabuzda, D. C. 2009, ApJL, 702, L148
[8] Contopoulos, I., & Kazanas, D. 1998, ApJ, 508, 859
[9] Contopoulos, I., Kazanas D., & Christodoulou, D. M. 2006, ApJ, 652, 1451
[10] Contopoulos, J. 1995, ApJ, 450, 616
[11] Contopoulos, J. & Lovelace, R. V. E. 1994, ApJ, 429, 139
[12] Coughlan, C., Murphy, R., McEnery, K., Patrick, H., Hallahan, R., & Gabuzda, D. C. 2011, Proc. 10th EVN Symposium, in press
[13] Croke, S. M., O'Sullivan, S. P., & Gabuzda, D. C. 2010, MNRAS, 402, 259
[14] Eilek, J. A., & Owen, F. N. 2002, ApJ, 567, 202
[15] Feain, I. J,. Ekers, R. D., Murphy, T., et al. 2009, ApJ, 707, 114
[16] Feretti, L., Dallacasa, D., Govoni, F., Giovannini, G., Taylor, G. B., & Klein, U. 1999, A&A, 344, 472
[17] Ferreira J., Petrucci P.-O., Henri G., Saugé L., & Pelletier G. 2006, A&A, 447, 813
[18] Gaskell, C. M. 2009, NewAR, 53, 140
[19] Gaskell, C. M. 2010, ASPC, 427, 68
[20] Goodlet, J. A., Kaiser, C. R., Best, P. N., & Dennett-Thorpe, J. 2004, MNRAS, 347, 508
[21] Govoni, F., Murgia, M., Feretti, L., Giovannini, G., Dolag, K. & Taylor, G. B. 2006, A&A, 460, 425
[22] Govoni, F., Dolag, K., Murgia, M., Feretti, L., Schindler, S., Giovannini, G., Boschin, W., Vacca, V., & Bonafede A. 2010, A&A, 522, 105
[23] Guidetti, D., Murgia, M., Govoni, F., Parma, P., Gregorini, L., de Ruitar, H. R., Cameron, R. A., & Fanti, R. 2008, A&A, 483, 699
[24] Guidetti, D., Laing, R. A., Murgia, M., Govoni, F., Gregorini, L., & Parma P. 2010, A&A, 514, 50
[25] Hallahan, R., & Gabuzda, D. C. 2008, Proceedings of Science, http://pos.sissa.it/cgi-bin/reader/conf.cgi?confid=72
[26] Kharb, P., O'Dea, C. P., Baum, S. A., et al. 2008, ApJS, 174, 74
[27] Kellermann, K. I., Lister, M. L., Homan, D. C., et al. 2004, ApJ, 609, 539
[28] Königl, A. 2010, MNRAS, 407, L79
[29] Krasnopolsky, R., Li, Z.-Y., & Shang, H. 2011, ApJL, submitted
[30] Laing, R. A., Bridle, A. H., Parma, P., & Murgia, M. 2008, MNRAS, 391, 521
[31] Laing, R. A., Canvin, J. R., Bridle, A. H., & Hardcastle, M. J. 2006, MNRAS, 372, 510
[32] Laing, R. A., Canvin, J. R., Cotton, W. D., & Bridle, A. H. 2006, MNRAS, 368, 48
[33] Lister, M. L. & Homan, D. C. 2005, AJ, 130, 1389
[34] Murgia, M., Govoni, F., Feretti, L., Giovannini, G., Dallacasa, D., Fanti, R., Taylor, G. B., & Dolag, K. 2004, A&A, 424, 429
[35] Pelletier, G., Sol, H., & Asseo, E. 1988, Phys. Rev. A, 38, 2552
[36] Reichstein, A., Gabuzda, D. C., Coughlan, C., & Murphy, R. 2011, Proc. 25th Symposium on Relativistic Astrophysics, Proceedings of Science, in press
[37] Spruit, H. C. 2010, LNP, 794, 233
[38] Taylor, G. B., Govoni, F., Allen, S. W., & Fabian, A. C. 2001, MNRAS, 326, 2
[39] Taylor, G. B., & Zavala, R. 2010, ApJ, 722, L183
[40] Venturi, T., & Taylor, G. B. 1999, AJ, 118, 1931

MEM and CLEAN Imaging of VLBA Polarisation Observations of Compact Active Galactic Nuclei

Colm Coughlan
Department of Physics, University College Cork, Ireland
E-mail: colmcoughlanirl@gmail.com

Denise Gabuzda
Department of Physics, University College Cork, Ireland
E-mail: d.gabuzda@ucc.ie

Abstract. The Maximum Entropy Method (MEM) for the deconvolution of radio interferometry images is mathematically well based and presents a number of advantages over the usual CLEAN deconvolution, such as appreciably higher resolution. The application of MEM for polarisation imaging remains relatively little studied. CLEAN and MEM intensity and polarisation techniques are discussed in application to recently obtained 18cm VLBA polarisation data for a sample of Active Galactic Nuclei.

1. Imaging Techniques in Radio Astronomy

Consider a radio source which gives rise to a brightness distribution, $I(x, y)$, in the radio sky of the Earth. There exists a Fourier transform (FT) relationship between the true brightness distribution and the complex visibility function $V(u, v)$ measured by a VLBI array as a function of the baseline vector.

$$V(u, v) = F.T.(I(x, y)) \tag{1}$$

It is not a simple case of taking the inverse Fourier transform of the visbilities to return to the true brightness distribution as, due to the limited uv coverage offered by Earth rotation synthesis VLBI, most components of the visibility function are not measured. Mathematically this corresponds to multiplying the total visibility by a sampling function, $S(u, v)$, which eliminates most of the visibilities and leaves the observer with only $V_{sampled}(u, v)$. This means that, even before noise is taken into account, any attempt to recover the sky brightness distribution from the measured uv visibilites is a mathematically ill-posed problem.

$$V_{sampled}(u, v) = V(u, v)S(u, v) + Noise \tag{2}$$

Deconvolution algorithms are used to attempt to interpolate the information contained in the measured visibilities to reconstruct the unmeasured visibilities and get as true an image of the sky brightness distribution as possible. There is no single "best" deconvolution algorithm, however the two most common are outlined below.

1.1. The CLEAN Algorithm
The CLEAN algorithm (Högbom, 1974) is a standard technique used to deconvolve the measured visibilities and attempt to recover the data from the unmeasured visibilities. First the measured visibilities are Fourier transformed to create a "dirty" map (containing only the information from the measured visibilities). This map is then described by a set of "CLEAN" components (δ functions). The set of "CLEAN" components is then convolved with the CLEAN beam (a Gaussian fit to the primary lobe of the point spread function) and the residual noise added in to give the final "clean" image. Images of Stokes I,Q and U parameters can be made independently using this technique.

1.2. The Maximum Entropy Method (MEM)
A second way to deconvolve the visibilities and recover some of the unmeasured data is to make a model intensity map of the source that satisfies specified mathematical criteria. This model can then be varied until the simulated visibilities agree with the actual data to within a specified limit, thus maximising the similarity of the model intensity and the true map. However, as the model is changed one cannot let the model visibilities converge to the actual measured visibilities, or the model would then simply yield the original "dirty" map. A regularising parameter is required to stop this convergence. In MEM, this parameter is the entropy of the image. Consider the following function:

$$J = H(I_m, P_m) - \alpha\chi_I^2(V_m, V_d) - \beta\chi_P^2(V_m, V_d) - conditions \quad (3)$$

where H is the entropy of the model map of the source and χ^2 is a measure of the difference between the model (subscript m) and observed (subscript d) visibilities (there are two such terms, one for intensity, $\chi_I^2(V_m, V_d)$, and a second for polarisation, $\chi_P^2(V_m, V_d)$. α and β are Lagrange parameters, and other conditions are also included which represent additional constraints, such as positivity of the Stokes I component. The optimal model of the source maximises the function J above. This results in a balance between entropy (representing noise, and the effect of unsampled visibilities) and fidelity to the observed data.

Multiple forms of entropy can be used; common choices include Shannon entropy (4), which is suitable for unpolarised emission, and the Gull and Skilling entropy (5), a generalised form of the Shannon entropy, suitable for polarised emission.

$$H = -\sum_k I_k log(\frac{I_k}{B_k}) \quad (4)$$

$$H = -\sum_k I_k(log(\frac{I_k}{eB_k}) + \frac{1+m_k}{2}log(\frac{1+m_k}{2}) + \frac{1-m_k}{2}log(\frac{1-m_k}{2})) \quad (5)$$

In these equations, the index k represents summation over pixels, and I_k and m_k are the intensity and fractional polarisation, respectively, at pixel k. B is a bias map, to which the solution defaults in the absence of data. See Holdaway [1] and Gull and Skilling [2] for more details about the MEM and different forms of polarisation entropy.

2. CLEAN and MEM : A comparision
Both CLEAN and MEM are widely used in image processing as deconvolution techniques, however each technique has specific strengths and weaknesses. CLEAN, while intuitive, does not have a firm mathematical footing, and it can be difficult to state the resolution of a CLEAN image exactly. MEM, while less intuitive, has a much better mathematical grounding, and has a resolution which can be demonstrated mathematically. As the deconvolution problem is

Journal of Physics: Conference Series **355** (2012) 012020 doi:10.1088/1742-6596/355/1/012020

inherantly ill-posed, neither technique is perfect - they cannot recreate the "true" image, and different types of artefacts occur in each technique. This means that regardless of the imaging technique used, the images produced must be interpreted before any conclusions can be made about features that may be present in them.

In both techniques, the resolution obtained is inversely proportional to the size of the longest baseline in the observing array. Due to the lack of firm mathematical foundations for the CLEAN algorithm, a conservative estimation of its resolution is usually taken to be the full width at half maximum of the CLEAN beam, which is a Gaussian fit to the primary lobe of the point spread function. Some information is also present on smaller scales, but can be difficult to interpret. MEM's resolution can be shown to be

$$x_{min} = \frac{1}{4u_{max}} \qquad (6) \qquad\qquad y_{min} = \frac{1}{4v_{max}} \qquad (7)$$

where x_{min} and y_{min} are the resolutions in radians in the image space, and u_{max} and v_{max} correspond to the length of the longest baseline in wavelengths in the visibility space [1]. This resolution is approximately four times smaller than the conservative estimate of the CLEAN algorithm.

3. Application of polarisation MEM to VLBI data

The standard radio astronomy software package Astronomy Image Processing System (AIPS) includes a task (VTESS) to conduct MEM deconvolution of intensity (Stokes I), but not polarisation (Stokes Q and U) images (see Cornwell and Evans [3] for details). The relation between the Stokes Q and U parameters and the polarised intensity and polarisation angle of a map is shown by equations (8) and (9).

$$P = \sqrt{Q^2 + U^2} \qquad (8)$$

$$\chi = \frac{1}{2} ArcTan(\frac{U}{Q}) \qquad (9)$$

We are in the process of investigating a number of ways of implementing fully polarised MEM to VLBI polarisation data. We have in the meantime produced a number of "proof of concept" MEM polarisation maps in AIPS by devising an algorithm to work around the limitations imposed by the AIPS MEM task (primarily the non-negativity requirement - whereas Stokes I must be positive, Stokes Q and U can be either positive or negative).

This algorithm involves using the initial (dirty) Q and U maps to define a binary mask which is $+1$ where the Q and U maps are positive, and -1 where the Q and U maps are negative. This mask is then multipled with the dirty Q and U maps to create a strictly positive map, which is then imaged using the AIPS VTESS task. The mask is then reapplied to the MEM model developed. A major limitation of this work-around is the assumption that the location of positive and negative pixels in the Q and U maps remain the same as in the original maps (only the amplitudes are allowed to vary).

4. Preliminary MEM Intensity and Polarisation VLBA Images

The prelimary images shown were obtained with the Very Long Baseline Array (see Coughlan et al [4]). Both the CLEAN and MEM images were made from the same uv data, which had been calibrated using standard methods in AIPS. The CLEAN map was convolved with the CLEAN beam and the MEM map convolved with a Gaussian beam corresponding to the resolution

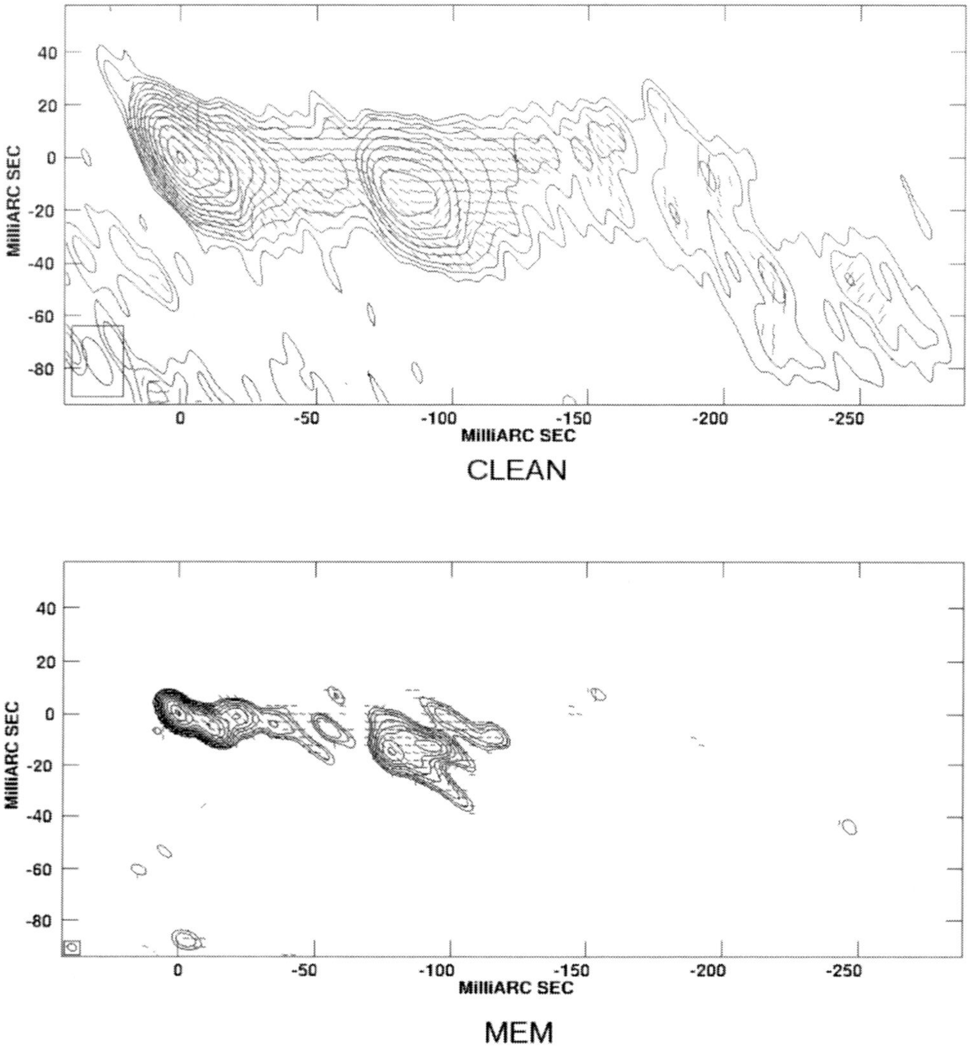

Figure 1. CLEAN and MEM images of J0433+0521 (3C120) at 1.36 GHz.

indicated by (6) and (7). In all cases, the MEM images lost some of the extended emission visible in the CLEAN images. However, the increased resolution of the MEM images provides fuller information about inner jet structure and morphology. The superimposed sticks indicate the local polarisation angles. In all cases there is good agreement between the MEM and CLEAN polarisation angles.

Journal of Physics: Conference Series **355** (2012) 012020 doi:10.1088/1742-6596/355/1/012020

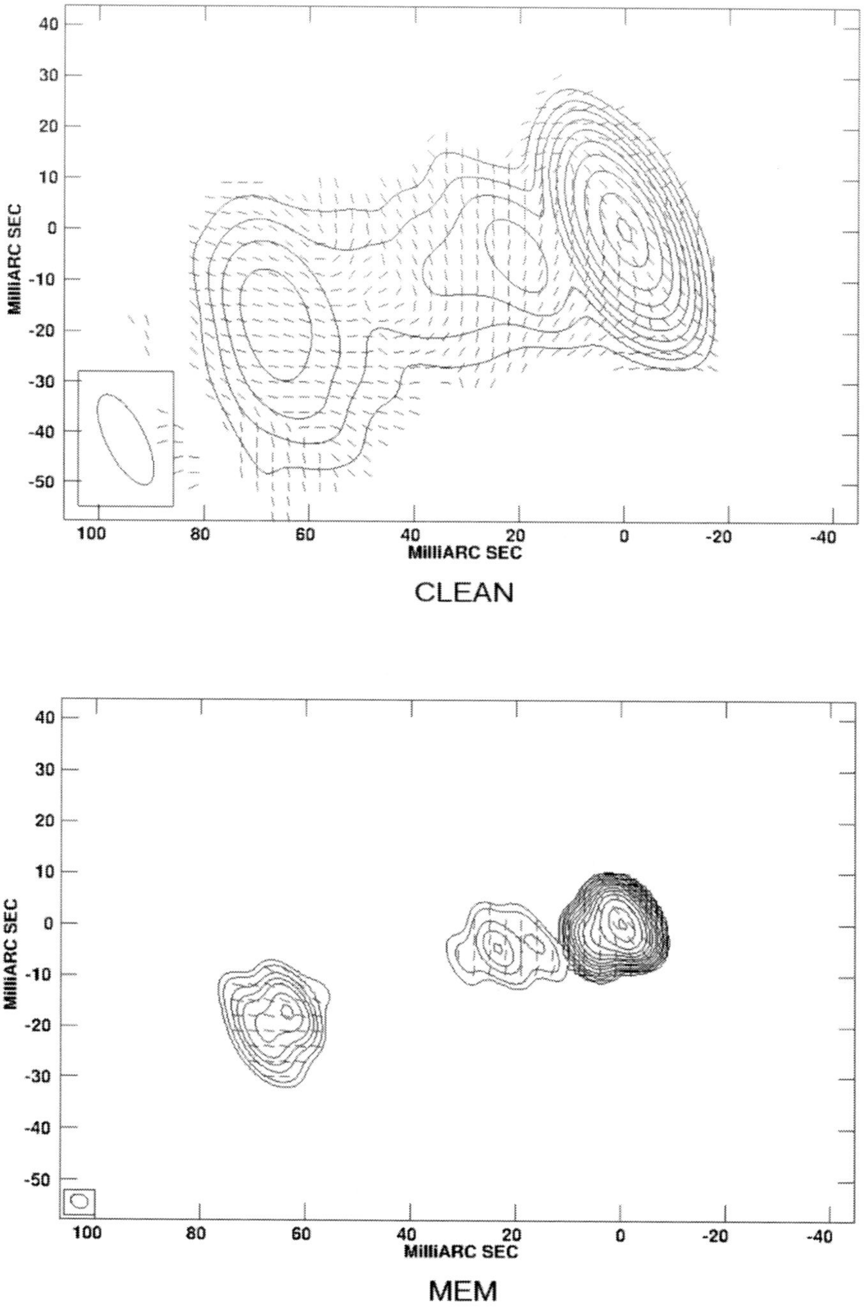

Figure 2. CLEAN and MEM images of J0750+1231 at 1.36 GHz.

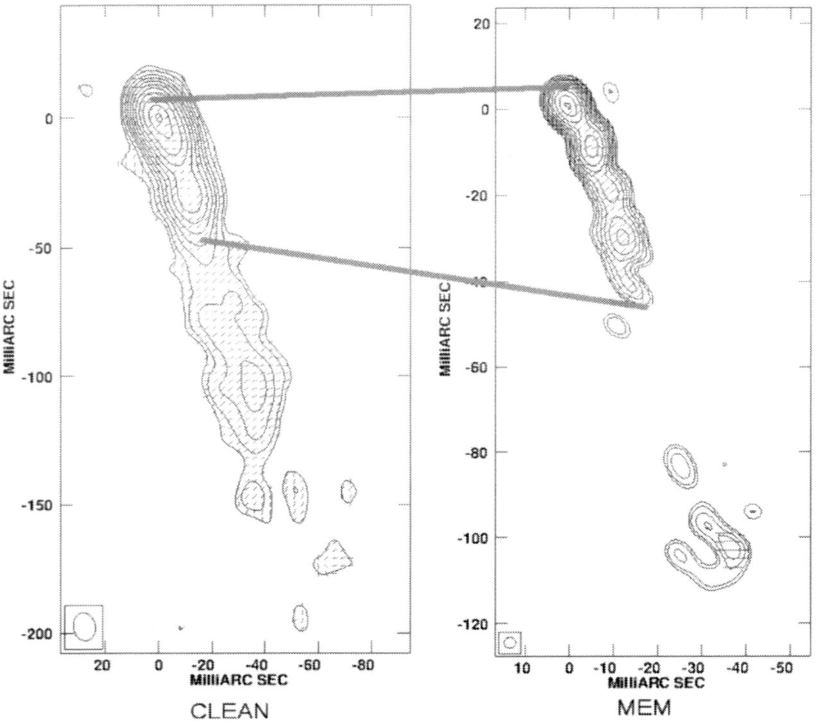

Figure 3. CLEAN and MEM images of J0841+7053 at 1.36 GHz.

5. Conclusion

Both CLEAN and MEM offer unique and complementery perspectives on the same uv visibility data. MEM appears less sensitive to low intensity emission, but provides higher resolution, allowing jet direction and morphology to be studied in more detail. There is good agreement between the CLEAN and approximate MEM polarisation maps we have constructed, suggesting that the approximations made in the work-around used to produce polarised MEM maps in AIPS have not adversely affected the polarisation. However, in order to generate a truely reliable MEM polarisation map a form of polarised entropy must be used and we are in the process of implementing the entropy (5) for this purpose. The synergy between CLEAN and MEM in studying polarised emission should then lead to a clearer picture of the polarisation intensity and direction along the jet. We also plan to investigate the construction of MEM spectral-index and Faraday-rotation maps based on multi-frequency intensity and polarisation VLBI maps.

Acknowlegements

This work was supported by the Irish Research Council for Science, Engineering and Technology (IRCSET).

References

[1] Holdaway M., 1990, Ph.D. Thesis, Brandeis University.
[2] Gull S.F., Skilling J., 1984, *Indirect Imaging*, ed. J.A. Roberts. Cambridge University Press p 267.
[3] Cornwell T.J., Evans K.F., 1985, *A&A*, **143**, 77-83.
[4] Coughlan et al, 2011, *Proc.* 10th *European VLBI Network Symposium*.

The Use of Faraday Rotation Sign Maps as a Diagnostic for Helical Jet Magnetic Fields

Andrea Reichstein and Denise Gabuzda
University College Cork, Ireland
E-mail: amn.reichstein@gmail.com, d.gabuzda@ucc.ie

Abstract. We present maps of the sign of the Faraday Rotation measure obtained from multi-frequency radio observations made with the Very Long Baseline Array (VLBA). The Active Galactic Nuclei (AGN) considered have B-field structures with a central "spine" of B-field orthogonal to the jet and/or a longitudinal B-field near one or both edges of the jet. This structure can plausibly be interpreted as being caused by a helical/toroidal jet magnetic field. Faraday Rotation is a rotation of the plane of polarization that occurs when the polarized radiation passes through a magnetized plasma. The sign of the RM is determined by the direction of the line-of-sight B-field in the region causing the Faraday Rotation, and an ordered toroidal or helical magnetic field associated with an AGN jet will thus produce a distinctive bilateral distribution of the RMs across the jet. We present and discuss RM-sign maps and their possible interpretation regarding the magnetic field geometries for several sources.

1. Introduction

When first discovered in the active galactic nucleus (AGN) 1055+018 [1], "spine–sheath" magnetic field structure was quite unexpected. It was suggested that, depending on the jet's angle to the line-of-sight, the image might be dominated by the "spine" or the "sheath", but wouldn't show both at once [1]. One possible explanation for an AGN with spine–sheath polarization structure is interaction of the jet with the surrounding medium, causing a deceleration of the plasma near the jet edges and stretching out the magnetic field along the jet, forming a shear layer. The polarization is strongest where the interaction is strongest. In the picture proposed in [1, 2], the orthogonal magnetic field along the central axis of the jet is caused by a series of transverse shocks. In the standard Blandford–Königl jet model [3], the transverse component of B-field decays less rapidly than the longitudinal field, so that the jet B-field will naturally tend to become more dominated by the toroidal component with distance from the jet base, providing another possible explanation for the spine of orthogonal B-field.

Another possible explanation of spine–sheath B-field structure is an overall helical magnetic field around the jet, where the projected toroidal component of the field is dominant at the jet axis, and the longitudinal field component becomes dominant at the jet edges [4, 5]. This explanation requires only one agent (the helical field), rather than two (the combined presence of a transverse or toroidal field and shear). Figure 1 shows how to explain a spine–sheath polarization structure with a helical magnetic field. To investigate this type of polarization structure, we are making use of the effect of Faraday Rotation.

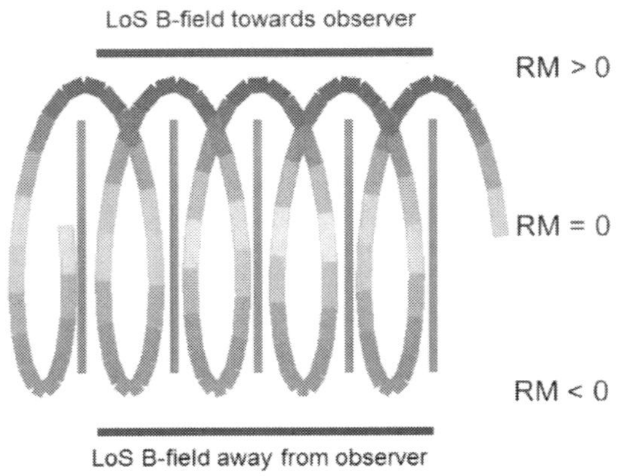

Figure 1. Schematic illustration how a helical magnetic field (its projection on the sky) can appear as a spine–sheath structure with the B field orthogonal to the jet axis in the center and longitudinal at the jet edges. In this case, the viewing angle is near 90° to the jet axis in the jet's rest frame; the rotation measure has opposite signs on either side of the jet, due to the different directions of the line-of-sight B field.

2. Faraday Rotation

When a linearly polarized electromagnetic wave travels through magnetized thermal plasma, the polarization angle χ rotates due to the different propagating speeds of left and right-circular polarized components of the wave. This is called Faraday Rotation. The amount of rotation is given by:

$$\chi = \chi_0 + RM\lambda^2 \qquad RM \propto \int n_e B \cdot dl$$

where χ_0 is the unrotated polarization angle, n_e is the electron density, $B \cdot dl$ picks out the line-of-sight magnetic field and λ is the observing wavelength. The coefficient of λ^2 is called the Rotation Measure (RM). Having simultaneous multifrequency observations, the RM can be easily determined by measuring the polarization angle at each wavelength and performing a linear fit on χ vs. λ^2. This also gives us the intrinsic (unrotated) polarization angle.

Systematic gradients in the RM across the jets of AGN can be interpreted as representing a systematic change in the line-of-sight component of a helical/toroidal B-field (Figure 1).

3. RM-sign maps

The idea behind RM-sign maps comes from theoretical simulations by Broderick & McKinney [6], who computed sub-parsec-scale B-field and RM distributions of 3D general-relativistic magnetohydrodynamic jets, which they then extrapolated to parsec (observable) scales (the computation time for their simulations to propagate to parsec scales would be prohibitive). They were able to reproduce some of the observed parsec scale RM morphologies, in particular transverse RM gradients produced by the helical field that was generated, and showed that most of the Faraday Rotation occurs in the outer layers of the jet. Even though the observed magnitude of the RM is strongly influenced by finite beam effects, the simulations still show clear gradients when convolved with large beams. These simulations confirmed that a helical B-field is generated and gives rise to RM gradients across the jet, resulting in a particular bilateral RM structure, which can be analysed with RM-sign maps. They are also a good tool to help distinguish whether the Faraday screen (the actual Faraday rotating material) is in the immediate vicinity of the jet or is a foreground screen somewhere between the source and the observer. If one is observing a random foreground screen there would be no bilateral structure correlated with the actual jet direction, but some random distribution of Faraday patches. If the jet is observed side-on in the rest frame of the jet, then the observed RM distribution would ideally be symmetric around a RM of 0 (see Figure 1), but in other cases, the RM distribution will be symmetric around some other value. RM-sign maps can thus be used as an indicator for a helical magnetic field along the jet, and can help exclude random foreground screens.

4. Observations and results

Twelve AGN showing evidence for spine–sheath polarization structures were observed at 4.6, 5.1, 7.9, 8.9, 12.9, and 15.4 GHz with the Very Long Baseline Array (VLBA) on the 27th of September 2007. The data were calibrated and imaged in the NRAO AIPS package using standard techniques. The EVPA calibration was carried out using integrated (VLA) polarization observations of some of the VLBA targets obtained close in time to the VLBA observations. The resulting χ vs. λ^2 plots are generally consistent with linear behaviour within the $1 - 2\sigma$ errors. In no case were rotations greater than 45° observed, consistent with Faraday rotation external to the source volume. Although some of the jets are not well resolved in the transverse direction, it can still be possible to reliably detect the presence of transverse polarization and RM structures, as shown by the simulated transverse profiles of Murphy & Gabuzda (these proceedings).

Analyzing the RM distributions, we have found some evidence for helical B-field structures in a number of these objects. As an example, Figure 2 shows a transverse RM gradient across the entire resolved jet region of 1633+382. The RM-sign map for this source features a clear bilateral structure correlated with the jet direction, consistent with a helical/toroidal magnetic field structure surrounding the jet. A partially transverse RM gradient was found across the

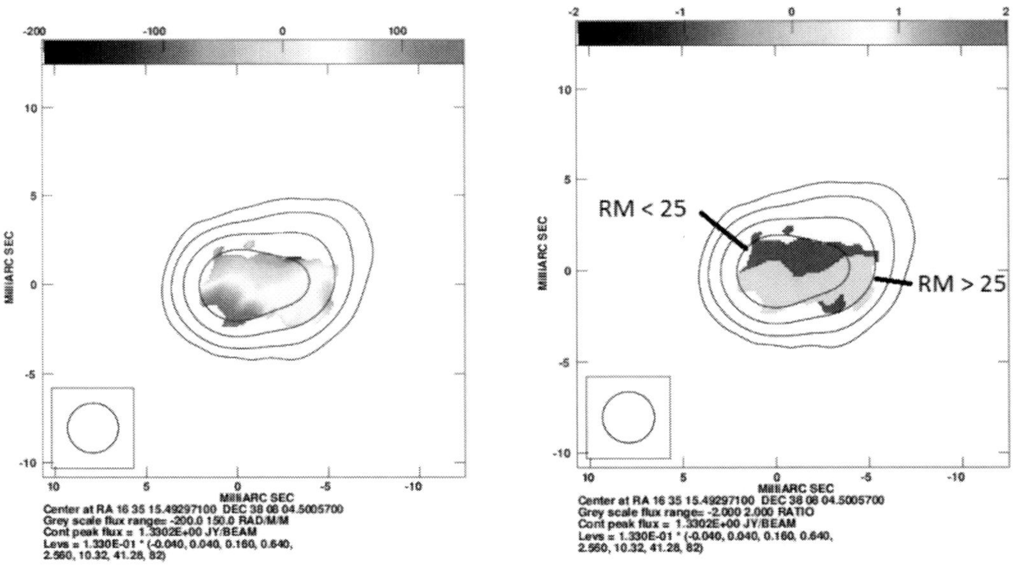

Figure 2. Left: RM map for 1633+382 using six frequencies from 4.6 to 15.4 GHz with contours of the 4.6 GHz total intensity map. Right: RM-sign map for 1633+382. The yellow and blue areas show RMs higher than and lower than 25 rad/m^2. It shows a bilateral structure correlated with the jet direction.

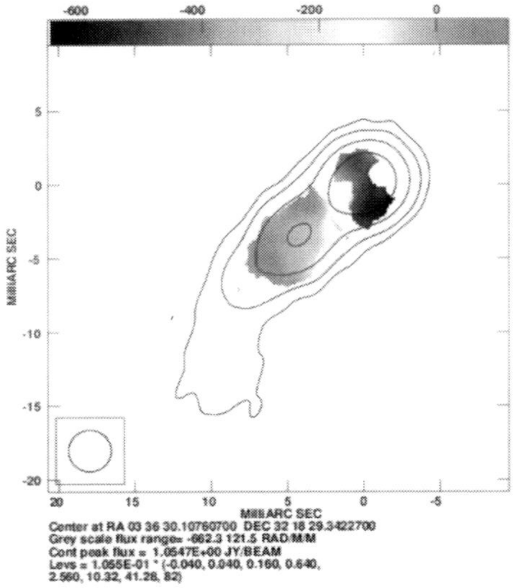
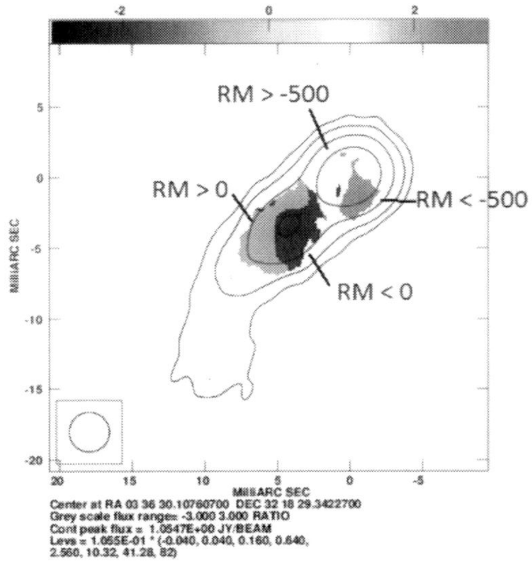

Figure 3. Left: RM map for 0333+321 using six frequencies from 4.6 to 15.4 GHz with contours of the 4.6 GHz total intensity map. Right: RM-sign map for 0333+321. The yellow and blue areas in the core indicate RMs higher than and lower than -500 rad/m^2. In the jet, the orange and dark blue regions show RMs higher than and lower than 0 rad/m^2. Both parts of the map shows a bilateral structure loosely correlated with the jet direction.

jet of 0333+321, see Figure 3 and also [7]. This result confirms the results of Asada et al. [8], who found a very similar RM gradient in this source. The RM-sign map for this source can be divided into two separate parts, one in the region of the core (roughly symmetrical around an RM of 500 rad/m^2) and the other one further out in the jet (roughly symmetric around an RM of zero). Even though we are looking at two separate parts of the jet with very different RM values, the direction of the gradient is roughly the same and the RM-sign maps in both parts are roughly correlated with the jet direction. Although there is clearly a transverse component to the RM gradient, the evidence for a helical B-field associated with this jet is weaker than for 1633+382, since the RM gradient is not as close to being transverse. The different RM values in the core region and jet could be due to somewhat different viewing angles and average electron densities in the core and jet. The RM map of 1150+812 (Figure 4) also shows some evidence for transverse gradients [7], however, the RM-sign map shows no consistent evidence for a correlation between the RM structure and the jet direction. The core region seems to show a tentatively transverse structure in the RM-sign map, but the RM-sign pattern in the jet shows no correlation with the jet direction. Thus, we do not find evidence for a bilateral RM-sign distribution indicative of a helical/toroidal jet B-field in this case.

5. Summary

Faraday Rotation gives information about the line-of-sight B-field. A simple explanation for transverse RM gradients and spine–sheath polarization structures is a helical magnetic field wrapped around the jet, whose changing line-of-sight B-field causes these gradients. In our images, we have found some evidence for transverse RM gradients in roughly half the analysed sources that show spine–sheath polarization structure [7]. The simulations of Broderick & McKinney [6] show that a helical B-field is generated by the jet launching, resulting in a

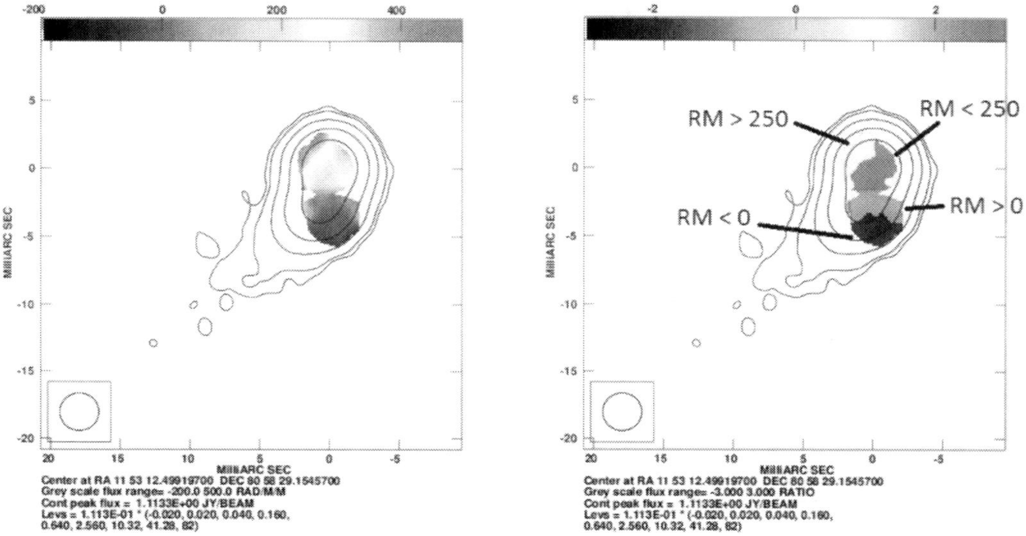

Figure 4. Left: RM map for 1150+812 using four frequencies from 4.6 to 8.9 GHz with contours of the 4.6 GHz total intensity map. Right: RM-sign map for 1150+812. The yellow area in the core shows RM higher than 250 rad/m^2, the blue area RM lower than 250 rad/m^2. Further out in the jet the orange region indicates RM higher than 0 rad/m^2, the dark blue area lower than 0 rad/m^2. The structure in the core region is roughly correlated with the jet direction, however the structure in the jet is not correlated with the jet direction.

particular bilateral RM structure, which can be analysed using RM-sign maps. Our results so far are varied. RM-sign maps constructed using our VLBA RM maps show such a bilateral structure for 1633+382, providing supporting evidence that its jet carries a helical B-field. The RM and RM-sign structure in 0333+321 are bilateral, but somewhat oblique, making this a weaker candidate for a jet with a helical B-field. The RM-sign map for 1150+812 does not show a clear and consistent bilateral pattern.

Overall, RM-sign maps can potentially provide a useful diagnostic for the presence of helical jet B-fields. However, there is always some probability that RM structures could appear to be correlated with the jet direction purely by chance. This makes it important to try to correlate the presence of bilateral RM-sign structures with other possible indicators of helical jet B-fields, such as spine–sheath polarization structure. The construction of RM-sign maps for larger numbers of AGN jets will also help reveal whether trends indicative of helical B-fields are indeed present in a statistical sense.

6. References

[1] Attridge, J. M.,Roberts, D. H., & Wardle, J. F. C. 1999, ApJ, 518, 87
[2] Wardle, J. F. C.,Cawthorne, T. V.,Roberts, D. H., & Brown,L. F. 1994, ApJ, 437, 122
[3] Blandford, R. D. & Konigl, A. 1979, ApJ, 232, 34
[4] Lyutikov, M.,Pariev, V. I., & Gabuzda, D. C. 2005, MNRAS, 360, 869
[5] Pushkarev, A. B.,Gabuzda, D. C.,Vetukhnovskaya, Y. N., & Yakimov, V. E. 2005, MNRAS, 356, 859
[6] Broderick, A. E. & McKinney, J. C. 2010, ApJ, 725, 750B
[7] Reichstein, A. & Gabuzda, D. C. 2011, *ArXiv e-prints*, 1102, 702R
[8] Asada, K.,Inoue, M.,Nakamura, M.,Kameno, S.,Nagai, H. 2008, ApJ, 682, 798

Faraday rotation in jets of AGN: the case of 3C 120

José L. Gómez[1], Mar Roca-Sogorb[1], Iván Agudo[1], Alan P. Marscher[2] and Svetlana G. Jorstad[2]

[1] Instituto de Astrofísica de Andalucía, CSIC, Glorieta de la Astronomía s/n, 1808 Granada, Spain.

[2] Institute for Astrophysical Research, Boston University, 725 Commonwealth Avenue, Boston, MA 02215, USA.

E-mail: jlgomez@iaa.es,mroca@iaa.es,iagudo@iaa.es,marscher@bu.edu,jorstad@bu.edu

Abstract. The source of Faraday rotation in the jet of the radio galaxy 3C 120 is analyzed through Very Long Baseline Array observations carried out between 1999 and 2007 at 15, 22 and 43 GHz. Uncorrelated changes in the linear polarization of the underlying jet emission and the Faraday rotation screen indicate that the emitting jet and the source of Faraday rotation are not closely connected physically and have different configurations for the magnetic field and/or kinematical properties. Furthermore, the existence of a region of enhanced rotation measure whose properties remain constant over three years requires a localized source of Faraday rotation, favoring a model in which a significant fraction of the rotation measure originates in foreground clouds.

1. Introduction

Helical magnetic fields are thought to play an important role in the formation and collimation of relativistic jets in active galactic nuclei (AGN) (e.g., [1–3]). The structure of the magnetic field in jets of AGN can be studied through polarimetric Very Long Baseline Interferometric (VLBI) observations, however, recent observations have revealed that the orientation of the polarization angle at parsec scales may be affected by Faraday rotation (e.g., [4, 5]). Therefore, the analysis of the Faraday rotation in jets of AGN is of special relevance to determine the actual orientation of the magnetic field structure on the plane of the sky.

Several possibilities have been suggested as the source of the Faraday rotation seen in jets of AGN. The first suggestion that it is produced in a sheath that is in close proximity with the jet was given by [6], based on the observation of a rotation measure (RM) gradient across the jet width in 3C 273. Such gradients are expected if the sheath is threaded by a helical magnetic field due to the change in the line of sight component of the magnetic field across the jet [7]. Further observations have also provide evidence for the existence of such gradients in other sources (e.g., [8–13]). However, [14] have raised concerns about the claimed transverse RM gradients by arguing that most of these observations lack the necessary resolution transverse to the jet. Additionally, there is strong observational evidence for the existence of foreground clouds (e.g., [15, 16]) that may contribute significantly to the observed RM [9, 17]. In particular, [9] present a study of the radio galaxy 3C 120 revealing the existence of a localized region of high RM between approximately 3 and 4 mas from the core. A smooth sheath around the jet in 3C 120 cannot produce a localized region of enhanced RM (although it successfully explains the observed gradients in RM and degree of polarization along and across the jet), so the RM was

Journal of Physics: Conference Series **355** (2012) 012022 doi:10.1088/1742-6596/355/1/012022

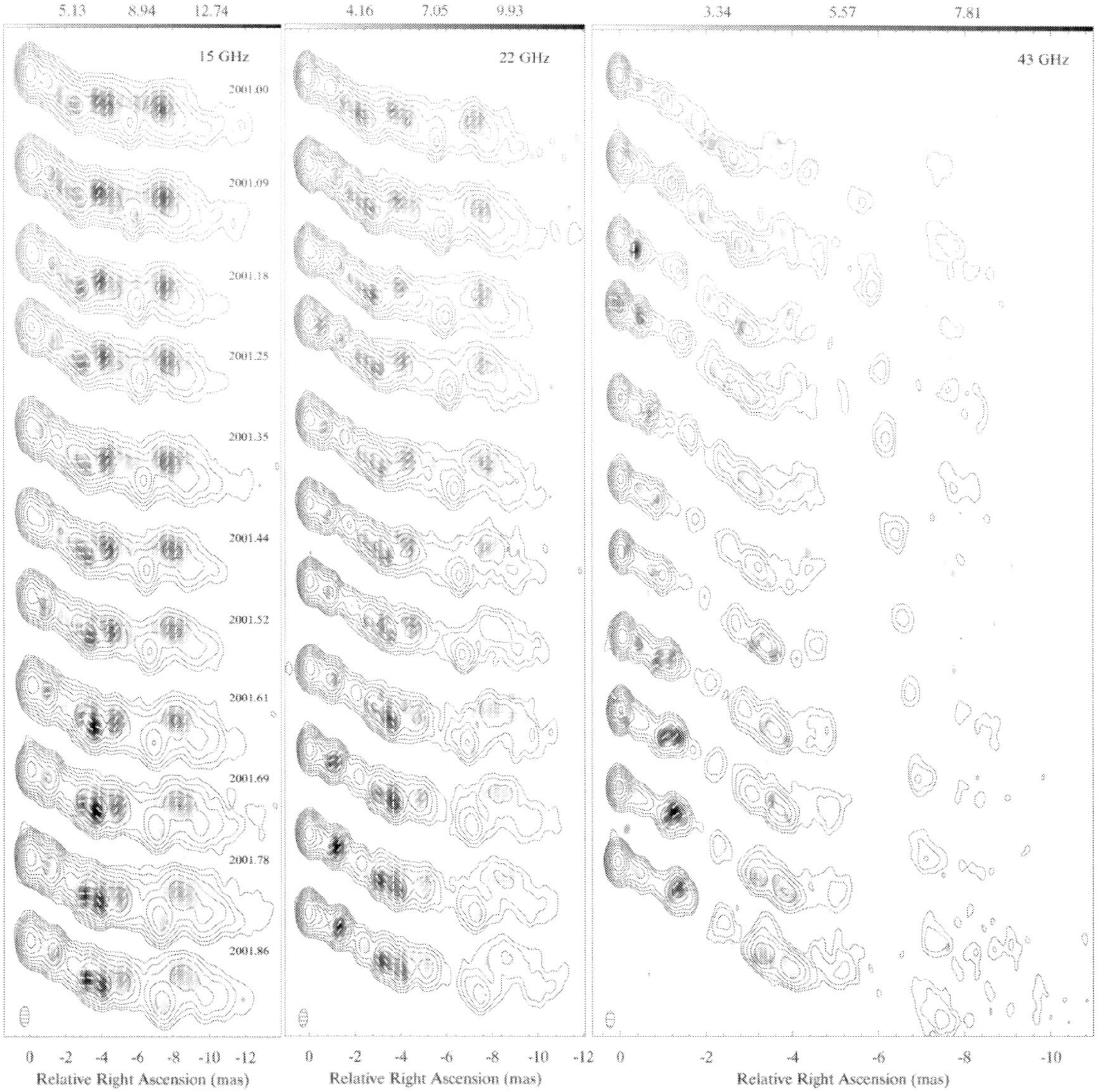

Figure 1. 15 (left), 22 (middle), and 43 GHz (right) VLBA images of 3C 120 in 2001. Total intensity contours are overlaid at 0.5 (0.8; 1.0), 1.1 (1.7; 2.2), 2.5 (3.6; 4.6), 5.6 (7.7; 9.9), 13 (16; 21), 28 (35; 45), 62 (74; 95), 138 (158; 203), 310 (337; 433), and 692 (716; 923) mJy beam^{-1} at 15 GHz (22; 43 GHz). Gray scale images show the linearly polarized intensity. Bars (of unit length) indicate the electric vector position angle, *uncorrected* for Faraday rotation. Note the different scale size used for each frequency.

assumed to originate in a foreground cloud, presumably interacting with the jet. In this paper we present further observations of the jet in the radio galaxy 3C 120 aimed to obtain a better understanding of the origin of the Faraday rotation seen in this source.

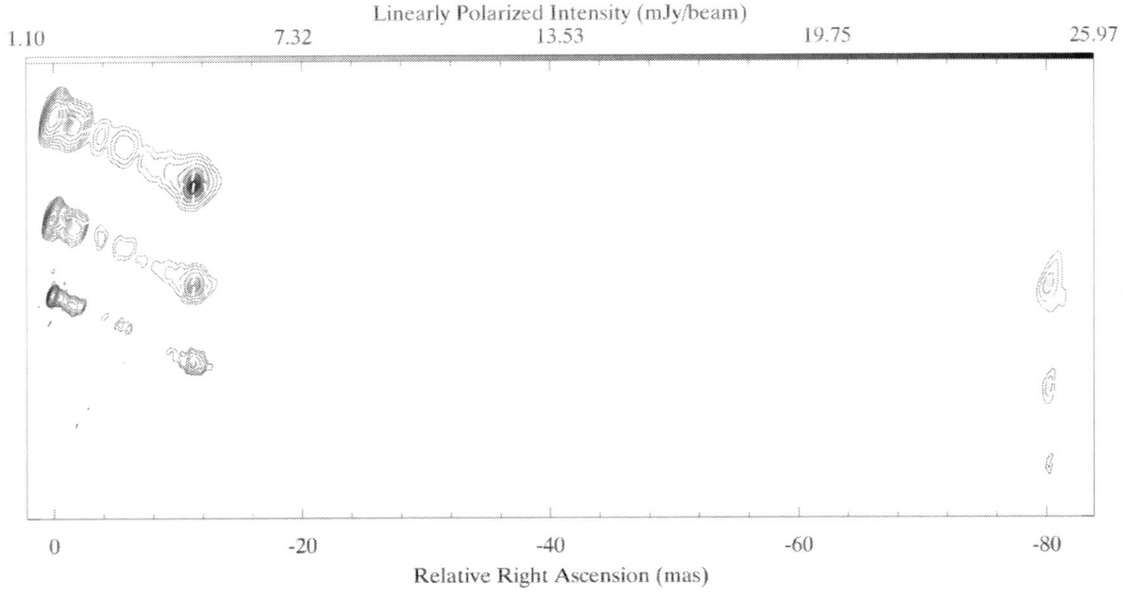

Figure 2. Same as Fig. 1 for observations in 2007 November 7 (2007.85) at 15, 22, and 43 GHz, from top to bottom, respectively. Contours are overlaid at 1.1, 2.4, 5.3, 11, 25, 53, 115, 249, 538, and 1164 mJy beam^{-1}.

2. Observations

We present Very Long Baseline Array (VLBA) polarimetric observations of the jet in 3C 120 carried out in 2001 and 2007 at the standard frequencies of 15, 22, and 43 GHz. Observations during 2001 cover a total of 12 monthly epochs, while those in 2007 were performed on November 7. Reduction of the data was performed with the AIPS software in the usual manner [18]. Opacity corrections at 22 and 43 GHz were introduced by solving for receiver temperature and zenith opacity at each antenna. The feed D-terms (instrumental polarization) were found to be very consistent over all observed sources and to remain stable across epochs during the 2001 observations.

The absolute phase offset between the right- and left-circularly polarized data, which determines the electric vector position angle (EVPA), was obtained by comparison of the integrated polarization of the VLBA images of several calibrators with VLA observations, as well as archival data from the UMRAO, MOJAVE, and NRAO long term monitoring programs. Estimated errors in the orientation of the EVPAs lie in the range of 5°-10°. After the initial reduction, the data were edited, self-calibrated, and imaged both in total and polarized intensity with a combination of AIPS and DIFMAP. For further details we refer the reader to [9, 19].

The total and linearly polarized intensity images corresponding to the 2001 monitoring program and those taken during 2007 are shown in Figs. 1 and 2, respectively. These images show a rich structure in both total and linearly polarized intensity even at the shortest wavelengths, being the emission at 80 mas particularly remarkable, as reported by [20].

2.1. Faraday rotation images

The analysis of the sequence of Faraday rotation images obtained from the 2001 observations (see [9]) reveals the presence of an RM screen in 3C 120 whose properties remain constant over one year, allowing the determination of the mean RM image shown in Fig. 3. The Faraday

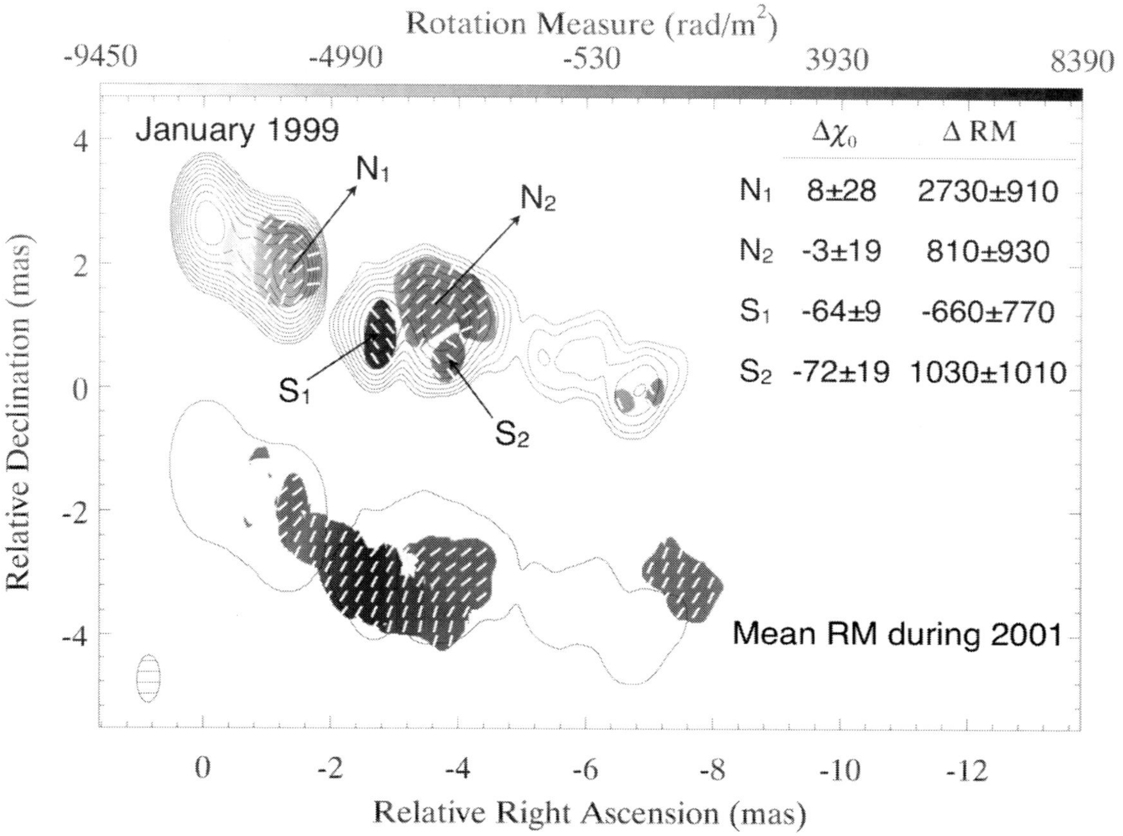

Figure 3. (top) Rotation measure map of 3C 120 in 10 January 1999 from VLBA observations at 22 and 43 GHz. Total intensity 22 GHz contours are overlaid at 3.6, 5.9, 9.8, 16, 27, 44, 73, 120, 198, 327 and 540 mJy beam^{-1}. (bottom) Map of the mean value of the rotation measure for observations during 2001 (see also [9]), shown for comparison. Bars indicate the RM-corrected EVPAs. The lowest contour of the January 1999 image is also shown for an easier comparison. Inset table shows the mean differences between epochs.

screen displays a localized region of enhanced RM between approximately 3 and 4 mas from the core, with a peak of $\sim$ 6000 rad m^{-2}, as well as gradients across and along the jet. The RM image obtained from observations in 1999 January 10 (see [21]) is also shown in Fig. 3.

The rotation measure image for observations in 2007 combining the 15, 22, 43 GHz data is shown in Fig. 4.

3. Uncorrelated changes in the emitting jet magnetic field and Faraday screen

In order to quantify any possible variation in the Faraday screen across epochs we have subtracted the rotation measure values of 1999 January from those of the mean RM map for 2001 and computed the mean value of the difference in the components (jet areas in the mean 2001 map) labeled in Fig. 3. Component S_1 is located at the localized region of high Faraday rotation, which is found to remain in the same jet area as in 2001, with very similar values of the rotation measure. However, the RM-corrected EVPAs show significant rotation, with a mean difference between both epochs of $-64 \pm 9°$. A similar situation is found for component S_2. On the other

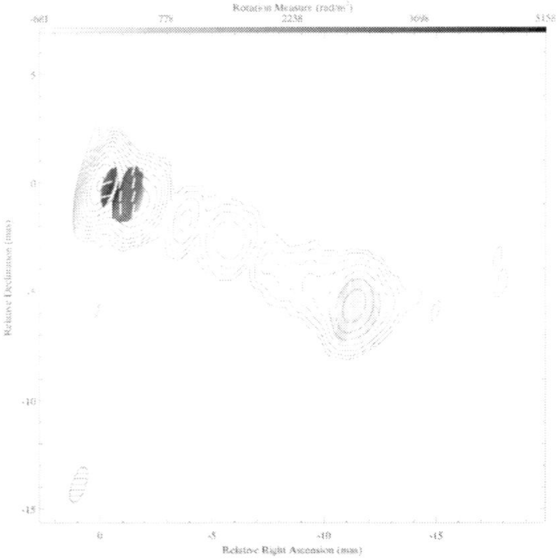

Figure 4. Rotation measure image of 3C 120 in 2007 between 15 and 43 GHz. Contours show the 15 GHz total intensity image. Bars indicate the RM-corrected EVPAs.

Figure 5. Total intensity VLBA images of 3C 120 for epochs 2001.00 and 2007.85 at 15 GHz. Both images have been convolved with the same beam of 1.14×0.54 mas at -1° (shown in the lower left corner) for easier comparison.

hand, component N_2 shows no significant changes in either the RM-corrected EVPAs or RM, with variations of $-3 \pm 19°$ and 810 ± 930 rad m^{-2}, respectively. For component N_1 we also find very similar RM-corrected EVPAs, with a variation of $8 \pm 28°$. However, the rotation measure in N_1 is observed to have changed significantly, with a difference between both epochs of 2730 ± 910 rad m^{-2}.

Therefore, we find uncorrelated changes in the linear polarization of the underlying jet emission and the Faraday rotation screen. While the RM remains constant in the outer jet (including the localized region of high RM) the RM-corrected EVPAs of two particular components (S_1 and S_2) rotate by almost 90°. On the other hand, in the innermost 2 mas the RM changes significantly but without variations in the RM-corrected EVPAs. These uncorrelated changes suggest that the emitting jet and source of RM are not closely connected physically. Furthermore, the existence of a three-year-long stationary region of enhanced RM requires a localized source of Faraday rotation, which favors a model in which a significant fraction of the Faraday rotation measure found in 3C 120 originates in foreground clouds, rather than in a sheath intimately associated with the emitting jet. In this case, Faraday rotation studies will provide valuable information about the ambient medium through which jets propagate, but will not be able to reveal further details about the emitting jet, such as the line-of-sight magnetic field, and hence to test whether they are threaded by helical magnetic fields.

4. No detection of the localized high RM region 2007: an effect of sampling

The RM map of Fig. 4 does *not* show the localized region of high RM that was found in previous observations. This can be understood by examining Fig. 5, which shows the superposition of the 15 GHz total intensity images at the 2001.00 and 2007.85 epochs, revealing a significant change in the jet geometry with time. The position angle of the innermost 3 mas structure has changed from $-126° \pm 2$ in 2001.00 to $-114° \pm 2$ in 2007.85. That is, the jet direction of ejection rotated to the north by 12° between these two epochs. As a consequence of this rotation the components in the 2007 jet do not travel across the localized region of enhanced RM, failing to reveal it, since it is only through the motion of superluminal components that we are able to map the jet polarization owing to the increase in energy density and magnetic field ordering that they produce.

The rotation of the direction of ejection is in agreement with the helical structure suggested by [22], and the proposed precessing/helical models of [23] and [24]. In particular, [23] estimated a precession period of 12.3 yr, predicting a swing to the north of the direction of ejection in the jet between our 2001 and 2007 epochs, as is indeed observed. Similar changes in the geometry of the innermost structure of the jet have also been found in the BL Lac object PKS 0735+178 related to changes in the overall activity of the source [25–29].

Acknowledgments

This research has been supported by the Spanish Ministry of Science and Innovation grants AYA2007-67627-C03-03 and AYA2010-14844, by the Regional Government of Andalucía (Spain) grant P09-FQM-4784, and by National Science Foundation grant AST-0907893. The VLBA is an instrument of the National Radio Astronomy Observatory, a facility of the National Science Foundation operated under cooperative agreement by Associated Universities, Inc.

References

[1] Komissarov S S, Barkov M V, Vlahakis N and Königl A 2007 *MNRAS* **380** 51
[2] Marscher A P, Jorstad S G, D'Arcangelo F D, Smith P S, Williams G G, Larionov V M, Oh H, Olmstead A R, Aller M F, Aller H D, McHardy I M, Lähteenmäki A, Tornikoski M, Valtaoja E, Hagen-Thorn V A, Kopatskaya E N, Gear W K, Tosti G, Kurtanidze O M, Nikolashvili M, Sigua L, Miller H R and Ryle W T 2008 *Nature* **452** 966
[3] McKinney J C and Blandford R D 2009 *MNRAS* **394** L126 (c) Journal compilation © 2009 RAS
[4] Taylor G B 1998 *ApJ* **506** 637
[5] Attridge J M, Wardle J F C and Homan D C 2005 *ApJ* **633** L85
[6] Asada K, Inoue M, Uchida Y, Kameno S, Fujisawa K, Iguchi S and Mutoh M 2002 *Publications of the Astronomical Society of Japan* **54** L39
[7] Laing R A 1981 *ApJ* **248** 87
[8] Gabuzda D C, Murray É and Cronin P 2004 *MNRAS* **351** L89
[9] Gómez J L, Marscher A P, Jorstad S G, Agudo I and Roca-Sogorb M 2008 *ApJ* **681** L69–L72
[10] Asada K, Inoue M, Nakamura M, Kameno S and Nagai H 2008 *ApJ* **682** 798
[11] O'Sullivan S P and Gabuzda D C 2009 *MNRAS* **393** 429
[12] Kharb P, Gabuzda D C, O'Dea C P, Shastri P and Baum S D 2009 *ApJ* **694** 1485
[13] Asada K, Nakamura M, Inoue M, Kameno S and Nagai H 2010 *ApJ* **720** 41
[14] Taylor G B and Zavala R 2010 *ApJ* **722** L183
[15] Oosterloo T A, Morganti R, Tzioumis A, Reynolds J, King E, McCulloch P M and Tsvetanov Z 2000 *AJ* **119** 2085
[16] Middelberg E, Agudo I, Roy A and Krichbaum T P 2007 *MNRAS* **377** 731
[17] Mantovani F, Rossetti A, Junor W, Saikia D J and Salter C J 2010 *A&A* **518** 33
[18] Leppanen K J, Zensus J A and Diamond P J 1995 *AJ* **110** 2479
[19] Gómez J L, Roca-Sogorb M, Agudo I, Marscher A P and Jorstad S G 2011 *ApJ* **733** 11
[20] Roca-Sogorb M, Gómez J L, Agudo I, Marscher A P and Jorstad S G 2010 *ApJ* **712** L160
[21] Gómez J L, Marscher A P, Alberdi A, Jorstad S G and García-Miró C 2000 *Science* **289** 2317–2320
[22] Gómez J L, Marscher A P, Alberdi A, Jorstad S G and Agudo I 2001 *ApJ* **561** L161–L164
[23] Caproni A and Abraham Z 2004 *MNRAS* **349** 1218
[24] Hardee P E, Walker R C and Gómez J L 2005 *ApJ* **620** 646–664
[25] Gómez J L, Marscher A P, Alberdi A and Gabuzda D C 1999 *ApJ* **519** 642–646
[26] Gómez J L, Guirado J C, Agudo I, Marscher A P, Alberdi A, Marcaide J M and Gabuzda D C 2001 *MNRAS* **328** 873–881
[27] Gabuzda D C, Gómez J L and Agudo I 2001 *MNRAS* **328** 719–725
[28] Agudo I, Gómez J L, Gabuzda D C, Marscher A P, Jorstad S G and Alberdi A 2006 *A&A* **453** 477–486
[29] Britzen S, Witzel A, Gong B P, Zhang J W, Krishna G, Goyal A, Aller M F, Aller H D and Zensus J A 2010 *A&A* **515** 105

Prominent polarized flares of the blazars AO 0235+164 and PKS 1510−089

Mahito Sasada

Department of Physical Science, Hiroshima University, Kagamiyama 1-3-1, Higashi-Hiroshima 739-8526, Japan

E-mail: `sasada@hep01.hepl.hiroshima-u.ac.jp`

Abstract. We report on multi-band photopolarimetric observations of the blazars AO 0235+164 and PKS 1510−089. These two blazars became in active states in 2008 and 2009, respectively. In these active states, prominent short-term flares were detected in both objects, having amplitudes of > 1 mag within 10 days. The $V-J$ color became bluer when the objects were brighter in these flares. On the other hand, the color of PKS 1510−089 exhibited a trend that it became redder when it was brighter, except for its prominent flare. This feature can be explained by the strong contribution of thermal emission from an accretion disk. The degree of polarization increased at the flares, and reached > 25 % at the maxima in AO 0235+164 and PKS 1510−089. We compare these flares with other short-term flares which were detected by our monitoring of 44 blazars. Those two flares had one of the largest variation amplitudes in both flux and degree of polarization. Furthermore, we found a significant positive correlation between the amplitudes of the flux and degree of polarization in the flares. It indicates that the short-term flares originate from the region where the magnetic field is aligned.

1. Introduction

There are two components in the emission from blazars [3]. The low energy component from radio to optical or X-ray bands is attributed to synchrotron radiation emitted by relativistic electrons in jets. The high energy component is likely to be inverse-Compton scattering radiation. The synchrotron emission from blazars is occasionally highly polarized in the optical band [2]. Since the polarization gives us a way to probe magnetic field of a nonthermal radiation source, observations of temporal variations of the polarization are important for study of the structure of a blazar jet. However, the temporal behavior of the polarization is known to be complex. The polarization varies in an erratic manner in most blazars (e.g. [2]). On the other hand, systematic variations have also been reported in several blazars, for example, rotations of the polarization (e.g. [11]; [19]).

It has been suggested that the erratic behavior of polarization is due to the composition of several polarization components [10]. From the data of OJ 287 from 2005 to 2009, Villforth, et al. have reported that its observed polarization behavior could be explained by separating the jet emission into two components: an optical polarization core and chaotic jet emission [25]. Uemura, et al. have separated a long-term trend from observed temporal variations of polarization in blazars using a Bayesian approach [24].

AO 0235+164 is one of the most famous blazars, showing violent variability whose amplitude is larger than 1 mag (e.g. [16]). The object showed large amplitude outbursts in past. Radio outbursts repeated quasi-regularly with a periodicity of ~5.7 years [12]. The object shows high polarization in the optical band, and the polarization parameters vary violently. The highest degree of polarization (PD) has been reported to be 43.9 % [7].

PKS 1510−089 is a radio-loud and highly polarized quasar. The object emits radiation from the radio to gamma-ray bands (e.g. [8]; [1]). The spectral energy distribution of the object has a bump structure over a synchrotron component in the ultraviolet (UV) band ([20]; [8]). The UV bump is thought to be thermal emission from an accretion disk. Polarization parameters of the object have also been varied. Marscher, et al. reported that the polarization rotated consecutively for 50±10 days [9].

In the above two blazars, we detected prominent short-term flares with a timescale of ~ 10 days, and found that they were associated with violent variations in polarization. We report their light curves, color and polarization variations, and discuss the characteristics in polarization variations of such short-term flares.

2. Observations

We performed monitoring of AO 0235+164 and PKS 1510−089 in the 2008–2009 seasons using TRISPEC attached to the Kanata 1.5-m telescope at Higashi-Hiroshima Observatory in Japan. TRISPEC has a CCD and two InSb arrays, which enable photopolarimetric observations in an optical and two NIR bands, simultaneously [26]. Unfortunately, the K_S-band readout system of TRISPEC was out of action. Thus, we obtained only V- and J-band data. A unit of the observing sequence consisted of successive exposures at four position angles of a half-wave plate; 0°, 45°, 22.°5 and 67.°5. A set of polarization parameters was derived from each set of the four exposures.

The integration time for an exposure depends on the sky condition and the brightness of the objects. Typical integration times were 120 and 108 sec in AO 0235+164, and 150 and 135 sec in PKS 1510−089 in the V and J bands, respectively.

All images were bias-subtracted and flat-fielded, and we performed an aperture photometry with a Java-based package. We performed a differential photometry with comparison stars taken in the same frames of AO 0235+164 and PKS 1510−089. The positions of the comparison stars are R.A.=$02^h38^m32^s.31$, Dec.=+16°35′59″.7 and R.A.=$15^h12^m53^s.19$, Dec.=−09°03′43″.6 (J2000.0), respectively. Their magnitudes are V=12.720, 13.282 and J=11.221, 12.205, respectively ([5]; [21]). The constancy of the comparison stars was checked by another neighbor stars in the same frames. The position of the check stars are R.A.=$02^h38^m36^s.70$, Dec.=+16°36′26″.6 and R.A.=$15^h12^m51^s.64$, Dec.=−09°05′24.″ (J2000.0), respectively. No significant flux variation was observed in the comparison stars over 0.16 and 0.07 mag in the V band during our observation period.

We confirmed that the instrumental polarization was smaller than 0.1 % in the V band using observations of unpolarized standard stars. We, thus, applied no correction for it. The zero point of the angle of polarization (PA) is corrected for in the equatorial coordinate system (measured from north to east) by observing the polarized stars, HD 19820 and HD 25443 [27].

3. Results

3.1. AO 0235+164

We had performed the photopolarimetric monitoring of AO 0235+164 from Aug. 12, 2008 to Feb. 18, 2009. Figure 1 shows the light curve of the object in the V band, $V-J$ color variation, and temporal variations of the polarization parameters in the V band. We also show relative magnitude, C1-C2, between the comparison and the check stars. In the top panel of figure 1, the object showed the violent variability. The object had been active for 153 days from JD 2454696, in which it was brighter than V = 18.0. The peak flux of the object was V=15.067±0.004 (JD 2454733). The faintest state of the object has been reported at V ~ 20 for 32 years from

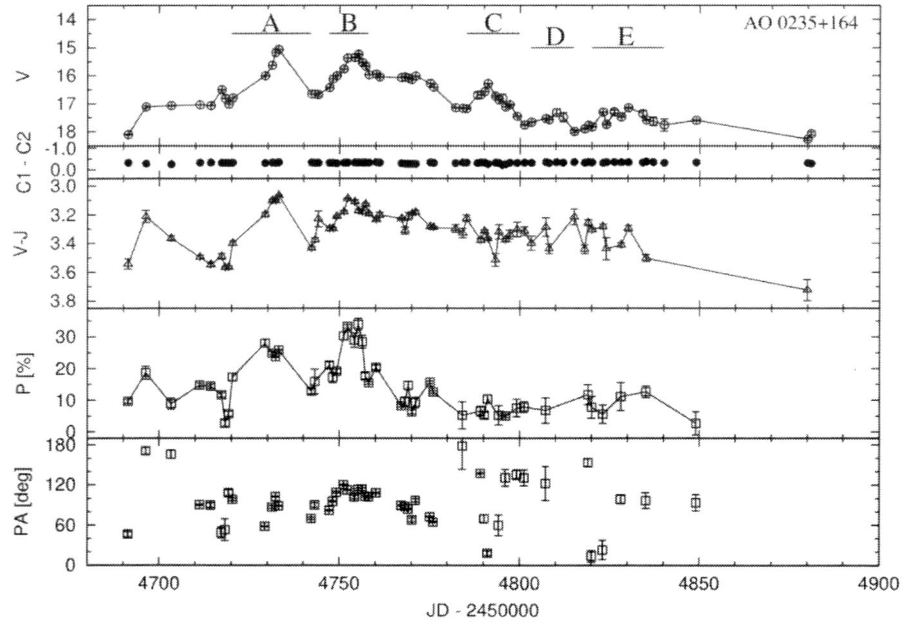

Figure 1. Light curve of AO 0235+164 in the V band, color variations of $V - J$, temporal variations of the PD and PA, % and degree. We also show the relative magnitude, C1-C2, between the comparison star (C1) and the check star (C2).

1975 to 2007 ([12]; [13]; [14]; [15]). Thus, the amplitude of the flux was about 5 mag from the faintest state. The light curve shows several short-term flares during the active state. We can see five flares labelled in figure 1 as "A" (from JD 2454720 to 2454742), "B" (from JD 2454747 to 2454758), "C" (from JD 2454785 to 2454801), "D" (from JD 2454803 to 2454815) and "E" (from JD 2454820 to 2454840).

The $V - J$ color was variable in our monitoring period, Δ $(V - J) \sim$ 0.5. In flares "A" and "B", the color became bluer. Hence, the object showed a bluer-when-brighter trend in these bright flares. The color also became bluer in JD 2454696, at the onset of the active state when the object was brightened rapidly. On the other hand, no clear bluer-when-brighter trend was associated with flares "C", "D" and "E". In the faintest state in JD 2454879, the object was the reddest.

The PD showed variable, and distributed from 0 % to ~30 %. In JD 2454755, the PD was the highest, 34±2 %. The PD was well correlated with the flux in flares "A" and "B". The peaks of the PD in flares "A" and "B" were higher than 25 %. However, the PD were not high during flares "C", "D" and "E", less than 16 %.

3.2. PKS 1510−089

We had also performed the photopolarimetric monitoring of PKS 1510−089 from Jan. 12 to Jul. 22 in 2009. Figure 2 shows the light curve of PKS 1510−089 in the V band, $V - J$ color variation, and temporal variations of the polarization parameters in the V band. The object has been brightened since JD 2454900, and a peak of the flux was once recorded in JD 2454928. After that, a prominent flare occurred around JD 2454960. We labelled this flare as "F" (from JD 2454954 to 2454966). In this flare, the peak of the flux reached about V=13.6 mag in

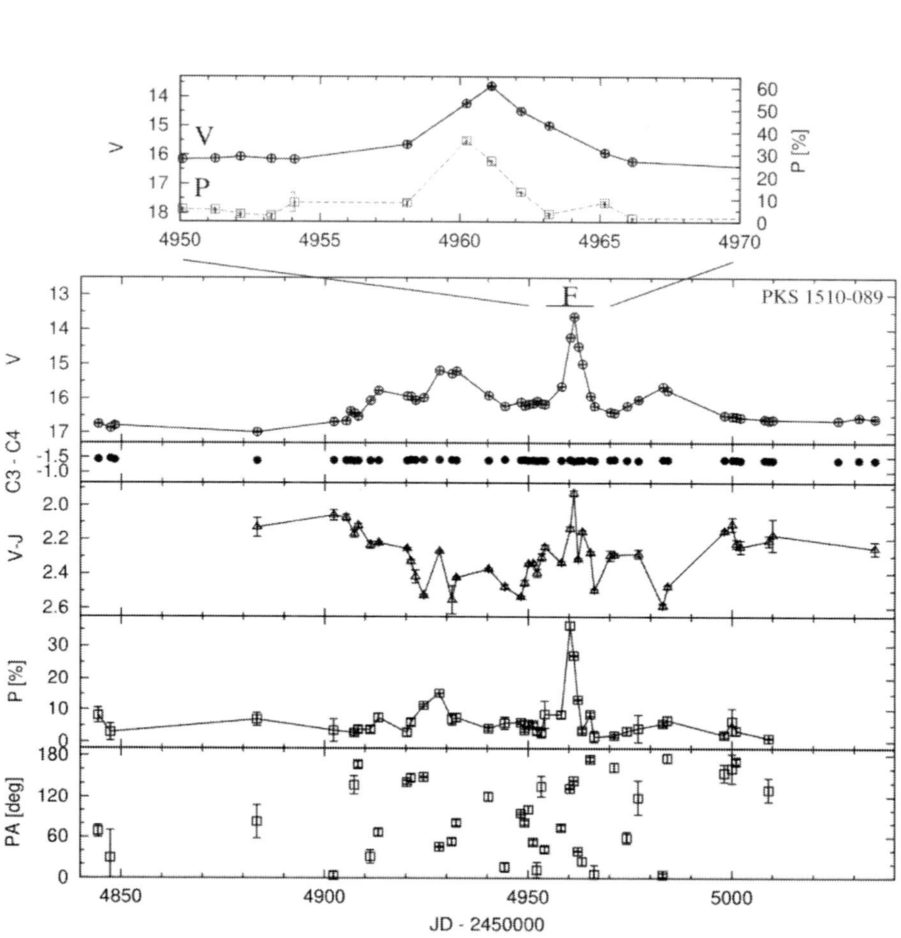

Figure 2. Light curve of PKS 1510−089 in the V band, color variations of $V - J$, temporal variations of the PD and PA, % and degree. The top panel shows the enlarged light curve and PD around flare "F". We also show the relative magnitude, C3-C4, between the comparison star (C3) and the check star (C4).

JD 2454961, and the amplitude of the magnitude between the peak and the bottom of the flare in JD 2454954 was ΔV=2.5 mag. After the flare peak, the flux was fallen down to V=16.2 mag in JD 2454966. Thus, the flux varied by 10 times within 10 days. The flux again rose after this flare, and the flux reached another peak at about V=15.7 mag in JD 2454966.

The $V - J$ color was about 2.1 in a faint state in JD 2454900. When the object became brighter, the $V - J$ color got redder to 2.5. This "redder-when-brighter" trend was saturated at about $V - J$=2.5. On the other hand, the color during the flare "F" showed a bluer-when-brighter trend. The $V - J$ color reached 1.9 in this flare.

Polarization parameters also varied in the whole observing period. The PA was distributed from 0 to 180°. The PD varied less than 10 % in a faint state from JD 2454844 to 2454911. The PD increased with the flux from JD 2454920, and reached 15 % at the maximum of a small flare in JD 2454928. In the rising phase of the flare "F", the PD rapidly increased, and reached its peak about 36 % in JD 2454960, in which the flux was still in a rising phase (see the top panel of figure 2). Thus, the peak of the PD was 1 day earlier than that of the flux. In the decay

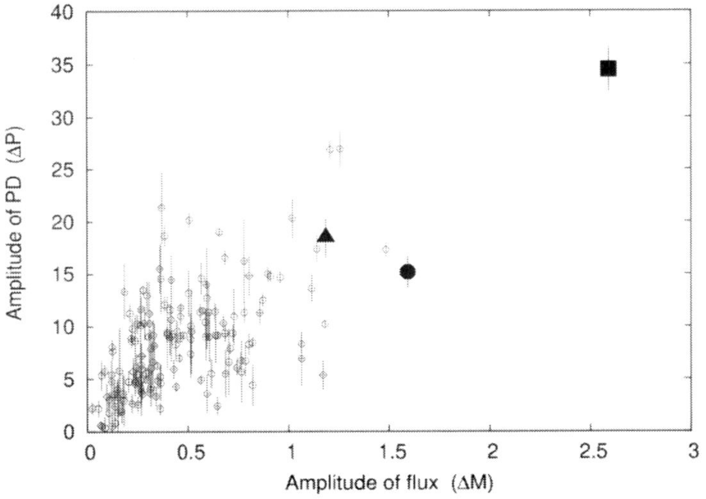

Figure 3. Correlation between the amplitudes of the flux, ΔM, and PD, ΔP, in short-term flares in blazars in the V band. Definitions of ΔM and ΔP are described in the text. The filled circle, triangle and square represent flares "A", "B" and "F", respectively.

phase of the flux, the PD had already been low, about 13 % in JD 2454962. After the flare, the PD did not vary more than 10 %.

4. Discussion

We reported that there were positive correlations between the flux and PD in both AO 0235+164 and PKS 1510−089. Such correlations between the flux and PD have been reported in other blazars in past studies (e.g. [22]; [23]; [18]), although the polarization behaviors have generally been reported to be erratic (e.g. [2]). We propose that short-term flares on a timescale of 10 days in blazars are associated with increases in PD, with changes of the PA of each flare. If a small flare occurs, this flare is buried by other radiation component, for example the long-term component as mentioned above, and then, the PD is hardly changed. However, a large amplitude flare may not be buried like flares "A", "B" and "F". In this situation, the observed PD should increase during the large flare. In fact, the observed PD during the large flares "A", "B" and "F" have been changed dramatically.

We investigated the correlation between amplitudes of the flux and PD during flares. We used the photopolarimetric data of 44 blazars which we obtained from 2008 to 2010 [6]. In this paper, we defined the flares and their amplitudes of the flux and PD as described below. First, we defined the peak of a flare as the observation point with the highest flux within ±10 days. Second, we can calculated the peak-to-valley amplitudes of the flux and PD in this 20 days period. Practically, the amplitude of the flux was calculated as $\Delta M = -(M_{\rm bri} - M_{\rm fai})$, where $M_{\rm bri}$ and $M_{\rm fai}$ are the brightest and faintest values of the V-band magnitude for the 20 days period. Similarly, the amplitude of PD is defined as $\Delta P = P_{\rm max} - P_{\rm min}$, where $P_{\rm max}$ and $P_{\rm min}$ are the maximum and minimum values of the PD for the 20 days period.

Figure 3 shows correlation between the amplitudes of the flux and PD of short-term flares selected with the above definition in the V band. There is a positive correlation, and the correlation coefficient is $0.66^{+0.08}_{-0.10}$. The filled circle, triangle and square represent flares "A", "B" and "F" of AO 0235+164 and PKS 1510−089, respectively. Those flares are one of the largest flux- and polarization-amplitude flares in our sample (figure 3). On the other hand, the correlation is weak in small flares; there are flares having a relatively large amplitude of flux, but a small amplitude of the PD, and vice versa. This is naturally expected since polarization variations of small flares are readily buried by another variation component in our picture.

Sasada, et al., reported that a microvariation of S5 0716+714 had a specific polarization component, and suggested that the microvariation originated from a small and local region where the magnetic field was aligned [17]. Our result indicates that the short-term flares on timescales of 10 days also originate from regions where the magnetic field is aligned.

5. Conclusion

We performed photometric and polarimetric monitoring of the blazars AO 0235+164 and PKS 1510−089, and observed their flares. The large flares "A", "B" and "F" exhibited the bluer-when-brighter trends. The $V-J$ color variation of PKS 1510−089 showed the redder-when-brighter trend, except for the period of flare "F". There were positive correlations between the flux and PD in both objects. We studied the correlation between the amplitudes of the flux and PD of short flares in 44 blazars. We found that there was a significant positive correlation between them. We propose that the short-term flares on timescales of 10 days originate from a region where the magnetic field is aligned. Observed polarization variations are occasionally erratic probably because polarization variations of small flares are readily buried by another variations having different PAs and/or timescales.

References

[1] Abdo, A. A., et al. 2010b, ApJ, 721, 1425
[2] Angel, J. R. P. & Stockman, H. S. 1980, A&A, 18, 321
[3] Fossati, G., Maraschi, L., Celloti, A., Comastri, A. & Ghisellini, G. 1998, MNRAS, 299, 433
[4] Fukugita, M., Shimasaku, K. & Ichikawa, T. 1995, PASP, 107, 945
[5] González − Pérez, J. N., Kidger, M. R. & Martín − Luis, F. 2001, AJ, 122, 2055
[6] Ikejiri, Y., et al. 2011, astro-ph/1105.0255
[7] Impey, C. D., Brand, P. W. J. L. & Tapia, S. 1982, MNRAS, 198, 1
[8] Kataoka, J., et al. 2008, ApJ, 672, 787
[9] Marscher, A. P., et al. 2010, ApJ, 710, L126
[10] Moore, R. L., Angel, J. R. P., Duerr, R., Lebofsky, M. J., Wisniewski, W. Z., Rieke, G. H., Axon, D. J., Bailey, J., Hough, J. M. & McGraw, J. T. 1982, ApJ, 260, 415
[11] Qian, S. J., Quirrenbach, A., Witzel, A., Krichbaum, T. P., Hummel, C. A. & Zensus, J. A. 1991, A&A, 241, 15
[12] Raiteri, C. M., et al. 2001, A&A, 377, 396
[13] Raiteri, C. M., et al. 2005, A&A, 438, 39
[14] Raiteri, C. M., et al. 2006, A&A, 459, 731
[15] Raiteri, C. M., et al. 2008, A&A, 480, 339
[16] Rieke, G. H., Grasdalen, G. L., Kinman, T. D., Hintzen, P., Wills, B. J. & Wills, D. 1976, Nature, 260, 754
[17] Sasada, M., et al. 2008, PASJ, 60, L37
[18] Sasada, M., et al. 2010, PASJ, 62, 645
[19] Sillanpää, A., Takalo, L. O., Nilsson, K. & Kikuchi, S. 1993, Ap&SS, 206, 55
[20] Singh, K. P., Shrader, C. R. & George, I. M. 1997, ApJ, 491, 515
[21] Skrutskie, M. F., et al. 2006, AJ, 131, 1163
[22] Smith, P. S., Balonek, T. J., Heckert, P. A. & Elston, R. 1986, ApJ, 305, 484
[23] Tosti, G., et al. 1998, A&A, 339, 41
[24] Uemura, M., et al. 2010, PASJ, 62, 69
[25] Villforth, C., et al. 2010, MNRAS, 402, 208
[26] Watanabe, M., et al. 2005, PASP, 117, 870
[27] Wolff, M. J., Nordsieck, K. H. & Nook,M. A. 1996, AJ, 111, 856

International Workshop on Beamed and Unbeamed Gamma-Rays from Galaxies IOP Publishing
Journal of Physics: Conference Series **355** (2012) 012024 doi:10.1088/1742-6596/355/1/012024

Do NLS1s have a beamed outflow? An unusual X-ray perspective for Mrk 766

Guido Risaliti

INAF-Arcetri observatory, Largo E. Fermi 5, I-50125 Firenze, ITALY, & Harvard-Smithsonian Ceter for Astrophysics, 60 Garden Street, Cambridge, MA, 02138, USA

E-mail: risaliti@arcetri.astro.it

Abstract. Recently, the detection of Gamma-ray emission from Narrow Line Seyfert 1s has suggested that these sources may emit beamed jets towards us. I report the results of a time-resolved spectral analysis of an XMM-Newton long observation of Mrk 766, showing that the X-ray source has been eclipsed several times by clouds with a cold, dense core, and a less dense, highly ionized tail. These clouds have blueshift velocities v> 10^4 km s^{-1} (as measured through He-like iron absorption lines), and are therefore part of a strong outflow. These results provide new important constraints on the geometry and structure of this source, and may be relevant to understand the peculiarity of NLS1s as a class of AGNs.

1. Introduction

X-ray absorption variability is a common feature in Active Galactic Nuclei (AGN). An analysis of a sample of nearby X-ray obscured AGN with multiple X-ray observations, performed a few years ago (Risaliti et al. 2002) revealed that column density (N_H) variations are almost ubiquitous in local Seyfert galaxies. More recent observations performed with *XMM-Newton*, *Chandra* and *Suzaku* further confirmed this finding. The physical implications of these measurements are that the circumnuclear X-ray absorber (or, at least, one component of it) must be clumpy, and located at sub-parsec distances from the central source. The comparison between different observations, typically performed at time distances of months-years, only provides upper limits to the intrinsic time scales of N_H variations. An improvement of these estimates can only be obtained through observational campaigns within a few weeks/days, and/or through the search for N_H variations within single long observations. Such short time-scale studies have been already performed for a handful sources: NGC 1365 (Risaliti et al. 2005, 2007, 2009), NGC 4388 (Elvis et al. 2004), NGC 4151 (Puccetti et al. 2007). In particular, in the case of the AGN in NGC 1365 we revealed extreme spectral changes, from Compton-thin (N_H in the range 10^{23} cm^{-2}) to reflection-dominated (N_H>10^{24} cm^{-2}) in time scales from a couple of days to ~10 hours. Such rapid events imply that the absorption is due to clouds with velocity v>10^3 km s^{-1}, at distances of the order of 10^4 gravitational radii (assuming that they are moving with Keplerian velocity around the central black hole). The physical size and density of the clouds are of the order of 10^{13} cm and 10^{10}-10^{11} cm^{-3}, respectively.[1] All these physical parameters are typical for Broad Line Region (BLR) clouds, strongly suggesting that the X-ray

[1] For a detailed derivation and discussion of these parameters, we refer to Risaliti et al. 2009. A rough estimate is also mentioned here in Section 2.

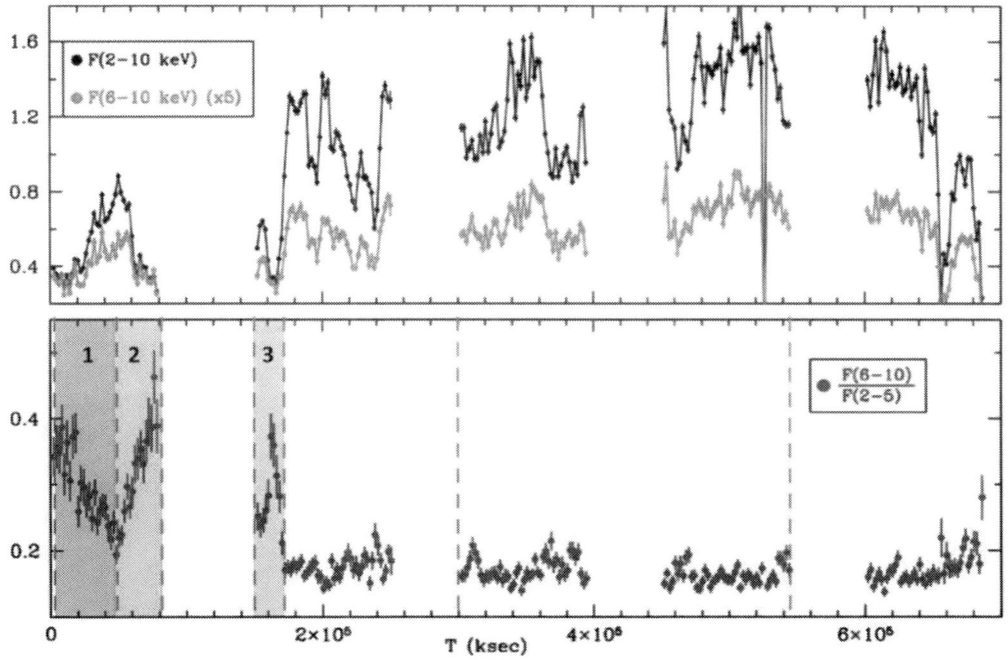

Figure 1. Flux (top) and hardness ratio (bottom) light curves from the *XMM-Newton* long observation of Mrk 766. The observation is made in five consecutive *XMM-Newton* orbits.

absorber and the clouds responsible for broad emission lines in the optical/UV are one and the same.

Here I present a special case of X-ray absorption variability of the narrow line Seyfert 1 galaxy Mrk 766. The occultations found in this source are particularly interesting because of the (on average) unobscured state of this source, which implies that the absorption by an intervening cloud has a large impact on the observed spectrum, which allows a more detailed, otherwise impossible analysis of the cloud structure.

2. Time-resolved analysis of the X-ray emission of Mrk 766

Mrk 766 has been observed by *XMM-Newton* for six consecutive orbit, for a total time of about 700 ks. We analyzed this source through a two-step method based on (1) the study of the hardness ratio light curve, and (2) the spectral analysis of the time intervals defined in the first step. This approach is illustrated in Fig. 1.

The upper panel of Fig. 1 shows the standard 2-10 keV flux light curve for this observation, with the well known strong variability on time scales of thousands of seconds, or even shorter. The lower panel shows the light curve of the (6-10 keV)/(2-5 keV) flux ratio. In general, this light curve shows much smaller variations, indicating that the continuum shape remains the same during most of the luminosity variations. However, clear exceptions are observed in at least three intervals, highlighted in Fig. 1. During these intervals it is possible that a cloud with N_H of the order of 10^{23} cm^{-2} has covered the central source, strongly decreasing the observed flux in the soft band, without affecting the hard band, and therefore increasing the observed hardness ratio.

Following this interpretation, in the first interval we should be observing a cloud uncovering the X-ray source (with the covering phase occurred before the beginning of the observation); in

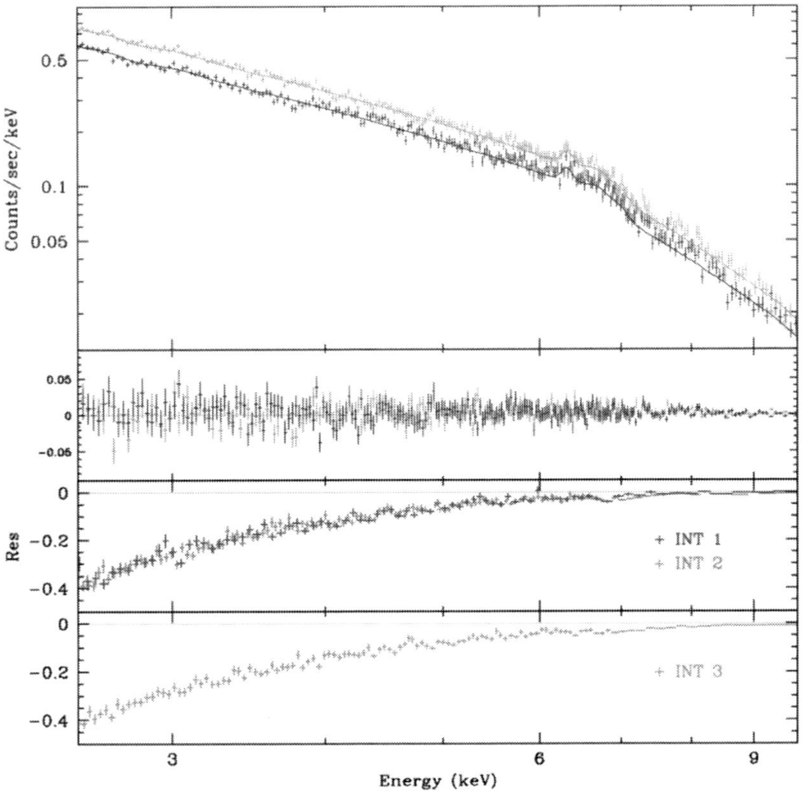

Figure 2. Results from the spectral analysis of the eclipses observed in Mrk 766. Top two panels: spectra, best fit model and residuals from the third and fourh orbit (Fig. 1), where no spectral changes are observed. Bottom panels: difference between the spectra in the three intervals with spectral variations (INT 1, 2 and 3, Fig. 1) and the best fit model for the third and fourth orbit (INT 4).

the second interval another cloud shoud be covering the X-ray source, with the uncovering phase not observed due to the "dead time" between two consecutive *XMM-Newton* orbits. Finally, in the third interval we should be observing the whole eclipse.

In order to check this scenario, we performed a complete analysis of the spectra obtained from the three highlighted intervals, and of those obtained from the third and fourth orbit, representing the standard spectral state of the source. In this analysis we allowed all the main spectral parameters of the model to vary among the different intervals. The results of the this study, illustrated in Fig. 2 are the following:

1) the 2-10 keV spectrum obtained from the third and fourth orbit (the "standard" state) is well reproduced by a typical model for type 1 AGNs, consisting of a power law, a reflection component and an iron emission line;

2) the variations observed in the first and second orbit can be completely reproduced through an obscuring cloud partially covering the X-ray source. Each of the three events (two in the first orbit, one in the second) are due to different clouds (with the first two possibly overlapping, see

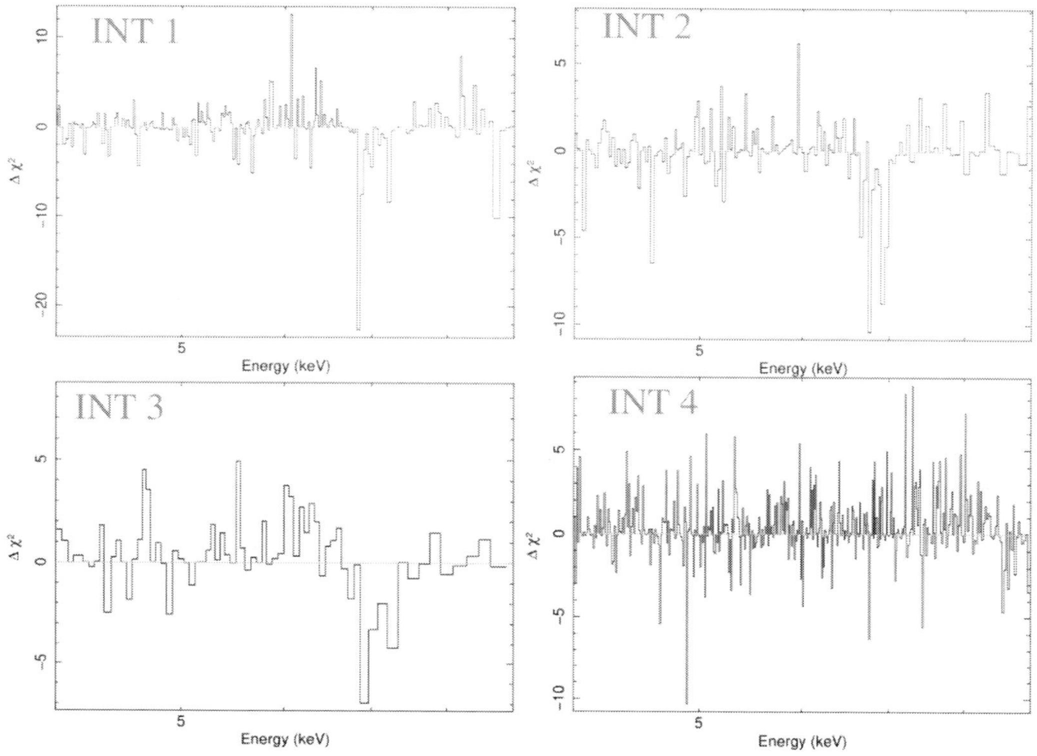

Figure 3. χ^2 residuals for the six time intervals corresponding to: the three occultation events (1, 2 and 3AC), the first interval after the third occultation (3D), the remainder of the second orbit (3EF) and the whole third and fourth orbits (4). The presence of iron absorption lines when the cold absorption cloud is also present, suggest the presence of a ionized tail.

Risaliti et al. 2011 for details), with column densities in the range 1-3$\times 10^{23}$ cm^{-2}.

3) The residuals of the best fits for all the three intervals with occultations show the presence of absorption lines in the 6.7-8 keV energy interval (Fig. 3), compatible with highly ionized iron (Fe XXV and Fe XXVI). A subsequent fit showed that these features are statistically highly significant, and are blueshifted by several thousands km/s. No such line is detected in any of the other spectral intervals, i.e. the highly ionized absorber is associated to the neutral clouds responsible for the continuum spectral variations. A new fit which allow for a free ionization state for the obscuring cloud shows that the two components must have a difference in ionization parameter by a factor of at least 50.

3. Conclusions

The analysis presented above shows that three occultations by BLR clouds happened during the long *XMM-Newton* observation of Mrk 766. At the same time of the "eclipses" due to the nearly neutral clouds, iron absorption lines due to highly ionized iron were also present, with outflowing velocities of several 10^3 km s^{-1} (up to 15,000 km s^{-1} for the third cloud). The different in ionization parameter between the two components is at least 50. This strongly suggests a cometary structure of the clouds (Fig. 4) with a high-density head, followed by a long, low density, highly ionized tail. In our model, the clouds have both an outflowing and a rotational (~Keplerian) component of the velocity. The rotational component is of the order of

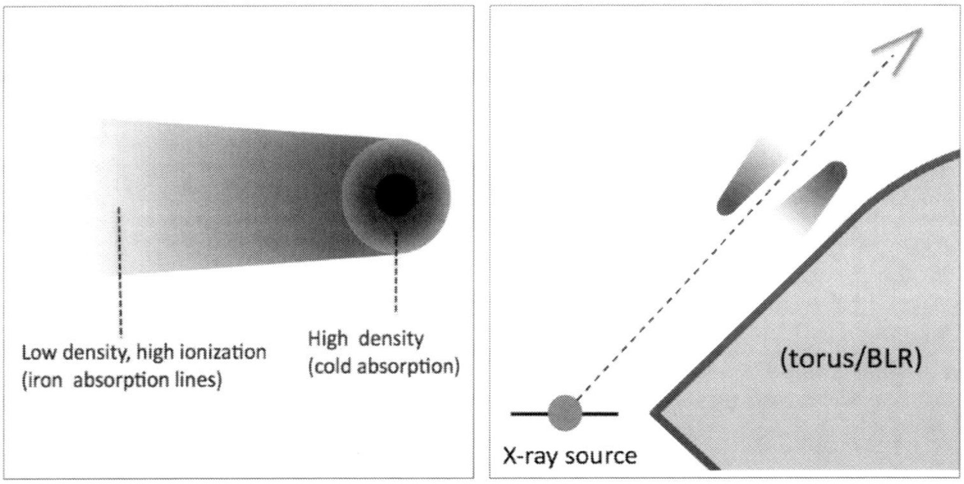

Figure 4. Scheme of the proposed structure of the absorber responsible for the observed occultations. Left: structure of an eclipsing cloud, as viewed from a direction orthogonal to the line of sight. Right: a possible structure of the circumnuclear gas in Mrk 766, as viewed from the disc/torus plane. The cloud tail can be either in front or behind the cold core with respect to the source.

several thousands km/s, in order to cover the X-ray source (whose size is assumed to be a few gravitational radii) in the observed eclipsing time.

This analysis is the first direct X-ray evidence of the presence of ionized material rapidly outflowing from the central source in a Narrow-Line Seyfert 1, and may be related to the presence to beamed outflows, as recently suggested to explain the radio and gamma-ray emission of several NLS1.

References

[1] Bianchi, S., Piconcelli, E., Chiaberge, M., Bailón, E. J., Matt, G., & Fiore, F. 2009, *ApJ*, 695, 781
[2] Elvis, M., Risaliti, G., Nicastro, F., Miller, J. M., Fiore, F., & Puccetti, S. 2004, *ApJ*, 615, L25
[3] Puccetti, S., Fiore, F., Risaliti, G., Capalbi, M., Elvis, M., & Nicastro, F. 2007, *MNRAS*, 377, 607
[4] Risaliti, G., Elvis, M., & Nicastro, F. 2002, *ApJ*, 571, 234
[5] Risaliti, G., Elvis, M., Fabbiano, G., Baldi, A., & Zezas, A. 2005, *ApJ*, 623, L93
[6] Risaliti, G., Elvis, M., Fabbiano, G., Baldi, A., Zezas, A., & Salvati, M. 2007, *ApJ*, 659, L111
[7] Risaliti, G., et al. 2009,*apj*, 696, 160
[8] Risaliti G., Nardini E., Salvati M., Elvis M., Fabbiano G., Maiolino R., Pietrini P., Torricelli-Ciamponi G., 2011, MNRAS, 410, 1027

Possible γ-ray Signals from Low-luminosity Active Galactic Nuclei

Hajime Takami

Max Planck Institute for Physics, Föhringer Ring 6, 80805 Munich, Germany

E-mail: takami@mpp.mpg.de

Abstract. Although it has been believed that both emission from an accretion disk and from a jet contributes to the spectral energy distribution of a low-luminosity active galactic nucleus (LLAGN), the respective contribution of them is still an open question. γ-ray signals from LLAGNs can constrain a jet component because a jet component is expected to associate with γ-rays by emission mechanisms similar to powerful blazars and radio galaxies, and then play an important role to understand the physics of disk-jet connection. Here, we demonstrate the γ-ray signals on the assumption that a jet component dominates in X-ray bands, and discuss possible parameter sets on an emission region and the detectability of the γ-rays. In some reasonable cases, Cherenkov Telescope Array can detect γ-rays from a nearby LLAGN.

1. Introduction

An active galactic nucleus (AGN) is a powerful object in the Universe, which is powered by gravitational energy through the accretion of surrounding gas onto a supermassive black hole (SMBH) located at the center of the accretion system. The accretion paradigm has been extensively applied to extremely powerful objects such as quasars and radio galaxies for extragalactic objects, but recently the idea that all the galaxies have their own SMBHs at their centers has been commonplace [1] and optical spectroscopic surveys have revealed that a significant fraction of nearby galaxies have active nuclei which are much less luminous than the powerful objects ($< 10^{40}$ erg s^{-1} in nuclear Hα luminosity; low-luminosity AGNs (LLAGNs)) [2]. The Eddington ratio of LLAGNs are much lower than the canonical value ($\sim$ 10 per cent) and reach $\sim 10^{-8}$ in some cases [3].

The low radiative efficiency and the weak feature of the big blue bump [4] indicate optically thin, radiatively inefficient accretion flows (RIAFs) in LLAGNs. Advection-dominated accretion flow (ADAF) models [5, 6], which are a kind of RIAF models, have successfully reproduced the spectral energy distribution (SED) of Sagittarius A* which is a SMBH located at the center of our Galaxy [7, 8, 9]. Recently, ADAF models have been applied to nearby LLAGNs to investigate the physics of the accretion disk of LLAGNs. However, it is a difficult task to measure radiation from the core of a LLAGN because very good angular resolution is required to avoid possible contamination from stellar and dust emission. This fact prevents us from obtaining good multi-wavelength SED and from understanding the nature of LLAGNs. Radio observations are a powerful tool, they have revealed a compact core morphology [10, 11] and have implied the existence of an ADAF component [12]. Recent observations have revealed that a ADAF component is not enough to reproduce total radio emission from a core, which indicates the contribution of jets [13, 14, 15]. These observations imply the coexistence of both disk and jet components in emission from LLAGNs. Recently, Ref. [16] demonstrated the SED modeling of several observed LLAGNs by using both the components and showed that observed X-ray data can be reproduced by either disk-dominated or jet-dominated radiation models without inconsistency because of sparse observational data. The respective contribution of them is connected with the understanding of the nature of the accretion disk of LLAGNs.

A possible solution to understand the origin of the X-rays from LLAGNs is to observe γ-rays. If the X-rays are the synchrotron radiation of relativistic electrons in a jet, the electrons upscatter the synchrotron photons (synchrotron self-compton) or ambient photons (external compton) and produce γ-rays, similarly to the cases of blazars and/or radio galaxies. Thus, γ-rays can be a hint of the jet component of radiation from LLAGNs.

Another motivation to consider γ-rays from LLAGNs is the commonality of γ-ray emission mechanisms among AGNs. Recent progress on Imaging Atmospheric Cherenkov Telescopes (IACTs) have detected a lot of very-high-energy (VHE) γ-ray emitted blazars and radio galaxies. Many BL Lac objects have been detected in VHE energies, which are associated with Fanaroff-Riley 1 galaxies based on a unification hypothesis [17], and their accretion disks are thought to be RIAFs. Thus, accretion physics is similar between VHE-detected BL Lac objects and LLAGNs. How common are γ-ray emission mechanisms among AGNs?

In this paper, we demonstrate possible γ-ray signals from a LLAGN, NGC 4278, as a representative on the assumption that a jet component dominates in X-ray bands by fitting observed SED in the framework of a synchrotron self-compton (SSC) model. We also discuss the detectability of the γ-ray signals by γ-ray observatories.

2. Model

We consider an one-zone SSC model for LLAGNs. Electrons are accelerated in a jet via particle acceleration mechanisms, such as shock acceleration [18], and are assumed to have a smoothed broken power-law spectrum,

$$\frac{dn_e}{d\gamma} = n_0 \gamma^{-s_1} \left(1 + \frac{\gamma}{\gamma_{\rm br}}\right)^{s_1 - s_2} \quad (\gamma_{\rm min} \le \gamma \le \gamma_{\rm max}), \tag{1}$$

where n_0, γ, $\gamma_{\rm min}$, $\gamma_{\rm max}$, and $\gamma_{\rm br}$ are the normalization factor of the electron number density in an emission blob, the Lorentz factor of electrons, the minimum of γ, the maximum of γ, and γ at a spectral break, respectively. $\gamma_{\rm br}$ and $\gamma_{\rm max}$ are determined by the cooling time-scale of electrons as discussed below, while $\gamma_{\rm min}$ is fixed at 1.

$\gamma_{\rm max}$ can be estimated by comparing the time-scale to accelerate electrons with the shortest energy-loss time-scale. The acceleration time-scale is $\tau_{\rm acc} = \theta_F r_g/c = 6 \times 10^2 \theta_{F,2} \gamma_7 {B_{-1}}^{-1}$ s, where r_g and c is the Larmor radius of an electron and the speed of light, respectively. $\theta_{F,2} = \theta_F/10^2$, $\gamma_7 = \gamma/10^7$, and $B_{-1} = B/10^{-1}$ G. We take θ_F to be a constant, which is equivalent to assuming the diffusion coefficient for accelerated particles to be proportional to the Bohm diffusion coefficient. $\theta_F \ge 10$ is a fairly conservative value, but $\theta_F \sim 1$ is also possible for mildly relativistic shocks [19].

The accelerated electrons suffer from energy-loss via synchrotron radiation and inverse compton scattering (ICS). The cooling time-scale of synchrotron radiation is $\tau_{\rm syn} = 3m_e c/4\sigma_T \gamma U_B = 8 \times 10^3 {\gamma_7}^{-1} {B_{-1}}^{-2}$ s, where m_e, $\sigma_{\rm T}$, and $U_B = B^2/8\pi$ are the electron mass, the cross-section of Thomson scattering and the energy density of magnetic field, respectively. The cooling time-scale of ICS has the same form, but U_B is replaced by $U_{\rm rad}$ in the Thomson regime, which is the energy density of radiation. A process with larger energy density dominantly contributes to the energy-loss of electrons. Practically, we don't know $U_{\rm rad}$ until finishing calculations. Thus, firstly we calculate SED on the assumption that U_B is dominant, and then check $U_{\rm rad}$. In the cases of the small size of an emission blob, $U_{\rm rad}$ could be larger than U_B, but these cases are not favored by several observational aspects as seen afterwards. Consequently, $\gamma_{\rm max}$ can be estimated by $\tau_{\rm acc} = \tau_{\rm syn}$,

$$\gamma_{\rm max} = \left(\frac{6\pi e}{\theta_F \sigma_T B}\right)^{-1/2} = 4 \times 10^7 \theta_{F,2}{}^{-1/2} B_{-1}{}^{-1/2}. \quad (2)$$

Note that this $\gamma_{\rm max}$ always satisfies the Hillas criterion that r_g should be smaller than R [20] in all the cases treated in this paper.

The cooling of electrons via synchrotron radiation makes a power-law index steeper by 1, $s_2 = s_1 + 1$ above a characteristic energy $\gamma_{\rm br}$, assuming that accelerated electrons are continuously provided into the blob. Since the electrons lose their energies in the blob, this characteristic energy can be estimated by $\tau_{\rm syn} < \tau_{\rm esc}$, where $\tau_{\rm esc} = 3R/c = 1 \times 10^6 R_{16}$ s, is the time-scale to escape electrons from the blob. Here, we assume that electrons escape from the blob by the velocity of $c/3$, which corresponds to the velocity of downstream fluid in a shock system under the strong shock limit of a relativistic shock. Thus,

$$\gamma_{\rm br} = \frac{2\pi m_e c^2}{\sigma_{\rm T} R B^2} = 8 \times 10^4 R_{16}{}^{-1} B_{-1}{}^{-2}. \quad (3)$$

We estimate the SEDs of a jet component of NGC 4278 in the overall energy range based on the SSC model under the electron spectrum modeled above. This model has five free parameters: $s_1(s_2)$, B, R, n_0, and δ, where $\delta = [\Gamma(1-\beta\cos\theta)]^{-1}$ is so-called Doppler factor of the blob. For convenience, we change a parameter n_0 into the ratio of electron energy density in the blob to U_B, $\eta = U_e/U_B$, where

$$U_e = m_e c^2 \int_{\gamma_{\rm min}}^{\gamma_{\rm max}} d\gamma \gamma \frac{dn_e}{d\gamma}. \quad (4)$$

Here, $\gamma_{\rm min}$ is assumed to be 1.

In order to demonstrate γ-ray spectra of LLAGNs, this study focuses on a high state of NGC 4278. This object is a LLAGN located at 16.7 Mpc from the Earth [21]. This source is sometimes categorized as a LINER or a radio-loud LLAGN. Recent observations in radio bands have revealed a two-sided relativistic parsec scale jet ($\beta \sim 0.75$) closely aligned to the line of sight ($2^\circ \le \theta \le 4^\circ$), where β and θ are the velocity in the unit of speed of light and the viewing angle of the jet, respectively [22]. These observables lead to the relativistic Doppler factor of the parsec scale jet of 2.6, which is much smaller than that of γ-ray blazars. In addition, there are several observational results with good resolution in optical and X-ray bands.

The size of an emission blob R can be constrained by several discussions. Monthly time-scale variability in X-ray bands [23] limits the blob size as

$$R \le \frac{\delta c \Delta t_{\rm obs}}{1+z} = 3 \times 10^{17} \delta_3 \Delta t_{\rm obs,-1} \ \ {\rm cm}, \quad (5)$$

where $\delta_3 = \delta/3$ and $\Delta t_{\rm obs,-1} = \Delta t_{\rm obs}/10^{-1}$ yr is the variability time-scale. On the other hand, the emission blob is natural not to be smaller than the Schwarzschild radius of the central black hole. Since the black hole mass of this object is $M_{\rm BH} \sim 3 \times 10^8 M_\odot$ [24, 25], the blob size is also limited as

$$R \ge \frac{2GM_{\rm BH}}{c^2} = 9 \times 10^{13} M_{\odot,8.5} \ \ [{\rm cm}], \quad (6)$$

where $M_{\odot,8.5} = M_{\rm BH}/10^{8.5} M_\odot$, and G and $M_\cdot$ are the gravitational constant and the solar mass, respectively. Thus, we consider $10^{14} \le R \le 10^{17}$ cm throughout the paper.

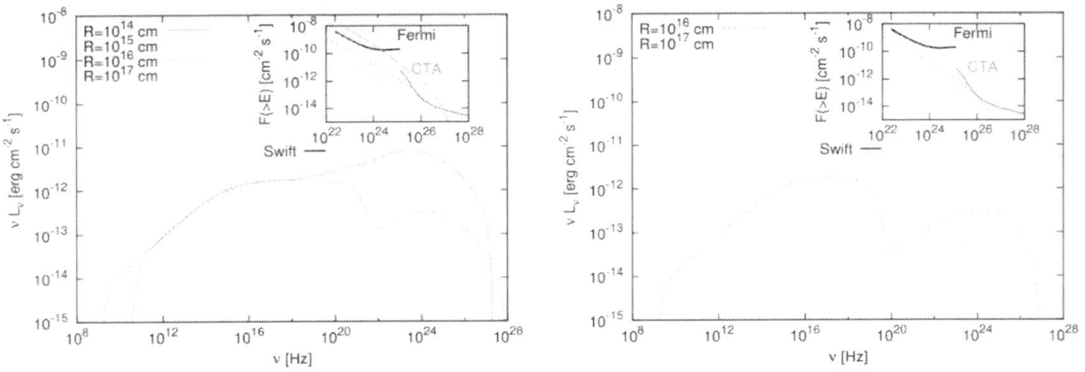

Figure 1. SEDs predicted on the assumptions of $\delta = 2.6$, and $\theta_F = 10^2$ (*left*) and $\theta_F = 10^4$ (*right*). The other parameters are listed in Table 1. A point in a near UV band is assumed to be the contribution of the disk of this system and is not used for the spectral fitting. The sensitivity limit of the *Swift*/BAT 54-month sky survey [26] is also shown. In addition, the corresponding integral fluxes, $F(> E)$, are shown with the integral sensitivity curves of the *Fermi*/LAT (5σ, 1 yr) [27] and CTA (5σ, 100h) [28] in the small panel.

3. Possible γ-ray Spectrum

Fig. 1 shows SEDs calculated on the assumptions of $\delta = 2.6$, and $\theta_F = 10^2$ (*left*) and $\theta_F = 10^4$ (*right*). $s_2 \sim 3$ is implied in the X-ray data. The absorption of high-energy γ-rays by extragalactic background light is taken into account by using a model of Ref. [29]. The other parameters used to fit observed data are listed in Table 1. $\delta = 2.6$ corresponds to a Doppler factor of radio jets estimated from radio images [22]. The sensitivity limit of the *Swift*/Burst Alert Telescope (BAT) 54-month sky survey [26] is also shown. In the small panel inside the figure, integral fluxes corresponding to the SEDs with the integral sensitivity curve of the *Fermi*Large Area Telescope (LAT) (5σ, 1 yr) [27] and the integral sensitivity goal of the Cherenkov Telescope Array (CTA) (5σ, 100 h) are shown [28].

The left panel demonstrates that a smaller blob predicts larger γ-ray flux. Since the case of a smaller blob requires the larger number density of synchrotron photons, which are seed photons for ICS, to reproduce the observed flux up to X-rays, larger γ-ray flux is realized.

However, the small size of the blob has several problems. Firstly, radiation energy density dominates compared to the energy density of magnetic field in the blob in the cases of $R = 10^{14}$ and $R = 10^{15}$ cm. In these cases, since $\gamma_{\rm max}$ and $\gamma_{\rm br}$ are determined by ICS, Eqs.2 and 3 are not valid. However, even including the effects of ICS, the model cannot reproduce the observed data self-consistently on the assumption of $\delta = 2.6$. $U_{\rm rad}$ much larger than $U_{\rm B}$ makes the spectral break frequency of synchrotron radiation be lower. In order to keep $\nu_{\rm br}$ to be $\sim 10^{15}$ Hz, $U_{\rm rad}$ should be reduced by decreasing n_e or B. Both of these decreases the flux of synchrotron radiation. In the former case, B is required to be larger to compensate the decrease of synchrotron radiation, but this leads to a break frequency lower than that calculated in Fig.1. Although η needs to be larger to reproduce the observed data in the latter case, the total $U_{\rm rad}$ is not reduced because ICS flux is proportional to ${n_e}^2$. The second reason is another spectral break due to synchrotron self-absorption. A smaller R produces the spectral break at a higher frequency, which is $\sim 10^{11}$ Hz and $\sim 10^{10}$ Hz for $R = 10^{14}$ and $R = 10^{15}$ cm, respectively. In these cases, radio emission from the blob does not dominate at ~ 5 GHz. For well-observed blazars, radio emission at 5 GHz is already optically thick and therefore radio emission from the blob is not dominant (e.g., Refs.[30, 31]). It is not clear at present whether this feature is

R [cm]	B [G]	η	B [G]	η
10^{14}	3.72	140.0	5.86	0.25
10^{15}	0.80	37.0	1.26	0.06
10^{16}	0.165	10.0	0.272	0.014
10^{17}	0.035	2.5	0.058	0.003

Table 1. Parameters required for spectral fits in Figs. 1 ($\delta = 2.6$; *left*) and 2 ($\delta = 10$; *right*).

common in radio-loud LLAGNs. However, the tight correlation between radio luminosity at 5 GHz and X-ray luminosity implies that the origin of these radiations is the same [32]. Based on this implication, $R \leq 10^{15.5}$ cm is not favored for $\delta = 2.6$.

The integral flux for $R = 10^{16}$ and 10^{17} cm are lower than the integral sensitivity curve of Fermi LAT. This is consistent with the fact that NGC 4278 is not listed in Fermi LAT 1yr source catalog [33]. Unfortunately, even 10 years observations do not reach the prediction for $R = 10^{16}$ cm, assuming that the sensitivity curve of Fermi LAT is roughly scaled inversely proportional to the square root of the total observation time. The integral flux is also compared with the integral sensitivity goal of CTA. The integral flux for $R = 10^{16}$ cm is above the CTA sensitivity at ~ 100 GeV. Thus, CTA can test the jet dominance of the total emission in this source in the case of $R \sim 10^{16}$ cm.

In order to check the uncertainty of θ_F, $\theta = 10^4$ are considered in the right panel. Since the cases of $R = 10^{14}$ and 10^{15} cm were ruled out, only SEDs for $R = 10^{16}$ and 10^{17} cm are shown in the figure. Following the change of θ_F, the spectral edge of synchrotron radiation is shifted to lower energy by a factor of 10^2. Also, the maximum energy of γ-rays decreases, but this effect is smaller than the synchrotron edge because of the Klein-Nishina effect. Thus, the detectability of the γ-rays by CTA for $R = 10^{16}$ is not largely changed.

Finally, we discuss SEDs in the case of δ larger than 2.6. Fig.2 shows SEDs calculated on the assumptions of $\delta = 10$ and $\theta_F = 10^4$. The other parameters are listed in Table 1. Since the high Doppler factor boosts the flux of radiation, the energy density of radiation much lower than that in the case of Fig.1 is enough to reproduce the observed flux up to X-rays. Thus, the energy density of magnetic field in the blob dominates compared to that of radiation even for $R = 10^{14}$ cm, but the cases of $R \leq 10^{15.5}$ cm are not favored due to synchrotron self-absorption again. The smaller energy densities of radiation and electrons leads to much smaller ICS components than those in Fig.1. Thus, even long-term observations by Fermi LAT and CTA could not confirm these γ-rays.

4. Summary

We have discussed γ-ray emission from LLAGN jets in the framework of a SSC model on the assumption of radio and X-rays are dominantly produced from jets. The γ-rays are a direct probe of a jet component in radio to X-ray bands without contamination from the other components, although the predicted flux is not large. Several observational results allowed us to constrain physical parameters in the emission region in jets. In the case of a Doppler factor as low as that of parsec scale jets and $R \sim 10^{16}$ cm, CTA may detect the γ-rays in the near future and test the jet dominance of radiation from LLAGNs. The determination of the respective contribution of disk and jet components will gives us a hint of a physical connection between a disk and relativistic jet in LLAGNs. For more details on this articles, see Ref. [34].

References

[1] Kormendy J and Richstone D 1995 *Ann. Rev. Astron. Astrophys.* **33** 581–624
[2] Ho L C, Filippenko A V and Sargent W L W 1997 *Astrophys. J. Suppl.* **112** 315–390

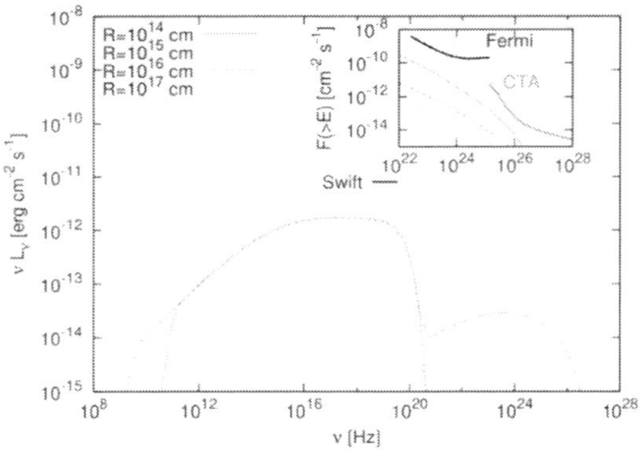

Figure 2. Same as the right panel of Fig. 1, but the cases of $\delta = 10$ and $\theta_F = 10^4$.

[3] Ho L C 2009 *Astrophys. J.* **699** 626–637
[4] Ho L C 2008 *Ann. Rev. Astron. Astrophys.* **46** 475–539
[5] Abramowicz M A, Chen X, Kato S, Lasota J and Regev O 1995 *Astrophys. J. Lett.* **438** L37–L39
[6] Narayan R and Yi I 1995 *Astrophys. J.* **444** 231–243
[7] Narayan R, Yi I and Mahadevan R 1995 *Nature* **374** 623–625
[8] Manmoto T, Mineshige S and Kusunose M 1997 *Astrophys. J.* **489** 791–803
[9] Mahadevan R 1998 *Nature* **394** 651–653
[10] Ho L C and Peng C Y 2001 *Astrophys. J.* **555** 650–662
[11] Nagar N M, Falcke H, Wilson A S and Ulvestad J S 2002 *Astron. & Astrophys.* **392** 53–82
[12] Doi A, Kameno S, Kohno K, Nakanishi K and Inoue M 2005 *Mon. Not. Roy. Astron. Soc.* **363** 692–704
[13] Ulvestad J S and Ho L C 2001 *Astrophys. J.* **562** L133–L136
[14] Anderson J M, Ulvestad J S and Ho L C 2004 *Astrophys. J.* **603** 42–50
[15] Wu Q and Cao X 2005 *Astrophys. J.* **621** 130–138
[16] Nemmen R S, Storchi-Bergmann T, Eracleous M and Yuan F 2010 *IAU Symposium* (*IAU Symposium* vol 267) pp 313–318
[17] Urry C M and Padovani P 1995 *Publ. Astron. Soc. Pacific* **107** 803–845
[18] Blandford R and Eichler D 1987 *Phys. Rep.* **154** 1–75
[19] Rachen J P and Meszaros P 1998 *Phys. Rev.* **D58** 123005
[20] Hillas A M 1984 *Ann. Rev. Astron. Astrophys.* **22** 425–444
[21] Tonry J L, Dressler A, Blakeslee J P, Ajhar E A, Fletcher A B, Luppino G A, Metzger M R and Moore C B 2001 *Astrophys. J.* **546** 681–693
[22] Giroletti M, Taylor G B and Giovannini G 2005 *Astrophys. J.* **622** 178–186
[23] Younes G, Porquet D, Sabra B, Grosso N, Reeves J N and Allen M G 2010 *Astron. & Astrophys.* **517** A33
[24] Wang T and Zhang X 2003 *Mon. Not. Roy. Astron. Soc.* **340** 793–798
[25] Chiaberge M, Capetti A and Macchetto F D 2005 *Astrophys. J.* **625** 716–726
[26] Cusumano G, La Parola V, Segreto A, Ferrigno C, Maselli A, Sbarufatti B, Romano P, Chincarini G, Giommi P, Masetti N, Moretti A, Parisi P and Tagliaferri G 2010 *Astron. & Astrophys.* **524** A64
[27] Atwood W B, Abdo A A, Ackermann M, Althouse W, Anderson B, Axelsson M, Baldini L, Ballet J, Band D L, Barbiellini G and et al 2009 *Astrophys. J.* **697** 1071–1102
[28] CTA Consortium T 2010 *ArXiv e-prints* (*Preprint* 1008.3703)
[29] Franceschini A, Rodighiero G and Vaccari M 2008 *Astron. & Astrophys.* **487** 837–852
[30] Anderhub H, Antonelli L A, Antoranz P, Backes M, Baixeras C, Balestra S, Barrio J A, Bastieri D, Becerra González J, Becker J K, and et al 2009 *Astrophys. J.* **705** 1624–1631
[31] Aharonian F, Akhperjanian A G, Anton G, Barres de Almeida U, Bazer-Bachi A R, Becherini Y, Behera B, Bernlöhr K, Boisson C, Bochow A and et al 2009 *Astrophys. J. Lett.* **696** L150–L155
[32] Panessa F, Barcons X, Bassani L, Cappi M, Carrera F J, Ho L C and Pellegrini S 2007 *Astron. & Astrophys.* **467** 519–527
[33] Abdo A A, Ackermann M, Ajello M, Allafort A, Antolini E, Atwood W B, Axelsson M, Baldini L, Ballet J, Barbiellini G and et al 2010 *Astrophys. J. Supp.* **188** 405–436
[34] Takami H 2011 *Mon. Not. Roy. Astron. Soc.* **413** 1845–1851

Journal of Physics: Conference Series **355** (2012) 012026 doi:10.1088/1742-6596/355/1/012026

Gamma-ray Spectra and the Extragalactic Background Light

Rudy Gilmore[1,*], **Rachel Somerville**[2,3], **Joel Primack**[4], **Alberto Domínguez**[5,6,7]

[1] SISSA Astrophysics Sector, 265 Bonomea, 34136 Trieste, Italy
[2] Space Telescope Science Institute, Baltimore, MD
[3] Department of Physics and Astronomy, Johns Hopkins University, Baltimore, MD 21218
[4] University of California, Santa Cruz, CA 95064 USA
[5] Visiting researcher at the Santa Cruz Institute for Particle Physics (SCIPP), University of California, Santa Cruz, CA 95064, USA
[6] Instituto de Astrofísica de Andalucía, CSIC, Apdo. Correos 3004, E-18080 Granada, Spain
[7] Departamento de Física Atómica, Molecular y Nuclear, Universidad de Sevilla, Apdo. Correos 1065, E-41080 Sevilla, Spain

E-mail: *rgilmore@sissa.it

Abstract. Attenuation of very high-energy gamma rays by pair-production with UV, optical and IR extragalactic background light (EBL) photons provides a link between the history of galaxy formation and gamma-ray astronomy. We present results from our latest semi-analytic models (SAMs), which employ the main ingredients thought to be important to galaxy formation and evolution, as well as an improved method for reprocessing of starlight by dust to mid- and far-IR wavelengths. These SAMs are based upon a hierarchical structural formation scenario, and are successful in reproducing a large variety of observational constraints such as number counts, redshift-dependent luminosity and mass functions, and color bimodality. Our fiducial model is based upon a WMAP5 cosmology and estimates the dust emission spectral energy distribution using templates based on Spitzer data. This model predicts a background flux considerably lower than found in optical and near-IR measurements that rely on subtraction of zodiacal and galactic foregrounds, and near the lower bounds set by number counts of resolvable sources at many wavelengths. We show predictions for the effect on extragalactic gamma-ray observations, and conclude with a discussion of the implications of our work and how the science of gamma-ray astronomy will continue to help constrain cosmology.

1. Introduction

The extragalactic background light (EBL) consists of photons emitted by galaxies over the history of the universe. Interactions between EBL photons and gamma rays can obscure gamma-rays sources, and produce a link between the history of galaxy formation and VHE (very high-energy) astrophysics. Direct measurement of the EBL (e.g. [1]) is difficult, and because the EBL is produced across cosmological time from all types of galaxies, sophisticated modeling is required to understand the buildup of this photon population. Modeling the sources of the EBL has been done historically using a variety of techniques, including evolution of galaxy properties that are either inferred over some range in wavelength [2, 3, 4] or directly observed in galaxy surveys [5, hereafter D11], backwards evolution models which assume that the redshift dependent evolution of galaxies can be found by applying a prescription to the local population [6, 7], or forward-evolution techniques, which model the growth of galaxies forward in time [8, 9, 10]. We present here an overview of results from our latest semi-analytic models [10, 11, 12] (SAMs), which employ the main ingredients thought to be important to galaxy formation and evolution, as well as an improved model for reprocessing of starlight by dust to mid- and far-IR wavelengths. Our semi-analytic method has the advantage of producing the evolution seen in galaxy properties in a natural way by modeling the population forward in time from the high-redshift early universe.

2. Methods

The SAMs used here are based on a ΛCDM cosmology [13, 14], in which galaxies form and develop in an underlying system of merging and growing dark matter halos [15]. Our assumed cosmology is based upon WMAP5 parameters [16]. A full description of the model is available in refs. [10] and [12]; the former includes a comprehensive comparison with available astrophysical data. Star formation in our model occurs in two modes, quiescent star formation in isolated galaxies and merger-driven starbursts. The chemical enrichment and star formation history of each galaxy are used to predict the stellar emission spectrum. We have adopted the stellar population models of [17] in this work, and a Chabrier [18] initial mass function.

The amount, composition, and evolution of dust in galaxies plays a crucial role in determining the overall spectrum of the EBL. Our model uses a self-consistent approach to the absorption and reemission of starlight by dust, with all absorbed radiative energy reemitted in the IR using spectral templates that are determined by the total galaxy IR luminosity. The absorption model is based on the 2-component prescription of [19]. We introduce an additional redshift-dependent parameter in our fiducial model to avoid overproducing dust in high-redshift galaxies and improve agreement with observed UV luminosity functions. IR spectra of reemitted light are modeled using the IR templates of [20], which are based on observations of local IR-luminous galaxies and utilize data from all three instruments on the Spitzer Space Telescope. Being observationally-based, these templates suffer from AGN contamination, and being based on low-redshift observations they may not correctly represent the properties of high-redshift star-forming galaxies, which produce a substantial fraction of the IR background. For comparison, we also show a model that is lacking the evolutionary term in the dust prescription, and uses dust reemission templates from the older work of [21]. This is to facilitate comparison with a scheme that more closely resembles the SAM results of [8, 9]. This model is designated DGS99 in the following figures.

3. Results

3.1. Astrophysical Results

In the left-hand panel of Figure 1, we show the local background flux, as a function of wavelength, as well as two older SAM predictions and the observational model of D11. All these models predict a background flux at UV to near-IR wavelengths that is near the level set by galaxy number counts ([22, 23, 24]). In the IR, and particularly around the far-IR peak at $\sim$ 100 μm, there is a considerable amount of divergence between the models. There is an especially striking discrepancy between the predictions of our fiducial model and the observational model of D11, which predicts a factor $\sim$ 2 higher contribution to the far-IR from rapidly star-forming galaxies. The DGS99 model shows a somewhat different shape in the mid- and far-IR, where the templates of [21], used in our previous work, have a substantially different spectral shape than the newer Spitzer-based templates of [20]; more discussion of this comparison can be found in [10]. The SAM prediction of [8] utilized a less-sophisticated dust absorption model in its galaxy treatment, and consequently underpredicts the background at wavelengths longer than 8 μm.

The right-hand side of Figure 1 shows how the emissivity of galaxies evolves as a function of redshift. In the high-redshift UV, our fiducial model is found to match the data of [25] and

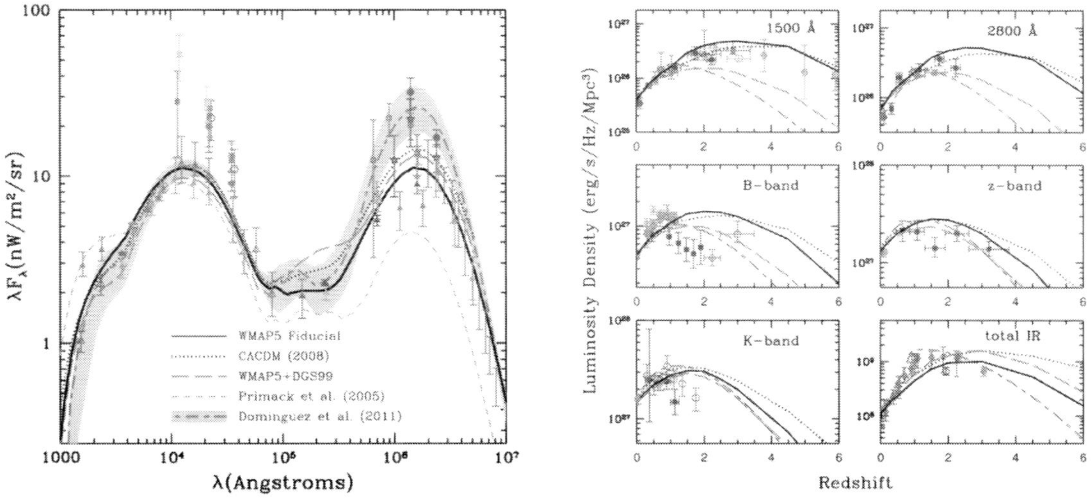

Figure 1. Left: The EBL predicted in our fiducial SAM and the 'DGS99' model. Shown for comparison are the models of [9] (CLCDM), the observational model of D11 with its associated error band, and [8]. Upward-pointing arrows are lower limits on the EBL from number counts seen in galaxy surveys, while other symbols are direct flux measurements. A complete list of the references for these data points can be found in [11].
Right: The evolving luminosity density in the universe at 5 wavelengths, and for all reemitted light in the IR. Note that for the latter, the y-axis has units of solar luminosities per Mpc^3. Line types here are as in the opposite panel. As before, observational references for this plot can be found in in [11].

others better than the DGS99 or D11 models, though this contribution has a minimal impact on the local background flux. The difference between the DGS99 and fiducial models is due entirely to the differing treatments of dust described in the last section. In the IR, the D11 model predicts a total flux that rises considerably faster with redshift than in any of our semi-analytic calculations. Several authors (e.g. [26, 27]) seem to support this rapid rise in IR output, and matching this fast evolution is an ongoing challenge for our semi-analytic models.

3.2. Gamma-ray Attenuation

Photon-photon interactions with sufficient center-of-mass energy can create electron-positron pairs. For multi-GeV and TeV gamma rays, interactions with UV and IR background photons can create an optically-thick barrier to passage over cosmological distances [28, 29]. This effect has been used by a variety of authors in recent years to constrain the EBL and disfavor or exclude specific models [30, 31, 32, 33, 34]. In the left-hand panel of Figure 2, we show the impact of gamma-ray attenuation as a function of observed energy for the models of the previous section. Increasing distance causes absorption features to increase in magnitude and appear at lower energies. The plateau seen between 1 and 10 TeV at low redshift is a product of the mid-IR valley in the EBL spectrum, and the large photon density in the far-IR peak produces a cutoff at higher energy. The fiducial SAM model predicts less attenuation at multi-TeV energies than the best-fit prediction of the D11 observational model, as opacity at these energies is mainly due to mid- to far-IR photons. At lower sub-TeV energies, where gamma rays have only enough energy to scatter off of UV and optical background photons, the situation is reversed and our fiducial model produces more attenuation than either D11 or our DGS99 model. UV emission is

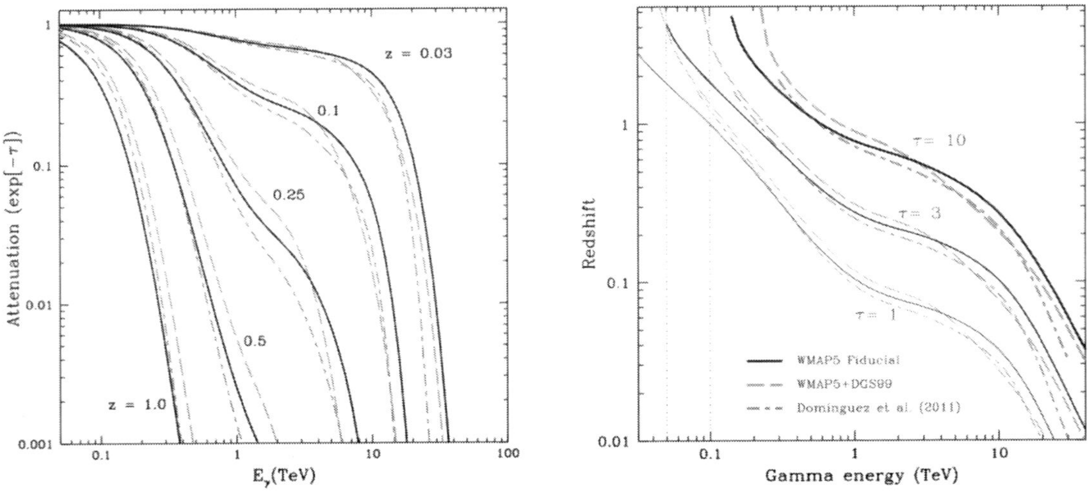

Figure 2. Left: The attenuation $e^{-\tau}$ of gamma rays vs. gamma-ray energy, for sources at $z = 0.03$, 0.1, 0.25, and 0.5, and 1. Results are compared for our fiducial (solid) and DGS99 (dashed blue) models, as well as the model of D11 (red dash-dotted).
Right: The gamma-ray attenuation edges for the fiducial (solid black) and DGS99 (dashed blue) models and model of D11 (red dash-dotted). The curves show the redshift at which the pair-production optical depth τ reaches the indicated value for a particular observed gamma-ray energy. The triplets of curves from lower left to upper right are the contours for $\tau = 1$, 3, and 10. We have included thin dotted lines to guide the eye at 50 and 100 GeV.

more suppressed by dust in these cases, and the UV EBL does not evolve as strongly in redshift. The right-hand panel of this plot is the 'attenuation edge', meaning the source redshift above which gamma rays suffer more attenuation than a given value. Contours here represent limits of $\tau = 1$, 3 and 10, corresponding respectively to modifications of the intrinsic flux by factors of 0.37, 0.05, and 0.000045. The thin dotted vertical lines on this plot at 50 and 100 GeV show the typical energy thresholds of current-generation ground-based atmospheric Cherenkov telescopes. The H.E.S.S. and VERITAS telescope arrays have typically claimed thresholds of $\sim$ 100 GeV, and the MAGIC telescope has viewed sources at energies as low as 50-60 GeV. The proposed CTA experiment will have a threshold below 50 GeV [35]. Our results suggest that the EBL will create a significant barrier to observations in the lower part of the CTA energy range only for source redshifts well above unity.

Figure 3, left panel, shows how the spectra of known gamma-ray blazars observed by ground-based Cherenkov telescopes are affected by the cumulative impact of our predicted EBL, including the evolution in redshift. As discussed in [31], a standard assumption in blazar emission scenarios is that the VHE spectra cannot be harder than $\Gamma = -1.5$, with $dN/dE \propto E^{-\Gamma}$. In the figure, the hardening of spectra resulting from de-convolving with our assumed EBL models is shown compared with the blazar redshift. More details on this and a data table with blazar information can be found in [11]. We find that our models are consistent with this standard assumption.

4. Discussion

The EBL presents one of the primary barriers to extragalactic gamma-ray astronomy with ground-based instruments. Our determination of a fairly low extragalactic background across

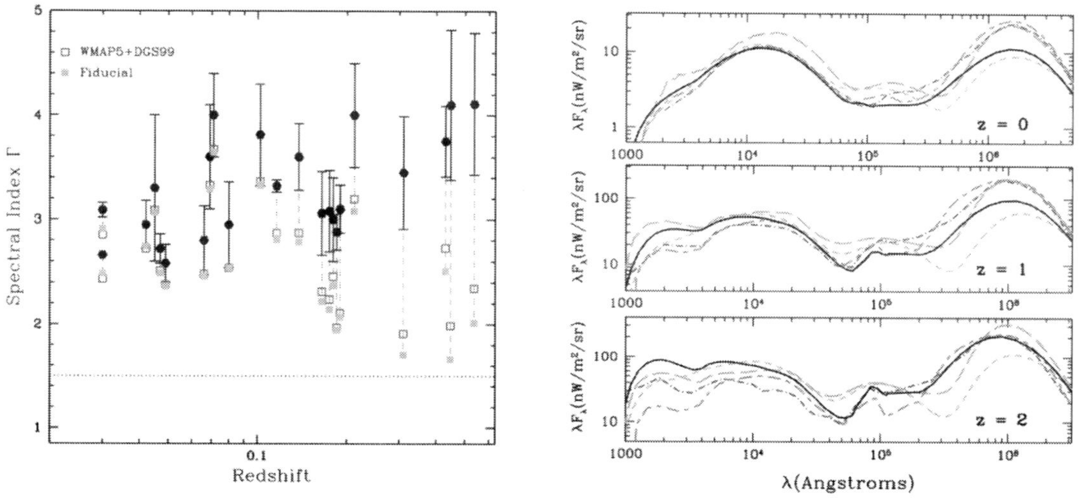

Figure 3. Left: Spectral corrections for our two semi-analytic EBL predictions to the measured spectral indices of gamma-ray blazars. The measured spectral index (Γ; $dN/dE \propto E^{-\Gamma}$) and redshift of each blazar is shown as a black hexagon with error bars, with the index corrected via the fiducial EBL shown as a solid orange point, and that corrected by the DGS99 model as an open blue point. The horizontal dotted line shows $\Gamma = 1.5$, which is typically taken as the hardest spectrum possible under usual assumptions. Some points have been shifted sideways slightly for readability.
Right: Our EBL predictions compared with several recent models from the literature. The solid black line shows the proper flux density from our WMAP5 model in the local universe and at $z = 1$ and $z = 2$. Other lines are from [7] (dashed-dotted blue), the best-fit model of [2] (long-dashed green), D11 (long-short dashed red), and model 'C' from [4] (dashed orange).

the optical and near- to mid-IR, supported by convergence with alternative methods such as [7] and D11, is an optimistic prediction for the future of the field. The weight of this evidence also increasingly points to an EBL that is well-known over this wavelength range, at a level near that of resolved number counts, though many questions remain about its redshift evolution due to uncertainty in high-redshift data.

In the right-hand side of Figure 3, we show a full comparison with several other recent EBL models in the local universe, and at redshifts 1 and 2. While these models tend to agree well in local flux predictions across the UV to near-IR peak, there is a considerable discrepancy in how flux at these wavelengths evolves to higher redshift. Models which make very similar predictions at $z = 0$ can have diverging predictions at $z > 0$, due to differing treatments of galaxy evolution. Studying the attenuation of extragalactic gamma rays is one way to constrain these differing predictions. TeV sources have now been detected above redshift 0.5 by ground-based telescopes [33], and at these distances the differences in evolution between EBL models become important. Fermi LAT has detected multi-GeV blazars and GRBs at much higher redshifts [36, 37]. While the limits from the first-year of LAT observations [34] only constrain extreme scenarios, more data from Fermi, combined with upgraded ground-based telescopes or future proposed experiments such as CTA may put interesting constraints on background evolution.

In the far-IR, a factor ~ 3 uncertainty exists in the absolute level of the EBL. As our treatment of these wavelengths is still rather primitive, we hope to upgrade our models in the future to better replicate the SEDs of high redshift star-forming galaxies which produce a significant fraction of the long-wavelength background, and test our predictions against new Herschel data. Gathering more multi-TeV data on nearby blazars such as Mkn 421 ($z = 0.031$) and Mkn 501 ($z = 0.034$) could also help constrain the normalization of the far-IR peak, by exploring the details of the cutoff that the EBL induces on the spectra at 10 to 30 TeV.

References

[1] Hauser M G and Dwek E 2001 ARA&A **39** 249–307
[2] Kneiske T M, Bretz T, Mannheim K and Hartmann D H 2004 A&A **413** 807–815
[3] Razzaque S, Dermer C D and Finke J D 2009 ApJ **697** 483–492 (*Preprint* 0807.4294)
[4] Finke J D, Razzaque S and Dermer C D 2010 ApJ **712** 238–249 (*Preprint* 0905.1115)
[5] Domínguez A, Primack J R, Rosario D J, Prada F, Gilmore R C, Faber S M, Koo D C, Somerville R S, Pérez-Torres M A and Pérez-González P 2011 MNRAS **410** 2556–2578 (*Preprint* 1007.1459)
[6] Stecker F W, Malkan M A and Scully S T 2006 ApJ **648** 774–783
[7] Franceschini A, Rodighiero G and Vaccari M 2008 A&A **487** 837–852 (*Preprint* 0805.1841)
[8] Primack J R, Bullock J S and Somerville R S 2005 *High Energy Gamma-Ray Astronomy* (*American Institute of Physics Conference Series* vol 745) ed Aharonian F A, Völk H J and Horns D pp 23–33
[9] Primack J R, Gilmore R C and Somerville R S 2008 *American Institute of Physics Conference Series* (*American Institute of Physics Conference Series* vol 1085) ed Aharonian F A, Hofmann W and Rieger F pp 71–82
[10] Somerville R S, Gilmore R C, Primack J R and Dominguez A 2011 *Preprint* 1104.0669
[11] Gilmore R C, Somerville R S, Primack J R and Domínguez A 2011 *Preprint* 1104.0671
[12] Somerville R S, Hopkins P F, Cox T J, Robertson B E and Hernquist L 2008 MNRAS **391** 481–506 (*Preprint* 0808.1227)
[13] White S D M and Rees M J 1978 MNRAS **183** 341
[14] Blumenthal G R, Fabee S M, Primack J R, Rees M J 1984 *Nature* **311** 517
[15] Somerville R S and Primack J R 1999 MNRAS **310** 1087–1110
[16] Komatsu E, Dunkley J, Nolta M R, Bennett C L, Gold B, Hinshaw G, Jarosik N, Larson D, Limon M, Page L, Spergel D N, Halpern M, Hill R S, Kogut A, Meyer S S, Tucker G S, Weiland J L, Wollack E, Wright E L 2008 ApJS **180** 330–376
[17] Bruzual G and Charlot S 2003 MNRAS **344** 1000–1028 (*Preprint* astro-ph/0309134)
[18] Chabrier G 2003 PASP **115** 763–795 (*Preprint* astro-ph/0304382)
[19] Charlot S and Fall S M 2000 ApJ **539** 718–731
[20] Rieke G H, Alonso-Herrero A, Weiner B J, Pérez-González P G, Blaylock M, Donley J L and Marcillac D 2009 ApJ **692** 556–573 (*Preprint* 0810.4150)
[21] Devriendt J E G, Guiderdoni B and Sadat R 1999 A&A **350** 381–398
[22] Madau P and Pozzetti L 2000 MNRAS **312** L9–L15
[23] Keenan R C, Barger A J, Cowie L L and Wang W 2010 ApJ **723** 40–46 (*Preprint* 1008.4216)
[24] Dolch T and Ferguson H 2011 *in prep*
[25] Bouwens R J, Illingworth G D, Franx M and Ford H 2007 ApJ **670** 928–958 (*Preprint* 0707.2080)
[26] Rodighiero G, Vaccari M, Franceschini A, Tresse L, Le Fevre O, Le Brun V, Mancini C, Matute I, Cimatti A, Marchetti L and Ilbert O 2010 A&A **515** A8 (*Preprint* 0910.5649)
[27] Le Floc'h E, Papovich C, Dole H, Bell E F, Lagache G, Rieke G H, Egami E, Pérez-González P G, Alonso-Herrero A, Rieke M J, Blaylock M, Engelbracht C W and Gordon K D 2005 ApJ **632** 169–190 (*Preprint* astro-ph/0506462)
[28] Fazio G G and Stecker F W 1970 Nature **226** 135
[29] Madau P and Phinney E S 1996 ApJ **456** 124
[30] Dwek E and Krennrich F 2005 ApJ **618** 657–674
[31] Aharonian F, Akhperjanian A G, Bazer-Bachi A R, Beilicke M, Benbow W, Berge D, Bernlöhr K, Boisson C, Bolz O, Borrel V and Braun I 2006 Nature **440** 1018–1021
[32] Mazin D and Raue M 2007 A&A **471** 439–452
[33] Albert J, Aliu E, Anderhub H, Antonelli L A, Antoranz P, Backes M, Baixeras C, Barrio J A, Bartko H, Bastieri D and Becker J K 2008 *Science* **320** 1752
[34] Abdo A A, Ackermann M, Ajello M, Allafort A, Atwood W B, Baldini L, Ballet J, Barbiellini G, Baring M G, Bastieri D, Baughman B M, Bechtol K, Bellazzini R, Berenji B and Bhat P N 2010 ApJ **723** 1082–1096
[35] The CTA Consortium 2010 (*Preprint* 1008.3703)
[36] Abdo A A, Ackermann M, Arimoto M, Asano K, Atwood W B, Axelsson M, Baldini L, Ballet J and Band D L 2009 *Science* **323** 1688
[37] Abdo A A 2009 (*Preprint* 0902.1559)

International Workshop on Beamed and Unbeamed Gamma-Rays from Galaxies IOP Publishing
Journal of Physics: Conference Series **355** (2012) 012027 doi:10.1088/1742-6596/355/1/012027

Influence of the cosmological evolution of the EBL on the limits derived from VHE gamma-ray blazar observations

Manuel Meyer[1], **Martin Raue**[1] **and Daniel Mazin**[2]

[1]Institut für Experimentalphysik, University of Hamburg, Luruper Chaussee 149, 22767 Hamburg, Germany

[2]Institut de Fisica d'Altes Energies (IFAE), Edifici Cn. Universitat Autonoma de Barcelona, 08193 Bellaterra, Spain

E-mail: `manuel.meyer@physik.uni-hamburg.de`

Abstract. Very high energy (VHE) γ-rays from distant sources such as blazars are attenuated due to the interaction with low energy photons of the extragalactic background light (EBL) in the ultraviolet to infrared wavelength band. Thus, the EBL leaves an imprint on the observed energy spectra of such sources. With assumptions about the source physics it is possible to derive limits on the EBL photon density. However, the EBL evolves with redshift a fact that often been neglected in the deviations of the limits. Here, the influence of evolution is examined by implementing it in a phenomenological way with four VHE spectra and compared to previous results.

1. Introduction

The extragalactic background light (EBL) is the diffuse isotropic background radiation field that ranges from ultraviolet / optical to far infrared wavelengths. The two main contributions to this radiation field are emitted starlight integrated over all epochs and the starlight absorbed and re-emitted by dust in galaxies [1, for a review]. Further possible components that add to the total EBL intensity are active galactic nuclei (AGN) [2], the very first (population III) stars [3] or even exotic candidates such as dark matter powered stars [4]. Knowledge of the spectral energy distribution (SED) of the EBL allows for insight of the cosmic star and galaxy formation rates as well as the amount of dust contained in galaxies. The EBL is however difficult to measure directly especially in the infrared due to foreground contamination [5].

With the advent of imaging air Cherenkov telescopes (IACTs) that are able to observe very high energy (energy $E \gtrsim 100$ GeV, VHE) γ-rays from distant sources it is possible to place upper limits on the intensity of the EBL under the assumption of a certain intrinsic source spectrum. VHE γ-rays can interact with EBL photons and produce electron-positron pairs. These γ-rays are thus removed from the photon beam and initial photon flux of the source, $\mathrm{d}N_{\mathrm{int}}/\mathrm{d}E$, is attenuated. This is commonly expressed as

$$\frac{\mathrm{d}N_{\mathrm{obs}}}{\mathrm{d}E} = \frac{\mathrm{d}N_{\mathrm{int}}}{\mathrm{d}E} \times \exp\left[-\tau_{\gamma\gamma}(E, z)\right], \quad (1)$$

where the observed spectrum is denoted by $\mathrm{d}N_{\mathrm{obs}}/\mathrm{d}E$. The parameter $\tau_{\gamma\gamma}$ is called the optical depth and is a three-fold integral over the cosmological distance to the source at redshift z_0, the cosine of the angle of incidence of the photons μ, and the energy ϵ of the background photons [6],

$$\tau_{\gamma\gamma}(E, z_0) = \int_0^{z_0} \mathrm{d}\ell(z) \int_{-1}^{+1} \mathrm{d}\mu \frac{1-\mu}{2} \int_{\epsilon'_{\mathrm{thr}}}^{\infty} \mathrm{d}\epsilon' n_{\mathrm{EBL}}(\epsilon', z) \sigma_{\gamma\gamma}(E', \epsilon', \mu). \tag{2}$$

The comoving photon number density of the EBL in the energy interval ϵ' and $\mathrm{d}\epsilon'$ at redshift z enters as n_{EBL} and the cross section for pair production as $\sigma_{\gamma\gamma}$. The threshold energy for pair production is given by $\epsilon'_{\mathrm{thr}} = \epsilon_{\mathrm{thr}}(E', \mu)$ with $E' = E(1 + z)$. The cross section is strongly peaked at a wavelength $\lambda = 1.24(E/\mathrm{TeV})$ µm, making the EBL the dominant background for the attenuation of VHE photons.

Several authors have used VHE spectra measured by IACTs to constrain the EBL [7, 8, 9, 10]. In [7] the authors test a great variety of shapes of the EBL intensity. For each shape they calculate the absorption and correct the observed VHE spectra accordingly. Subsequently, they compare the obtained intrinsic spectra with theoretical predictions, i.e. if the intrinsic spectra are described with power laws, $\mathrm{d}N_{\mathrm{int}}/\mathrm{d}E \propto E^{-\Gamma}$, with $\Gamma < 1.5$ or if the spectra are best described with an exponential pile up at the highest energies. If one of the two conditions is true the corresponding EBL shape is excluded. The envelope shape of allowed shapes gives the upper limit of the EBL intensity.

This article follows the same approach but its goal is to examine the consequences on the EBL upper limits if the evolution is not neglected but implemented in a phenomenological way outlined by [11]. However, the value of $\Gamma = 1.5$ as a limit is somewhat under debate. Harder spectra, i.e. spectral indices $\Gamma < 1.5$, can be achieved e.g. in the case of internal photon absorption [12], proton-synchrotron radiation [13, 14] or narrow electron injection spectra [15, 16, 17]. In this work, the focus lies on the influence of the evolution with redshift on upper limits of the EBL intensity. Thus, the limiting value of 1.5 is chosen for simplicity and to compare the results to [7].

2. Method of constraining the EBL density with VHE observations

2.1. Calculating the EBL photon density

Following [7], the EBL SED νI_ν (in nW m^{-2} sr^{-1}) is described by splines,

$$\nu I_\nu = \sum_{i=0}^{k} w_i s_{i,p}(\epsilon) \tag{3}$$

with the B-Splines $s_{i,p}$,

$$s_{i,0}(\epsilon) = \begin{cases} 1 & \epsilon_i \leq \epsilon < \epsilon_{i+1} \text{ and } \epsilon_i < \epsilon_{i+1}, \\ 0 & \text{otherwise,} \end{cases} \tag{4}$$

$$s_{i,p}(\epsilon) = \frac{\epsilon - \epsilon_i}{\epsilon_{i+1} - \epsilon_i} s_{i,p-1}(\epsilon) + \frac{\epsilon_{i+p+1} - \epsilon}{\epsilon_{i+p+1} - \epsilon_{i+p}} s_{i+1,p-1}(\epsilon), \tag{5}$$

and the order p is set to two. Each spline is defined by the choice of k supporting points in the $(\lambda, \nu I_\nu)$-plane, where λ, $i = 1, \ldots, k$ are the knot points and w_i are the weights of the spline. For the knot points 16 equidistant points in $\log_{10}(\lambda)$ between 0.1 and 1000 µm are chosen and 12 equidistant points of the EBL density in $\log_{10}(\nu I_\nu)$ between 0.06 and 100 nW m^2 sr^{-1} are set as weights. It should be stressed that the splines will in general not run through each weight and knot. The grid defined by the selection of weights and knots is limited by an envelope shape, which is shown in Figure 1. The minimum allowed shape is set to reproduce the lower limits from the galaxy number counts from Spitzer [18] while the maximum allowed shape roughly follows the measurements and upper limits. In total this range of knots and weights allows for 518,400 different EBL shapes.

Assuming the EBL intensity is constant in redshift, the differential photon number density is given by [20]

$$n_{\mathrm{EBL}}(\epsilon, z) = \frac{4\pi}{c\epsilon^2}(1 + z)^2 \nu I_\nu(\epsilon = h\nu). \tag{6}$$

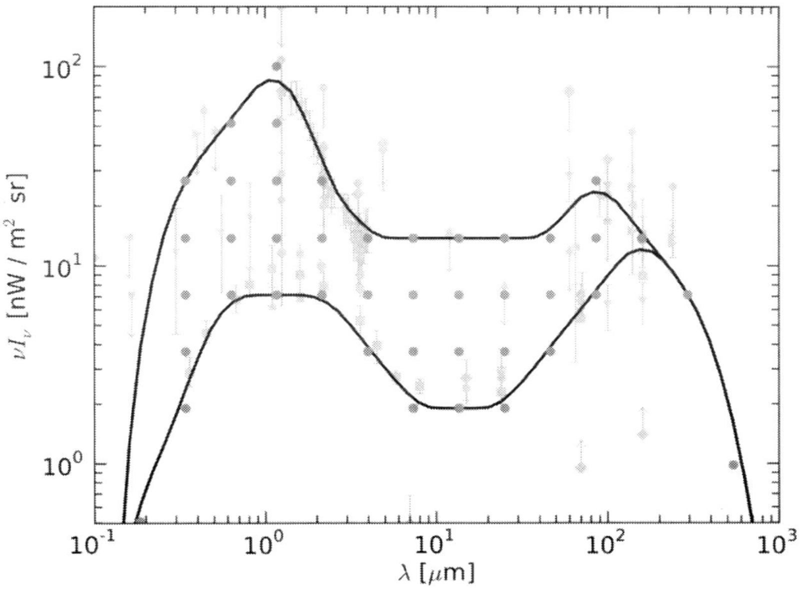

Figure 1. Grid points that are used to construct EBL shapes. The points are shown in red together with the minimal and maximal shape depicted as black lines. Direct measurements as well as lower and upper limits on the EBL are shown in gray [19].

However, the evolution of stars, dust and galaxies with redshift leads to an evolution of the EBL with redshift as well. This is also implemented in the modeling of the EBL (see e.g. [21, 20, 22]). Here, it is accounted for with the phenomenological ansatz of [11]. The cosmological photon number density scaling is changed from $n_{\mathrm{EBL}} \propto (1+z)^2$ to $n_{\mathrm{EBL}} \propto (1+z)^{2-f_{\mathrm{evo}}}$. For a value of $f_{\mathrm{evo}} = 1.2$ a good agreement is found (the deviation in $\tau_{\gamma\gamma}$ is found to be $< 4\%$ for $z = 0.2$ and $< 10\%$ for $z = 0.5$) for redshifts $z \lesssim 0.7$ between this simplified approach and complete EBL model calculations [11]. On the other hand, neglecting the evolution entirely leads to an overestimation of the optical depth between 10% ($z = 0.2$) and 35% ($z = 0.5$). Thus, including the evolution will generally weaken the upper limits on the EBL energy density.

2.2. *Exclusion criteria*

The derivation of upper limits is limited here to the three spectra, namely measurements of the sources H1426-428 [23] (measured by HEGRA), 1ES1101-232 [24] (observed with H.E.S.S.) and 3C279 [25] (measured by MAGIC). If an EBL shape is excluded is determined in the same fashion as in [7]. For a particular EBL shape the intrinsic spectrum is calculated by inverting Eq. 1. Subsequently, analytical functions (given in Table 2 of [7]) are fitted to the data points. An EBL shape is excluded if one of the following conditions is met:

(i) The photon index (or one of the photon indices if the spectrum is best described with a (double) broken power law) Γ of the analytical function is smaller than 1.5,

$$\Gamma + \sigma_\Gamma + \sigma_{\mathrm{sys}} < 1.5, \tag{7}$$

where σ_Γ is the uncertainty associated with the fitting procedure and σ_{sys} is the systematic error of the photon index quoted in the corresponding reference.

(ii) The spectrum is best described with a function that contains an exponential pile up at the largest energies.

	Source			
	H1426+428	1ES1101-232	3C279	Combined
With evolution	13.91%	21.50%	58.78%	0.93%
W/o evolution	13.05%	17.78%	49.61%	0.60%

Table 1. Percentage of allowed shapes for each spectrum with and without evolution. The numbers show that if evolution is neglected up to 10% (3C279) more shapes are excluded.

The upper limit EBL shape is defined as the envelope shape of all allowed shapes.

3. Results

The upper limits are derived using the three VHE spectra from the sources H1426+428 [23], 1ES1101-232 [24] and 3C279 [25]. The first two spectra are included in the analysis of [7]. The sources are located at redshifts of 0.129, 0.186 and 0.536 respectively and thus a weakening of the upper limits due to the inclusion of the evolution of the EBL can be expected. This is seen in Figure 2 which shows a comparison of the limits with and without evolution together with the limits derived in [7]. For completeness, two EBL models are shown as well. The blue solid line indicates the limits with evolution which are indeed weaker than those derived without evolution (dark blue dashed curve). The differences are most prominent in the transition at near infrared wavelengths where 1ES1101-232 and 3C279 give the most constraining limits which can be inferred from the strongly peaked cross section for pair production and the energy ranges covered by the spectra. This is natural as the effect of the evolution becomes more pronounced for higher redshifts. Table 1 lists the ratio of the number of allowed shapes to the total number of shapes for each spectrum with and without evolution.

The deviations from the upper limits of [7] can be explained, on the one hand, by the slightly different choice of the grid for the splines. In the present analysis the grid is shifted towards lower values of the intensity along the y-axis (see Figure 1) in order to cover the lower limits from galaxy counts. The spectrum of Markarian 501 [26] which excludes most EBL shapes in [7] is not considered here since the evolution should only have a negligible impact on this spectrum as the source is relatively close by with $z = 0.034$. On the other hand, the previously unconsidered source 3C279 improves the upper limits at optical wavelengths.

4. Conclusions

In this paper the effect of the evolution of the EBL density with redshift on upper limits from VHE observations is studied. The evolution is implemented in the phenomenological way outlined in [11]. The expectation is confirmed that the evolution generally weakens the upper limits since the photon number density of the EBL is reduced. The result underlines the necessity to include this effect in upper-limit calculations which has not been done so far. Furthermore, new upper limits from the source 3C279 are presented that improve the upper limits of [7] in the optical by roughly a factor of 2.

Acknowledgements

The authors would like to thank the organizers of the conference *Beamed and unbeamed gamma rays from galaxies* 2011 for the opportunity to present the results of this work. This work was made possible with the support of the state excellence cluster "Cosmic radiation fields and search for dark matter" at the university of Hamburg.

References

[1] Hauser M G and Dwek E 2001 *Ann. Rev. Astron. Astrophys.* **39** 249–307 (*Preprint* `arXiv:astro-ph/0105539`)
[2] Matute I, La Franca F, Pozzi F, Gruppioni C, Lari C and Zamorani G 2006 *Astron. Astrophys.* **451** 443–456 (*Preprint* `arXiv:astro-ph/0601355`)

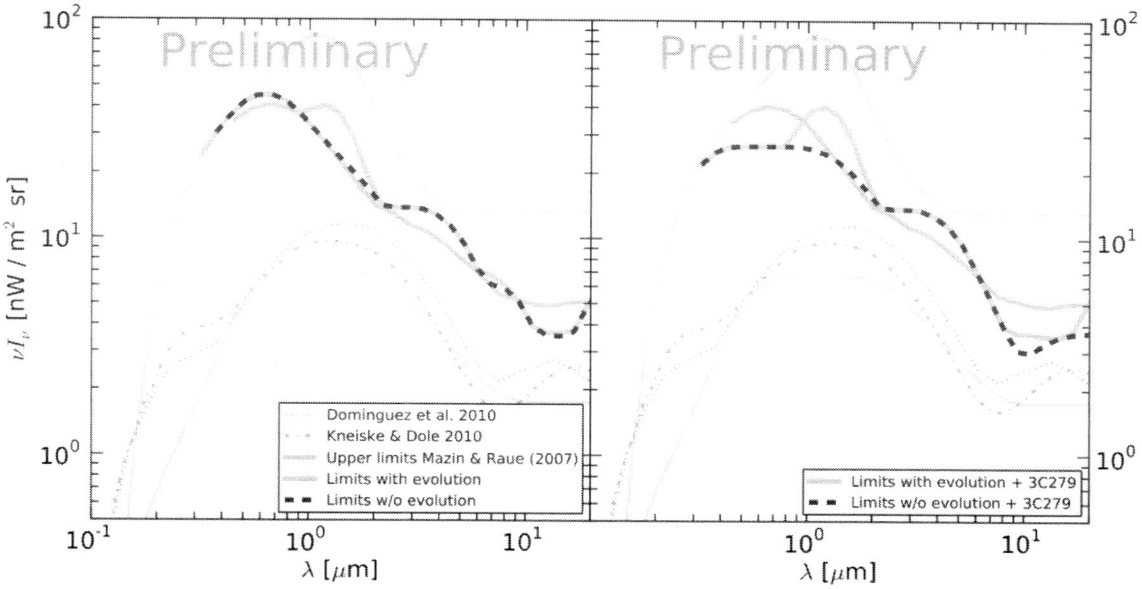

Figure 2. Upper limits with and without evolution. As a comparison the EBL models of [21] and [22] are shown together with the limits of [7].

[3] Raue M, Kneiske T and Mazin D 2009 *Astron. Astrophys.* **498** 25–35 (*Preprint* 0806.2574)
[4] Maurer A, Raue M, Kneiske T, Elsässer D, Hauschildt P and Horns D 2011 *in preperation*
[5] Hauser M G, Arendt R G, Kelsall T, Dwek E, Odegard N, Weiland J L, Freudenreich H T, Reach W T, Silverberg R F, Moseley S H, Pei Y C, Lubin P, Mather J C, Shafer R A, Smoot G F, Weiss R, Wilkinson D T and Wright E L 1998 *Astrophys. J.* **508** 25–43 (*Preprint* arXiv:astro-ph/9806167)
[6] Gould R J and Schréder G P 1967 *Physical Review* **155** 1404–1407
[7] Mazin D and Raue M 2007 *Astron. Astrophys.* **471** 439–452 (*Preprint* arXiv:astro-ph/0701694)
[8] Georganopoulos M, Finke J D and Reyes L C 2010 *Astrophys. J., Lett.* **714** L157–L161 (*Preprint* 1004.0017)
[9] Mankuzhiyil N, Persic M and Tavecchio F 2010 *Astrophys. J., Lett.* **715** L16–L20 (*Preprint* 1004.2032)
[10] Orr M R, Krennrich F and Dwek E 2011 *Astrophys. J.* **733** 77–+ (*Preprint* 1101.3498)
[11] Raue M and Mazin D 2008 *International Journal of Modern Physics D* **17** 1515–1520 (*Preprint* 0802.0129)
[12] Aharonian F A, Khangulyan D and Costamante L 2008 *Mon. Not. R. Astron. Soc.* **387** 1206–1214 (*Preprint* 0801.3198)
[13] Aharonian F A 2000 *New Astronomy* **5** 377–395 (*Preprint* arXiv:astro-ph/0003159)
[14] Zacharopoulou O, Khangulyan D, Aharonian F A and Costamante L 2011 *Astrophys. J.* **738** 157–+ (*Preprint* 1106.3129)
[15] Katarzyński K, Ghisellini G, Tavecchio F, Gracia J and Maraschi L 2006 *Mon. Not. R. Astron. Soc.* **368** L52–L56 (*Preprint* arXiv:astro-ph/0603030)
[16] Tavecchio F, Ghisellini G, Ghirlanda G, Costamante L and Franceschini A 2009 *Mon. Not. R. Astron. Soc.* **399** L59–L63 (*Preprint* 0905.0899)
[17] Lefa E, Rieger F M and Aharonian F 2011 *ArXiv e-prints* (*Preprint* 1106.4201)
[18] Fazio G G, Ashby M L N, Barmby P, Hora J L, Huang J S, Pahre M A, Wang Z, Willner S P, Arendt R G, Moseley S H, Brodwin M, Eisenhardt P, Stern D, Tollestrup E V and Wright E L 2004 *Astrophys. J., Suppl. Ser.* **154** 39–43 (*Preprint* arXiv:astro-ph/0405595)
[19] Raue M and Mazin D 2011 *ArXiv e-prints* (*Preprint* 1106.4384)
[20] Franceschini A, Rodighiero G and Vaccari M 2008 *Astron. Astrophys.* **487** 837–852 (*Preprint* 0805.1841)
[21] Domínguez A, Primack J R, Rosario D J, Prada F, Gilmore R C, Faber S M, Koo D C, Somerville R S, Pérez-Torres M A, Pérez-González P, Huang J S, Davis M, Guhathakurta P, Barmby P, Conselice C J, Lozano M, Newman J A and Cooper M C 2011 *Mon. Not. R. Astron. Soc.* **410** 2556–2578 (*Preprint* 1007.1459)
[22] Kneiske T M and Dole H 2010 *Astron. Astrophys.* **515** A19+ (*Preprint* 1001.2132)
[23] Aharonian (for the HEGRA Collaboration) F *et al.* 2003 *Astron. Astrophys.* **403** 523–528 (*Preprint* arXiv:astro-ph/0301437)
[24] Aharonian (for the HESS Collaboration) F *et al.* 2007 *Astron. Astrophys.* **470** 475–489 (*Preprint* 0705.2946)
[25] MAGIC Collaboration *et al.* 2008 *Science* **320** 1752– (*Preprint* 0807.2822)

[26] Aharonian (for the HEGRA Collaboration) F A *et al.* 1999 *Astron. Astrophys.* **349** 11–28 (*Preprint* `arXiv:astro-ph/9903386`)

Indications for a transparent universe at very high energies

Manuel Meyer[1] **and Dieter Horns**[1]

[1]Institut für Experimentalphysik, University of Hamburg, Luruper Chaussee 149, 22767 Hamburg, Germany

E-mail: manuel.meyer@physik.uni-hamburg.de

Abstract. The transparency of the universe for very high energy (VHE) photons is limited due to pair-production with low energy photons of the extra galactic background light (EBL) in the optical to infrared band. Here, we use 56 energy spectra from VHE emitting active galactic nuclei (AGN) from redshift 0.004 to 0.536 to search for signatures of deviations from the minimum expected opacity. A statistical study of the individual measurements reveals indications for an overcorrection of AGN spectra with current EBL models. Axion like particles are discussed as a possible explanation of the result.

1. Introduction

The opacity of the universe for very high energy (VHE) photons (with an energy $E \gtrsim 100$ GeV) arises through the pair production process in the interaction with background radiation fields, $\gamma_{\mathrm{VHE}} + \gamma_{\mathrm{bkg}} \rightarrow e^+ + e^-$. As a consequence, the intrinsic photon flux $\mathrm{d}N_{\mathrm{int}}/\mathrm{d}E$ of a cosmological source with redshift z is dimmed which is commonly expressed as

$$\frac{\mathrm{d}N_{\mathrm{obs}}}{\mathrm{d}E} = \frac{\mathrm{d}N_{\mathrm{int}}}{\mathrm{d}E} \times \exp\left[-\tau_{\gamma\gamma}(E, z)\right], \quad (1)$$

where the observed spectrum is denoted as $\mathrm{d}N_{\mathrm{obs}}/\mathrm{d}E$. The optical depth $\tau_{\gamma\gamma}$ is a threefold integral over the cross section for pair production and the number density of background photons over the cosmological distence, the cosine of the angle between the photon momenta, and the energy of the background photons. The cross section for the pair production strongly peaks at a wavelength $\lambda = 1.24(E/\mathrm{TeV})$ µm which makes the extragalactic background light (EBL) the radiation field responsible for the attenuation of VHE photon fluxes. The EBL ranges from the UV / optical to the far infra-red wavelength band and originates mainly from starlight integrated over all epochs and starlight that has been absorbed and re-emitted by dust in galaxies [1]. These two contributions lead to two peaks in the spectral energy distribution (SED) of the EBL: the first at around 1 µm for the emitted starlight and at around 100 µm for the re-emitted dust component. Direct measurements of the EBL are challenging due to the contamination with foreground emission such as the zodiacal light. Therefore, in the past years many models have emerged that try to describe the EBL [2, 3, 4, for some recent examples].

In grand unified theories that aim to combine the standard model (SM) of particle physics with gravity, effects can occur that alter the opacity of the universe. This is for example the case in certain quantum gravity theories that predict the breakdown of Lorentz invariance [5, 6] or in the presence of exotic particles like axions or axion like particles (ALPs) [7, 8, 9, 10, 11]. Here, the attention is focused on ALPs. Photons can convert into ALPs in magnetic fields that pervade the source, the Milky Way and presumably in the intergalactic medium. The effect of photon-ALPs oscillation is twofold. Fluxes of photons for which the universe appears optically thin, i.e. whose energies and source distances translate into small values of $\tau_{\gamma\gamma}$, are additionally dimmed as the photons convert to ALPs which are not detected on earth. With increasing $\tau_{\gamma\gamma}$ the photon-ALP conversion probability increases as well so that there is higher probability that photons oscillate into ALPs and back into photons. Consequently, the attenuation is decreased as ALPs are not subject to pair production. As a result, the correction for fluxes of photons for which the universe is optically thick is too large and a hardening of the intrinsic spectra is expected.

In this article we statistically investigate 55 VHE spectra from 25 sources obtained with the imaging air Cherenkov telescopes CAT, HEGRA, H.E.S.S., MAGIC, VERITAS and WHIPPLE for an overcorrection of the observed spectrum for high values of $\tau_{\gamma\gamma}$. This guarantees that our results are as independent from source physics as possible as $\tau_{\gamma\gamma}$ is a nontrivial combination of the energy of the γ-ray photon and the distance of the source. For example, high values of the optical depth may be the result of a distant source (like 3C 279) at several hundreds of GeV or for a relatively close source (like Markarian 501) measured at tens of TeV. The observed spectra are corrected for the EBL absorption by inverting Eq. 1 and using the lower limit EBL model by [3]. This lower limit model predicts an EBL which is just in agreement with the 1σ downward fluctuation of the galaxy number counts by *Spitzer* [12] and gives the lowest possible absorption. This ensures that our results indeed indicate a new phenomenon rather than reflecting the uncertainties of the EBL modeling.

The considered sources are all active galactic nuclei (AGN) with known redshift[1]. It follows from Eq. 1 and from the ignorance of the exact shape of the SED of the EBL that it is impossible to measure the intrinsic source spectrum directly. The measured photon indices[2] vary between ~ 1.4 and ~ 4 which makes it difficult to infer some generic intrinsic spectrum valid for all blazars. Furthermore, the source sample might suffer from observational bias, i.e. only certain sources are detected at large redshifts.

These difficulties stress the necessity that the statistical test introduced here is independent of any assumptions of the intrinsic VHE spectra and / or their distance alone. An additional advantage of the test presented in the next section is its independence of the statistical uncertainties of the single measurements.

2. A statistical test to search for an overcorrection of VHE spectra

As discussed in the introduction, an overcorrection of the observed spectra in the presence of ALPs is expected for high values of the optical depth. This corresponds to a combination of a high redshift of the source and / or a measurement at energies beyond ~ 10 TeV. Only a few sources are available where these conditions are met. As a compromise between statistics and the strength of the expected effect, we define optical thick as $\tau_{\gamma\gamma} \geq 2$ and optical thin as $\tau_{\gamma\gamma} < 2$. This value of the optical depth also ensures that the dimming due to pair production is not confused with the dimming of the initial flux because of photon-ALPs conversion. The latter can reduce the initial photon flux by maximally 1/3 [7] which already occurs for $\tau_{\gamma\gamma} \approx 0.4$.

Table 1. Analytical functions used to fit the data points of the spectra. The smoothing parameter f of the broken power law is fixed to 4 and the energy normalization E_0 is set to 1 TeV.

Function Name	$dN/dE(E)$	Fit parameters
Power law	$N_0(E/E_0)^{-\Gamma}$	N_0, Γ
Logarithmic parabola	$N_0(E/E_0)^{-\Gamma+\beta\ln(E/E_0)}$	N_0, Γ, β
Broken power law	$N_0(E/E_0)^{-\Gamma_1}[1 + (E/E_{\mathrm{break}})^f]^{(\Gamma_1-\Gamma_2)/f}$	$N_0, \Gamma_1, \Gamma_2, E_{\mathrm{break}}$

The statistical test is devised as follows. For a given VHE spectrum the data points are determined

[1] Sources for which the redshift is under ongoing discussion such as 3C 66A, PG 1553+113 and S 50716+714 are not considered either.

[2] Most observed VHE spectra are adequately described by power laws of the form $dN/dE \propto E^{-\Gamma}$, where Γ is called the photon index, see also Table 1.

Journal of Physics: Conference Series **355** (2012) 012028 doi:10.1088/1742-6596/355/1/012028

that correspond to an optical depth $\tau_{\gamma\gamma} < 1$. To these points a simple power law is fitted using a χ^2-minimization algorithm. If the resulting p-value of the fit is less than 5% a logarithmic parabola is fitted instead and if the p-value still remains below 5% a broken power law is used. The analytic functions are summarized in Table 1.

However, only for 5 spectra other functions than a power law are necessary to describe the data. These are spectra from Markarian 421 and Markarian 501 which are close sources with redshifts of 0.031 and 0.034, respectively, and one spectrum of PKS 2155-304. Already the observed Markarian spectra show curvature and since they are relatively close by many data points enter the fit. In the case that the spectrum only contains one data point with $\tau_{\gamma\gamma} < 1$ it is not considered, if it contains two data points the fit parameters of the power law are derived analytically.

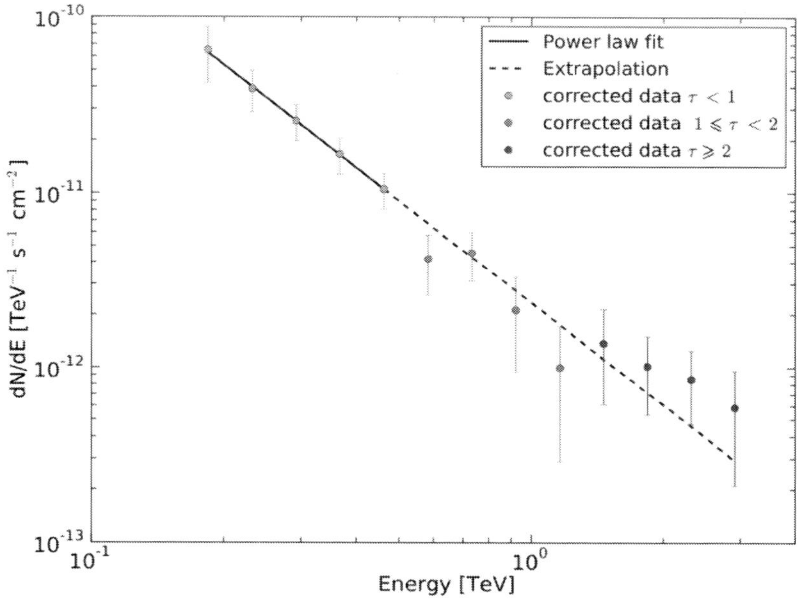

Figure 1. The spectrum of 1ES1101-232 measured by H.E.S.S. [13] with $1\,\sigma$ statistical errors on the flux. The first five points correspond to a $\tau_{\gamma\gamma}$ value smaller than one and thus enter into the power-law fit. The fit itself is shown as the black solid line. The fit is extrapolated to all other points (dashed line) and the ratios R_i are calculated according to Eq. 2. The ratios of the red points enter $\mathcal{S}_{\rm thin}$ while the ratios of the dark red points contribute to $\mathcal{S}_{\rm thick}$, see Eq. 3 and 4.

The functions determined this way are extrapolated to all remaining data points. The procedure is shown in Fig. 1 for the spectrum of 1ES1101-232 measured by H.E.S.S. [13]. The deviation of the i-th data point from the extrapolation is quantified by the ratio R_i,

$$R_i = \frac{\ln f^{\rm ext}(E_i) - \ln f_i^{\rm int}}{\ln f^{\rm ext}(E_i) + \ln f_i^{\rm int}}. \tag{2}$$

Here, $f^{\rm ext}$ denotes the extrapolation of the analytic description of the data points with $\tau_{\gamma\gamma} < 1$ and $f_i^{\rm int}$ is the shorthand notation for $\mathrm{d}N_i^{\rm int}/\mathrm{d}E$. If the corrected data point lies above the extrapolation, R will be positive since $f^{\rm ext}$ and $f^{\rm int}$ are always smaller than one. Simulations show that this definition of R gives the most sensitive result for the statistical test [14], opposed to e.g. a definition without the logarithm

[15]. For our purposes to find an overcorrection of the observed spectra, the choice of analytical functions can be considered conservative as they show less curvature than e.g. a power law modified with an exponential cut-off because the extrapolation of such a function could lead to an overestimation of R for the highest energy bins. The ratios are calculated for all considered VHE spectra[3] and two distributions of ratios are defined that correspond to optical thin and thick measurements, respectively,

$$\mathcal{S}_{\text{thin}} = \{R_i \mid 1 \leq \tau_{\gamma\gamma}(E_i, z) < 2\}, \quad (3)$$

$$\mathcal{S}_{\text{thick}} = \{R_i \mid \tau_{\gamma\gamma}(E_i, z) \geq 2\}. \quad (4)$$

These two distributions are compared with the Kolmogorov-Smirnov (KS) test [15, 16] under the null-hypothesis that the underlying probability distributions are equal.

3. Results

From the 56 spectra, only 26 (7) spectra have data points that correspond to an optical depth ≥ 1 (≥ 2). Figure 2 shows the result of the KS-test. The upper panel displays the optical depth for each data point of every spectrum plotted against the ratios R with their corresponding errors (only the errors of the individual flux measurements are considered). The red points contribute to the $\mathcal{S}_{\text{thick}}$ distribution whereas the blue points all correspond to an optical depth of $1 \leq \tau_{\gamma\gamma} < 2$ and are thus part of $\mathcal{S}_{\text{thin}}$. From the discussion in Section 2 it is clear that the number of points in $\mathcal{S}_{\text{thick}}$ (labeled N_2 in the figure) is smaller than in $\mathcal{S}_{\text{thin}}$ (labeled N_1) since less observations are available for such large attenuations. A trend is visible that observations at large $\tau_{\gamma\gamma}$ tend to show higher values of R as well. This is confirmed in the lower panel which depicts the corresponding cumulative distribution functions (CDFs). The green line simply visualizes that the analytical functions are a good description of the observations as the data points that enter the fit itself also enter into this CDF. The situation is different for the extrapolated data points. The blue line represents the CDF of $\mathcal{S}_{\text{thin}}$. About 60% of the ratios are below zero and not 50% as one would naively expect. The reason for this might be an intrinsic curvature of the spectrum or an underestimation of the EBL density for $1 \leq \tau_{\gamma\gamma} < 2$ by the lower limit model. In the case of the extrapolation to values that correspond $\tau_{\gamma\gamma} \geq 2$ the CDF (red line) show a systematic increase towards higher values of R, indicating a hardening of the intrinsic spectra which can be taken as evidence that the correction is indeed too strong. The maximum distance D between the CDFs of $\mathcal{S}_{\text{thin}}$ and $\mathcal{S}_{\text{thick}}$ is found to be $D \approx 0.62$ for which the KS-test gives a probability of $Q_{KS} \approx 2.81 \times 10^{-4}$ that the underlying probability distributions are equal. This corresponds to a significance of 3.45σ (one-sided confidence interval) that they are different from each other. A cross check with galactic sources shows that this result cannot be explained by instrumental effects such as an overestimation of the flux in the highest energy bins [14].

4. Conclusion

In this article a new approach based on the Kolmogorov-Smirnov test is presented that searches for systematic changes in VHE spectra corrected for the EBL absorption in the transition from $\tau_{\gamma\gamma}(E, z) < 2$ (optically thin) to $\tau_{\gamma\gamma}(E, z) \geq 2$ (optically thick). The approach is independent of the exact shape of the intrinsic VHE spectra and, thus, the obtained results do not suffer from possible observational bias, e.g. only certain types of sources are detected at high redshifts.

An indication with a significance of 3.45σ is found that optically thick measurements are over corrected by a lower limit EBL model. If the significance of this result is confirmed and improved with future observations[4] it demands for an explanation other than tuning the EBL density as the lower limit model already predicts a minimum attenuation that is in accordance with lower limit measurements of the EBL. The result is difficult to explain with source physics since many objects measured at

[3] For the entire VHE sample considered here, see [14] and references therein.

[4] There are $\mathcal{O}(10)$ new detections of VHE emitting AGN announced, see e.g. `http://tevcat.uchicago.edu`

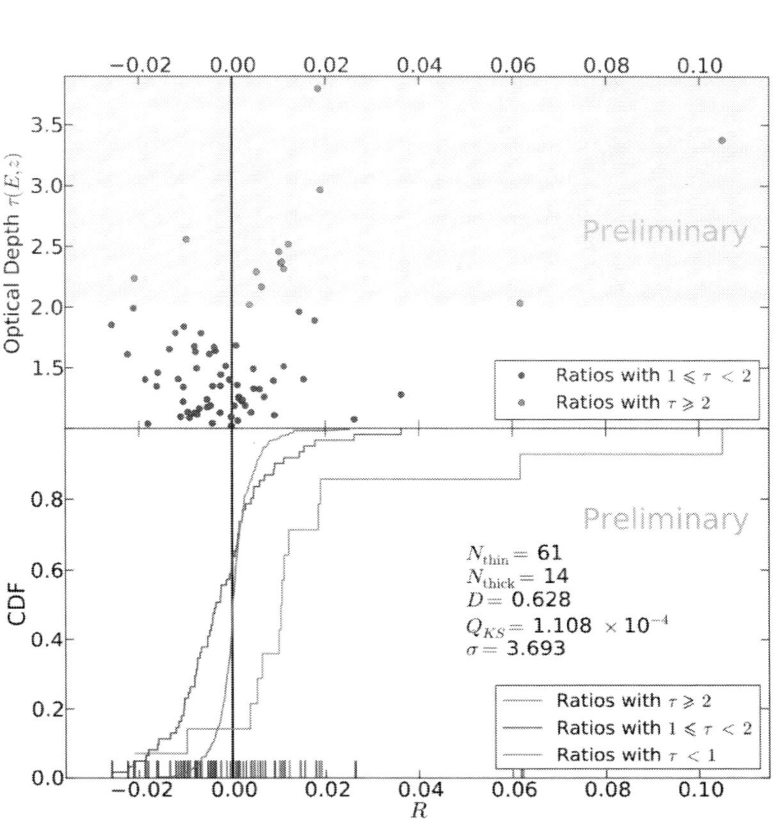

Figure 2. Result of the KS-test, see text for details.

different redshift and energy enter the test. For instance, an upturn of the intrinsic spectra at multiple TeV energies due to e.g. second order inverse Compton scattering [17], comptonization of cosmic microwave background (CMB) photons [18] or interactions of ultra-high energy cosmic rays with the CMB [19, 20, 21] is not able to explain our findings. One the one hand, such features could possibly enter the ratios of $\mathcal{S}_{\rm thin}$ and $\mathcal{S}_{\rm thick}$ since they are disinguishied by $\tau_{\gamma\gamma}$ and not by the energy E. On the other hand, distant sources like 3C 279 measured below 1 TeV also contribute to the result. One candidate for a way out might be given by photon-ALPs conversion as theory predicts a decrease of the attenuation for large values of the optical depth.

Acknowledgements

The authors would like to thank the organizers of the conference *Beamed and unbeamed gamma rays from galaxies* 2011 for the opportunity to present the results of this work. This work was made possible with the support of the LEXI young investigators group "Cosmic radiation fields and search for dark matter" at the university of Hamburg.

References

[1] Hauser M G and Dwek E 2001 *Ann. Rev. Astron. Astrophys.* **39** 249–307 (*Preprint* `arXiv:astro-ph/0105539`)
[2] Franceschini A, Rodighiero G and Vaccari M 2008 *Astron. Astrophys.* **487** 837–852 (*Preprint* `0805.1841`)

[3] Kneiske T M and Dole H 2010 *Astron. Astrophys.* **515** A19+ (*Preprint* 1001.2132)
[4] Domínguez A, Primack J R, Rosario D J *et al.* 2010 *Mon. Not. R. Astron. Soc.* 1665–+ (*Preprint* 1007.1459)
[5] Jacob U and Piran T 2008 *Phys. Rev. D* **78** 124010–+ (*Preprint* 0810.1318)
[6] Shao L and Ma B 2010 *Modern Physics Letters A* **25** 3251–3266 (*Preprint* 1007.2269)
[7] Mirizzi A, Raffelt G G and Serpico P D 2007 *Phys. Rev. D* **76** 023001–+ (*Preprint* 0704.3044)
[8] Hochmuth K A and Sigl G 2007 *Phys. Rev. D* **76** 123011–+ (*Preprint* 0708.1144)
[9] Hooper D and Serpico P D 2007 *Physical Review Letters* **99** 231102–+ (*Preprint* 0706.3203)
[10] Sánchez-Conde M A, Paneque D, Bloom E, Prada F and Domínguez A 2009 *Phys. Rev. D* **79** 123511–+ (*Preprint* 0905.3270)
[11] Mirizzi A and Montanino D 2009 *JCAP* **12** 4–+ (*Preprint* 0911.0015)
[12] Fazio G G, Ashby M L N, Barmby P, Hora J L, Huang J S, Pahre M A, Wang Z, Willner S P, Arendt R G, Moseley S H, Brodwin M, Eisenhardt P, Stern D, Tollestrup E V and Wright E L 2004 *Astrophys. J., Suppl. Ser.* **154** 39–43 (*Preprint* arXiv:astro-ph/0405595)
[13] Aharonian (for the HESS Collaboration) F *et al.* 2007 *Astron. Astrophys.* **470** 475–489 (*Preprint* 0705.2946)
[14] Meyer M and Horns D 2011 *in preparation*
[15] Meyer M and Horns D 2010 *Proceedings of Cosmic Radiation Fields: Sources in the early Universe (CRF 2010). November 9-12, 2010. Desy, Germany. Editors: Martin Raue (chair), Tanja Kneiske, Dieter Horns, Dominik Elsaesser, and Peter Hauschildt. Published online at http://pos.sissa.it/cgi-bin/reader/conf.cgi?confid=121, p.11* ed M Raue, T Kneiske, D Horns, D Elsaesser, & P Hauschildt pp 11–+
[16] Press W H 2002 *Numerical recipes in C++ : the art of scientific computing*
[17] Bloom S D and Marscher A P 1996 *Astrophys. J.* **461** 657–+
[18] Böttcher M, Dermer C D and Finke J D 2008 *Astrophys. J., Lett.* **679** L9–L12 (*Preprint* 0804.3515)
[19] Essey W and Kusenko A 2010 *Astroparticle Physics* **33** 81–85 (*Preprint* 0905.1162)
[20] Essey W, Kalashev O E, Kusenko A and Beacom J F 2010 *Physical Review Letters* **104** 141102–+ (*Preprint* 0912.3976)
[21] Essey W, Kalashev O, Kusenko A and Beacom J F 2011 *Astrophys. J.* **731** 51–+ (*Preprint* 1011.6340)

Journal of Physics: Conference Series **355** (2012) 012029 doi:10.1088/1742-6596/355/1/012029

Gamma-ray variability and correlation properties of blazars observed with Fermi LAT

S. Larsson on behalf of the Fermi LAT collaboration and many multiwavelength collaborators

Department of Astronomy, Stockholm University and The Oskar Klein Centre for Cosmoparticle Physics, Stockholm, Sweden

E-mail: `stefan@astro.su.se`

Abstract. The Fermi Large Area Telescope (Fermi LAT) provides long term systematic monitoring observations of the gamma-ray emission from blazars. The variability properties and the correlation with other wavelength bands are important clues for the evaluation of blazar models. We present results from timing and multiwavelength correlation analysis and discuss differences between blazar classes.

1. Introduction

More than 60 AGNs, almost all blazars, were detected by EGRET on Compton GRO, which established these sources as a powerful class of gamma-ray emitters [1]. In its first year of operation Fermi LAT has increased the number of known gamma-ray blazars by a factor of 10. Even more importantly the instrument is mapping the full sky every three hours, which allow regular monitoring of these sources on time scales from hours to years. These monitoring observations now form an important part of many ongoing efforts to study the variability and multiwavelength properties of blazars. Such observations contains information about the relative origin of different spectral components and about dynamical and radiation processes in blazar jets.

2. The Fermi Large Area Telescope

Fermi (the Fermi Gamma-ray Space Telescope) was launched in June 2008 from Cape Canaveral in Florida. The main instrument onboard Fermi is the LAT (Large Area Telescope) which is sensitive to gamma-rays in the energy range 30 MeV to 300 GeV [2]. With a much larger field of view and with a much improved energy resolution compared to its predecessors, Fermi is now providing unprecedented data for studies of active galactic nuclei and other gamma-ray sources. Fermi is operated primarily in a sky survey mode where the full sky is mapped every three hours. Since LAT has a very wide field of view (about 20% of the sky) the exposure on any particular sky position is large. Pointed observations are executed only to follow up GRBs or during other exceptional outbursts. The sky surveying and wide field of view of the instrument also mean that the potential for new, serendipitous discoveries is very large. The angular resolution, which increases with photon energy, is about 1 degree at 1 GeV, while source localizations are typically 0.1 degrees.

3. LBAS, 1LAC and 2LAC

Based on the first 3 months of observations a LAT bright-AGN Source list (LBAS) was produced by the Fermi LAT collaboration [3]. This list contains 106 high confidence associations (58 FSRQs, 42 BL Lacs, 2 radio galaxies and 4 of unknown type). These were sources with detection test statistic, TS > 100, corresponding approximately to 10σ detections. By comparison, the First LAT AGN catalog, 1LAC [4] based on 11 months of data included already 663 high confidence associations (sources with TS > 25, or approximately 5σ). In the recently released second AGN catalog, 2LAC, this number of sources was increased by 50%[5].

4. Blazar variability

Blazar light curves are often dominated by strong flares. The origin of these flares can e.g. be the bulk injection of new particles into the jet or strong internal shocks. If the variability is made up of shot pulses of different lengths and amplitudes produced by such events, the Power Density Spectrum (PDS) is determined by the shapes and amplitudes of the pulses and by correlations of their relative distribution in time. While pulse shapes might be associated with, e.g., cooling or light travel time scales, pulse correlations will contain information on the processes that are responsible for creating the pulses, which could be episodes of strong activity near the central black hole. By calculating the PDS, the Structure Function (SF), and other statistical mesures, we can test alternative mechanisms to produce the observed variability, keeping in mind that the same PDS or SF can be produced by more than one stochastic process.

The Fermi LAT data can be used to study variability on time scales of hours to years, to measure spectral variations and to catch flaring sources with high efficiency. An example of the capability to study flux and spectral variations on subday time scales is given by the analysis of the recent outburst of 3C454.3 [6].

For timing analysis the regular sampling provided by Fermi LAT is a great advantage. The initial aim of our study of Fermi blazars is a characterization the gamma-ray variability. The tools that have been used for this are,

- χ^2 and excess variance to quantify the presence of variability and its significance.
- Duty cycle analysis
- Auto correlation function (ACF)
- Structure function (SF)
- Power Density Spectra (PDS)
- Flare shape analysis

Even though the ACF, SF and PDS contain the same information, this information is expressed differently, which makes it useful to compute all three. The direct fitting of flare shapes is based on the assumption that individual flare pulses can be identified. The advantage with this approach is the possibility to extract light curve phase information, which is lost in the calculations of ACF/SF/PDS. This includes e.g. time asymmetry.

First results of our work, using 11 months of data and light curves with 3 to 7 days binning for the LBAS sources was recently published [7] and is partly summarized here.

4.1. Power Density Spectra

Power Density Spectra (PDS) of blazars have previously been studied primarily for optical and radio data and have then in general been described by a power law $P(\nu) \propto \nu^{-\alpha}$ with α typically in the range 1 to 2.

For the PDS analysis of Fermi blazars, light curves with 3 day binning was used for the 15 brightest LBAS sources (9 FSRQs and 6 BL Lacs). For a further 13 FSRQs with slightly lower flux, the binning was increased to 4 days. A standard Fourier transform was used to compute power density spectra (PDS) and the white noise level corresponding to the estimated measurement errors was subtracted.

The power spectral density is normalized to fractional variance per frequency unit ($rms^2\ I^{-2}\ Day^{-1}$, where I is the mean flux) and the PDS points are averaged in logarithmic frequency bins. An example PDS for 3C273 is shown in Figure 1 and the 1-week-binned light curve for the same source is presented in Figure 2.

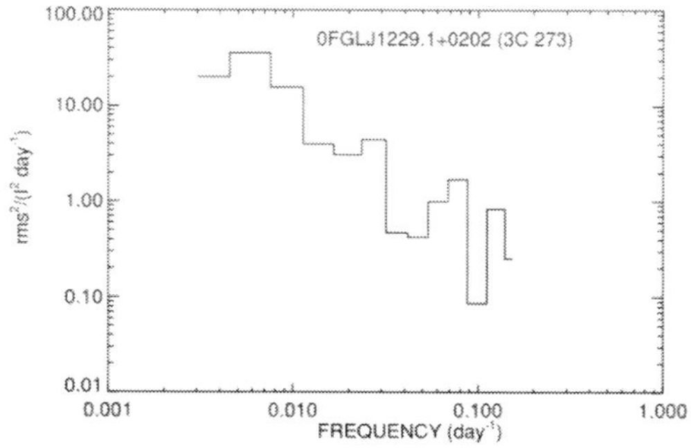

Figure 1. PDS of the 11 month Fermi light curve for 3C273. A 3 day binned light curve was used. The PDS is normalized to variance divided with the square of the mean flux and per frequency unit (from [7]).

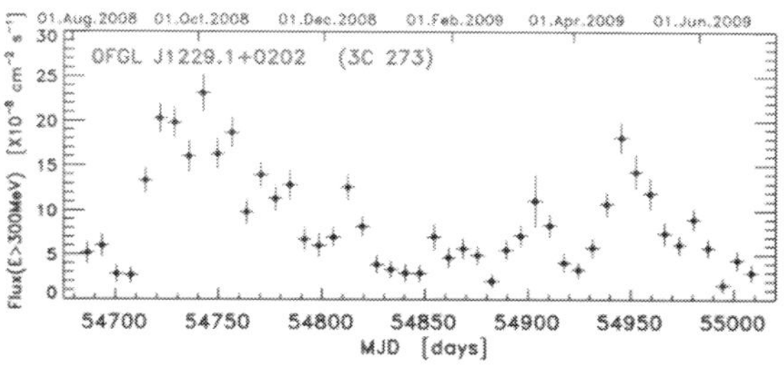

Figure 2. Fermi 11 month light curve (1 week binning) for 3C273 (from [7]).

There are a number of effects that can distort the shape of the computed PDS and cause it to differ from the 'true' intrinsic PDS describing the variability of the source. These effects are,

- Stochastic variability in a time limited observation.
- Effects of the observational data and the analysis (e.g. aliasing).
- Systematic errors
- Statistical (measurement) errors.

For the analysis presented here the last effect dominates at high frequencies and the first effect at the lower frequencies where source variance is much larger than the measurement noise. The white noise level used in the analysis has a significant effect on the slope deduced from the PDS fit. This means that the result is dependent of a reliable estimate of the measurement errors. These errors were therefore also checked by comparing some light curves with the corresponding ones obtained by direct aperture photometry, for which Poisson statistics is valid. This showed that the uncertainty in error estimates is not a significant problem for the brightest sources. For the less bright ones, including all the BL Lacs, this effect may however, introduce a systematic bias in the PDS slope. The effect was estimated by repeating the analysis for a range of possible white noise levels and also by analysis of light curves extracted with different time bins (from 1 to 7 days). To limit the effect of the white noise on the determination of the PDS slope, the fits only used frequencies up to 0.02 day^{-1}.

Observational and instrument systematics were investigated by analyzing pulsar light curves extracted from the 11 month data with the same procedure as for the blazars. The most prominent effect is a periodic modulation that is identified with the 54 day precession period of the Fermi satellite orbit. In the PDS for individual blazars this peak is often hidden by the stochastic variability but does show up when averaging the PDS of a number of sources. The frequency bin at this period was not used when PDS slopes were estimated.

With the available data the uncertainty in PDS shape for individual sources is still dominated by stochastic variability. To reduce this and the statistical fluctuations we can average the PDS for a group of sources under the assumption that the differences in PDS shape is small compared to the random fluctuations expected due to the action of the (presumed) underlying stochastic process. For the 9 brightest FSRQs this averaged PDS is shown in Figure 3. For frequencies below 0.017 we obtain a best fit slope of 1.4 $\pm$0.1. The same PDS is also shown in Figure 4 together with the corresponding averaged PDS for the 9 BL Lacs and the 13 moderately bright FSRQs. The differences in power-law slope are not statistically significant (errors are estimated from the scatter within each group of sources).

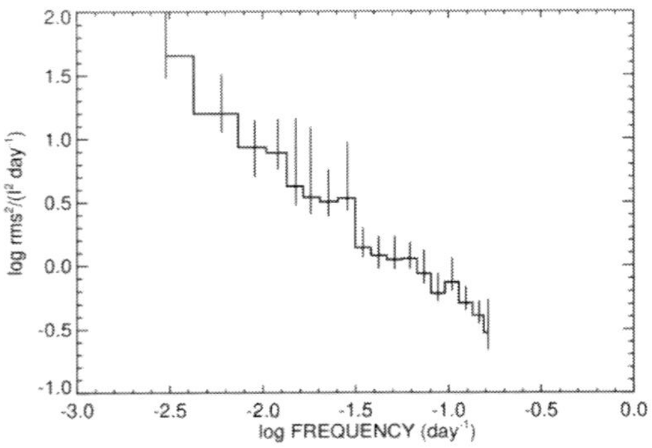

Figure 3. Average Power Density spectrum, PDS, for the 9 brightest FSRQs. White noise level based on light curve error estimates has been subtracted. The error bars are asymmetric 1 sigma errors of the mean. Our best fit estimate is a PDS slope of 1.4 $\pm$0.1 (from [7]).

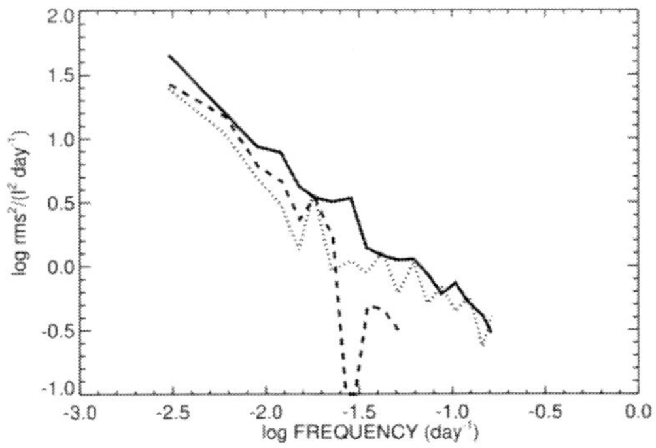

Figure 4. Comparison of the averaged PDS for three sets of sources, the 9 bright FSRQs from the upper plot (solid line), the 6 brightest BL Lac's (dotted line) and 13 additional FSRQs with TS > 1000 (dashed line). Best fit slope for the BL Lac and fainter FSRQs is 1.7 $\pm$0.3 and 1.5 $\pm$0.2 respectively (from [7]).

Since the number of sources is small and the length of observations relatively short it is too early to draw firm conclusions from these results. It is nonetheless interesting to note a difference between the low and the high synchrotron peaked BL Lacs (LSP and HSP resp). While the average slope for the BL Lacs is at least as steep as that of the FSRQs, the two HSPs, Mkr 421 and PKS2155-305, both have PDS slopes flatter than 1.0 (Figure 5). Mkr 421 is the brightest soft X-ray blazar and a PDS for long time scales at those energies can be computed from the ASM light curve. The PDS is well fitted by a power law with index close to -1, as shown in Figure 6.

When sources are in a high state it is possible to use light curves with smaller time bins and to compute a significant PDS to higher frequencies. In Figure 7 we have combined a PDS for the time around the December 2009 outburst of 3C454.3 with its 11 month LBAS PDS [8]. The fact that the two PDS join each other smoothly at the overlapping frequencies implies that the fractional variability during the bright outburst is similar to that of the longer time scale. The same conclusion was drawn when an intense flaring episode for 3C273 was compared to its 11 month LBAS PDS [9].

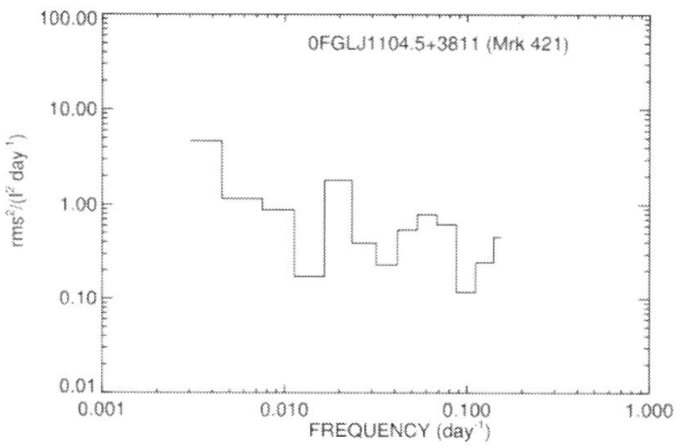

Figure 5. Gamma-ray PDS for the 11 month Fermi light curve of the HSP BL Lac Mkr 421 (from [7]).

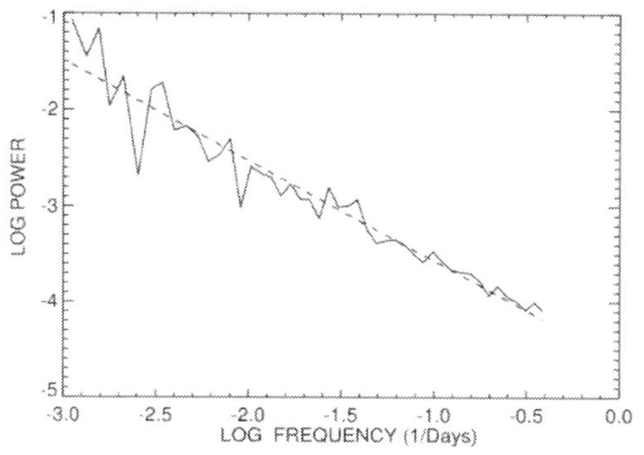

Figure 6. PDS of the soft X-ray ASM light curve for Mkr 421 (12 years of data). The PDS power law slope = -1.04 ±0.04.

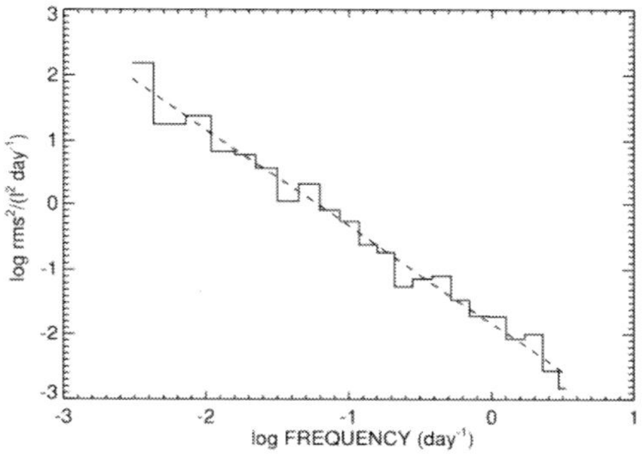

Figure 7. PDS for the FSRQ 3C454.3 combining the PDS for the 11 month LBAS data and that of the December 2009 outburst (from [8]).

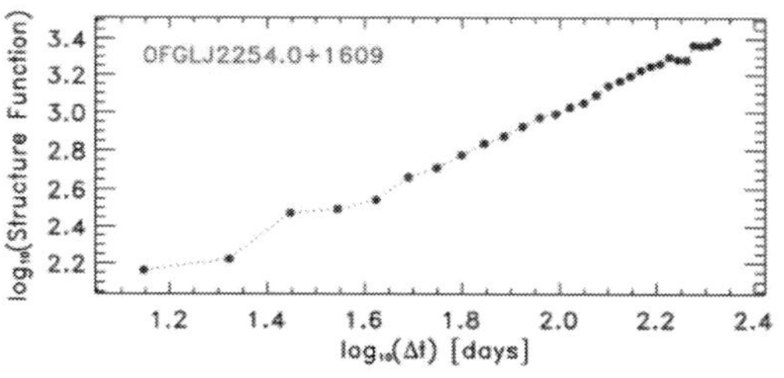

Figure 8. Structure function for the 1-week-binned LBAS 11 month light curve of 3C454.3 (from [7]).

4.2. Structure function analysis

The structure function is related to both the ACF and PDS. The slope of the structure function is related to α, the power law index of the PDS as, $S \propto (\Delta t)^{\alpha-1}$. Using 1-week-binned light curves, the structure function was computed for 56 LBAS sources and the corresponding α was estimated for each source. This is shown in Figure 8 for 3C454.3. For the whole sample the α values showed a wide distribution with a peak at about 1.3.

4.3. Summary of variability properties

The Fermi LAT data has allowed us to perform a systematic study of gamma-ray variability of blazars.

The main results are:

(i) 2/3 of the studied sources are variable. (This fraction is increasing with time.)

(ii) The time spent in high states are $< 1/4$ of the total observing time.

(iii) The relative variance is larger for FSRQs than for BL Lacs.

(iv) ACF time scales are $\approx$ 4 to >10 weeks.

(v) Typical PDS slopes are $\approx$ 1.3 to 1.5

(vi) No evidence for persistent characteristic time scale(s).

(vii) Flare profiles are on average symmetric

(viii) The fractional variability during outburst was found to be similar to its longer term mean in the two objects where this was studied.

5. Multiwavelength correlation

The relation between variability at different wavelengths can be investigated in a number of different ways. Ultimately one would like to describe and interpret the detailed time evolution of the full Spectral Energy Distribution (SED). In practice one is limited by statistics and by a limited coverage in wavelength and time. One of the most common tools to extract information is the Cross Correlation Function (CCF), which averages correlation information from a whole time series. The CCF and its discrete version, the Discrete Cross Correlation Function (DCCF), give the correlation between two light curves for different relative time shifts or lags. In the most simple case where one light curve is just a delayed version of the other, the correlation function will show a peak at the time lag corresponding to the time shift between the two bands. In reality the relation between the two bands may be more complex, including e.g. different time responses and time changes in the correlation properties.

In correlation analysis one should also keep in mind that a detrending will affect the variability timescales to which the analysis is sensitive. Without detrending the correlation is often dominated by large amplitude variations on longer time scales. With a detrending, e.g. by fitting and subtracting a low order polynomial, these variations are removed and variability on shorter time scales will dominate the CCF. An example of strong correlation between optical and gamma-ray variations on both long and short time scale is PKS 0537-441 as can be seen directly in the light curves of Figure 9. The DCCF for the same data is shown in Figure 10. By comparison the correlation amplitude for the detrended light curves is reduced from 0.9 to 0.6.

A more complex correlation behavior was seen in PKS 1510-089 [10]. Four flares of about one month each were observed in 2008 to 2009 and three of these had good coverage in both optical and gamma-rays. The total DCCF (Figure 11) has a number of correlation peaks. These are mostly caused by random correlations. Two correlation features however, appear to be persistent for all the flares. One is a correlation on short time scales, within 1 - 2 days of lag zero, The second is a strong peak at a lag of about 2 weeks (gamma-rays leading optical). A detailed comparison of the light curves show that the envelope (start and end) of each flare is about the same in the two bands but the shape of the flare is different. The ratio of gamma-ray to optical flux is higher in the beginning of the flare than towards the end. On top of this the flares exhibit strong variability on shorter time scales which is partly correlated and responsible for the correlation peak near zero lag.

6. Status and future prospects

The effort to characterize the time variability of gamma-ray emission from blazars is presently limited mainly by the length of available observations. Variability properties on short time scales can appear to be quite different from the longer term mean due to the stochastic nature of the variability. Likewise, the multiwavelength properties of flares are not constant. This is true not just when comparing different objects but also for different flares in the same object.

To reveal the nature of the stochastic process or processes driving the observed variability it is essential to collect long sequences of observations. Multiwavelength observations are just as important in order to identify physical processes and their connection to each other and to the central engine at the heart of the blazar and its jet. Fermi is playing a key role, not just by

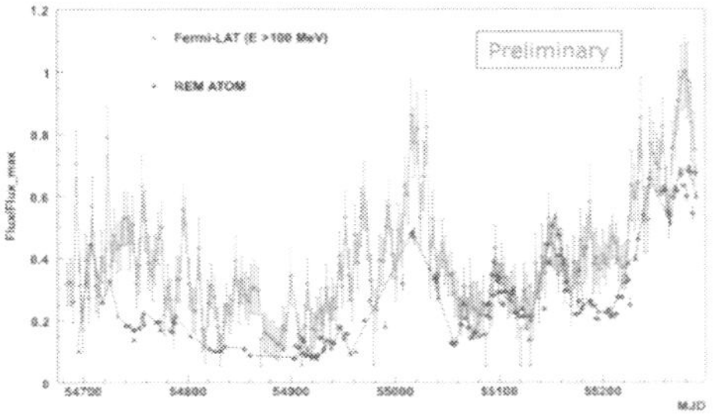

Figure 9. Comparison between γ-ray and R-band light curves for PKS 0537-441. The LAT (red dots) and R (blue triangles; ATOM and REM) light curves are normalized for comparison (in preparation).

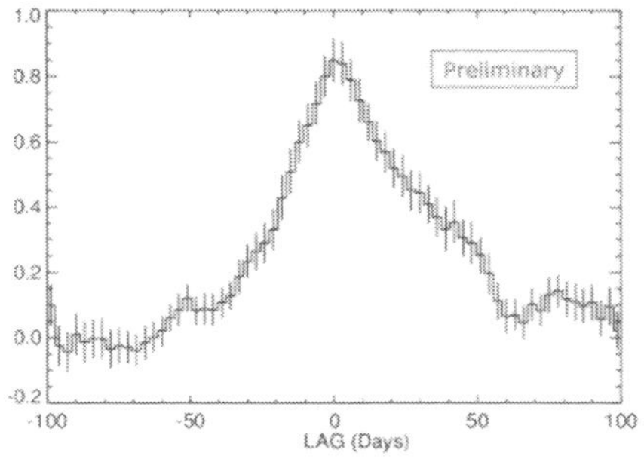

Figure 10. DCCF between the Gamma-ray and R band light curves in Figure 9.

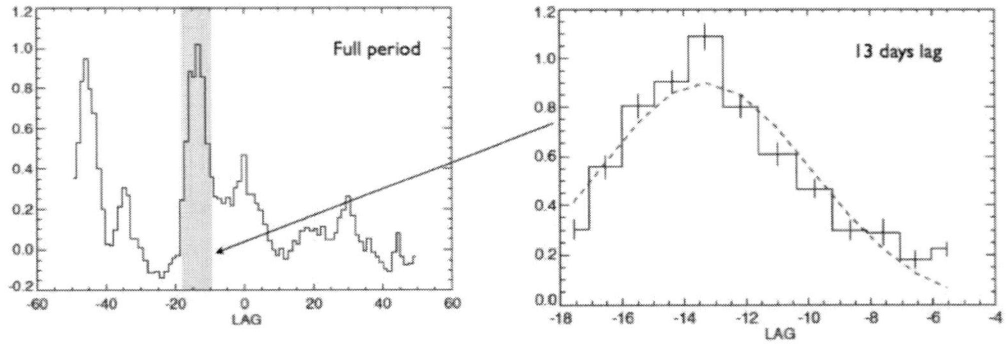

Figure 11. DCCF between gamma-rays and the R band for a period with three major flares in PKS1510-089 (from [10]). In a short light curve with just a few flares, there will also occur correlations at lags corresponding to the time between the flares. This is the cause of the peak at lag -50 days in this case.

the unique data it is providing but also as a catalyst and reference for all other observations of blazars. The present decade has all the prerequisites to become a golden age of blazar research.

7. Acknowledgments

The *Fermi* LAT Collaboration acknowledges support from a number of agencies and institutes for both development and the operation of the LAT as well as scientific data analysis. These include NASA and DOE in the United States, CEA/Irfu and IN2P3/CNRS in France, ASI and INFN in Italy, MEXT, KEK, and JAXA in Japan, and the K. A. Wallenberg Foundation, the Swedish Research Council and the National Space Board in Sweden. Additional support from INAF in Italy and CNES in France for science analysis during the operations phase is also gratefully acknowledged. Larsson is grateful to the Swedish National Space Board for funding his work in the Fermi project.

Journal of Physics: Conference Series **355** (2012) 012029 doi:10.1088/1742-6596/355/1/012029

8. References

[1] Hartmann, R.C., et al. 2001 *Astrophys. J.* **553**, 683
[2] Atwood, W. B., et al. 2009 *Astrophys. J.* **697** 1071
[3] Abdo, A, A., et al. 2009 *Astrophys. J.* **700** 597
[4] Abdo, A, A., et al. 2010 *Astrophys. J.* **715** 429
[5] Ackermann, M., et al. 2011, submitted to *Astrophys. J.*, *ArXiv*:1108.1420
[6] Abdo, A, A., et al. 2011 *Astrophys. J.* **733** L26
[7] Abdo, A, A., et al. 2010 *Astrophys. J.* **722** 520
[8] Abdo, A, A., et al. 2010 *Astrophys. J.* **721** 1383
[9] Abdo, A, A., et al. 2010 *Astrophys. J.* **714** L73
[10] Abdo, A, A., et al. 2010 *Astrophys. J.* **721** 1425

Fermi Gamma-ray Space Telescope Observations of the Gamma-ray Outburst from 3C454.3 in November 2010

B Lott[1], C D Dermer [2], L Escande[1] on behalf of the Fermi-LAT Collaboration

[1] Centre d'Études Nucléaires Bordeaux Gradignan, Universit de Bordeaux, CNRS/IN2P3, UMR 5797, Gradignan, 33175, France

E-mail: `lott@cenbg.in2p3.fr`

[2] Space Science Division, Naval Research Laboratory, Washington, DC 20375, USA

Abstract. The flat-spectrum radio quasar 3C454.3 underwent an extraordinary 5-day γ-ray outburst in November 2010 when the daily flux measured with the *Fermi* Large Area Telescope (LAT) at photon energies $E > 100\,\mathrm{MeV}$ reached $(66\pm2)\times10^{-6}$ph cm^{-2}s^{-1}. This is a factor of 3 higher than its previous maximum flux recorded in December 2009 and > 5 times brighter than the Vela pulsar, which is normally the brightest source in the γ-ray sky. The 3-hr peak flux was $(85\pm5)\times10^{-6}$ph cm^{-2}s^{-1}, corresponding to an apparent isotropic luminosity of $(2.1\pm0.2)\times10^{50}$ erg s^{-1}, the highest ever recorded for a blazar. In this talk, we investigate the features of this exceptional event in the γ-ray band of the *Fermi*-LAT.

1. Introduction

The radio source 3C454.3, a well-known flat-spectrum radio quasar (FSRQ) at redshift $z = 0.859$, has shown remarkably high activity since 2000. It has been particularly bright in the γ-ray band covered by AGILE and the *Fermi*-LAT, reaching a daily record flux level F[$E > 100\,MeV$] (F_{100} in units of 10^{-6}ph cm^{-2}s^{-1}) of 22 ± 1 in December 2009 [19, 6]. This high flux allowed detailed analysis to be performed, making it the best-studied blazar in the GeV band. Gamma-ray variability on timescales as short as a few hours [20] and a flux-doubling time scale of less than 3 hours for a short subflare on 2009 Dec 5 [6] have been reported. In the LAT energy band, 3C454.3 exhibits a spectrum with a clear departure from a power-law distribution, characterized by a break around 2 GeV [1]. This is found to be a common feature of bright γ-ray FSRQs [4]. Little variation of the break energy and spectra for large differences in flux states was observed for 3C454.3 [6]. From $\gamma\gamma$-opacity constraints, a minimum Doppler factor $\delta_{min} \approx 13$ was derived from the flux variability time [6], and highest-energy photon measurements, in accord with independent measurements of δ from superluminal motion observations [13].

In November 2010, the source displayed sustained activity at a flux of $F_{100} \approx 10$ for several days before showing a fast rise to record levels of $F_{100} = 55$, then rising to $F_{100} \approx 80$ (as measured over 6 hr-long periods). In this talk, the intraday variability and the associated spectral changes in the γ-ray band of 3C454.3 are presented.

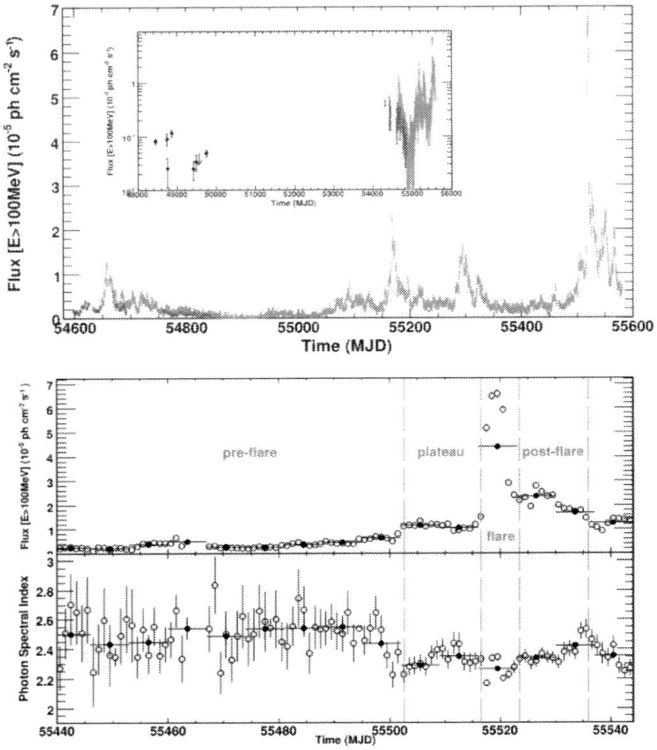

Figure 1. Top, main figure: Daily light curve of 3C454.3 measured with the Fermi-LAT since launch. Inset: Historical light curve. Black points are from EGRET [12] and blue points are from AGILE [19]. Bottom: Light curve of the flux F_{100} (top) and Γ (bottom) for a 103-day period including both the slowly increasing flux phase, the plateau, the flare and the post-flare. The open and filled symbols correspond to daily and weekly averaged quantities respectively. Error bars are statistical only.

2. Observations and analysis

We refer to [5] for details concerning the analysis. The data presented here are restricted to the 100 MeV–200 GeV range and were collected from, MJD55440 (2010 September 1) to MJD55543 (2010 December 13) in survey mode. Source variability was investigated by producing light curves with various time binnings (3 hours, 6 hours, 1 day, 1 week) and over different energy ranges (E>100 MeV, E>1 GeV, E=0.1-1 GeV).

3. Results

Figure 1 (top panel) showing the historical F_{100} light curve illustrates the spectacular rise in activity of 3C454.3 over the years. The second panel displays the light curves with time binnings of 1-day (open circles) and 1-week (filled circles) during the outburst period. A 13-day-long plateau precedes the 5-day-long flare, confirming the trend previously observed in the July 2008 and December 2009 flares, but it is longer in duration and higher in flux than those in previous flares [6]. The onset of this plateau is clearly accompanied by a weak but significant hardening of the spectrum (Γ changes from 2.50 ± 0.02 to 2.32 ± 0.03), as observed on a weekly time scale in the bottom panel of Figure 1. The daily flux decreases by a factor of about 3 in 4 days at the

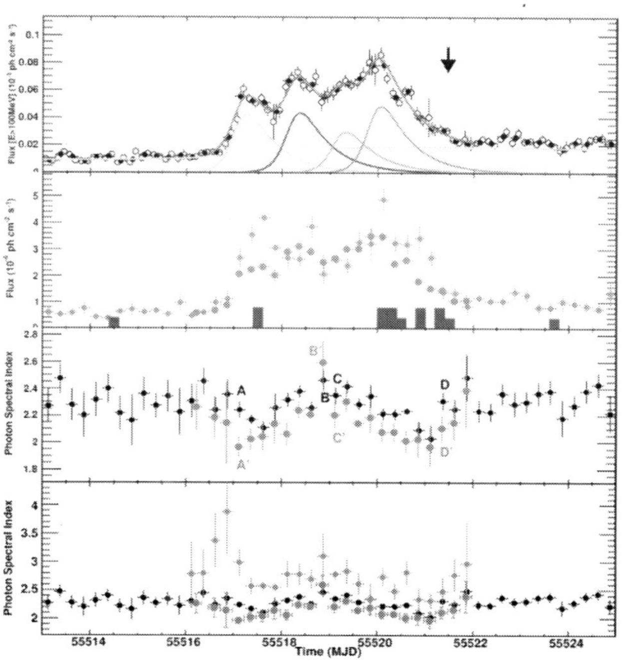

Figure 2. Top panel: light curve of the flux above 100 MeV. Open and filled symbols correspond to 3-hr and 6-hr averaged quantities respectively. The lines correspond to the results of a five-component fit (four subflares and a slowly-varying background) as described in the text, using the 6-hr data. The arrow depicts the detection time of the 31-GeV photon. Second panel: flux $F_{0.1-1\mathrm{GeV}}$, multiplied by a factor of 0.05 (red); flux $F_{1\mathrm{GeV}}$ (blue). The blue histogram represents the times of detection of $E > 10$ GeV photons. Third panel: Γ measured at E>100 MeV (black) and E=100 MeV-1 GeV (red). Bottom panel: Γ measured at E>100 MeV (black), E=100 MeV-1 GeV (red) and E>1 GeV (blue).

end of the flare. The flare is followed by a slowly decaying activity around $F_{100} = 20$. Different time periods labeled pre-flare, plateau, flare and post-flare in Figure 1 are considered in the following.

The light curves for F_{100} with 6-hr and 3-hr time binnings (focusing on the flare period) are shown in the upper panel of Figure 2. The F[$E > 1$ GeV] light curves with 6-hr and 3-hr time binnings are given in the second panel of Figure 2. The corresponding evolution of Γ is plotted in the third and bottom panels. As can be seen, the major flare lasts for about 5 days. In the F_{100} light curves, it seems to comprise four subflares, with the flux peaking during the last one. The rise time of the MJD55516.5 flare is 12 hours for a factor of 4 increase in flux (i.e., a doubling time of 6 hours). This is shorter than the previous fastest relative flux variation claimed in the GeV band for a major flare, which was from PKS 1502+106, when an increase by a factor of 3 in 12 hours (i.e., a doubling time of 7.5 hours) was found [3]. In the upper panel of Figure 2, a fit consisting of a slowly varying background and four faster temporally evolving components was performed between MJD55516.5 and MJD55522 for both the 6-hr and 3-hr light curves. Each component is assumed to be fit by a function of the form $F = 2F_0(e^{(t_0-t)/T_r} + e^{(t-t_0)/T_f})^{-1}$ [2], where T_r and T_f are the rising and falling times, respectively, and F_0 is the flux at t_0 representing approximately the flare amplitude. With the rise time T_r set equal for all subflares, and likewise for the fall time T_f, we find that $T_r = 4.5 \pm 1$hr and $T_f = 15 \pm 2$hr gives a good fit to both the

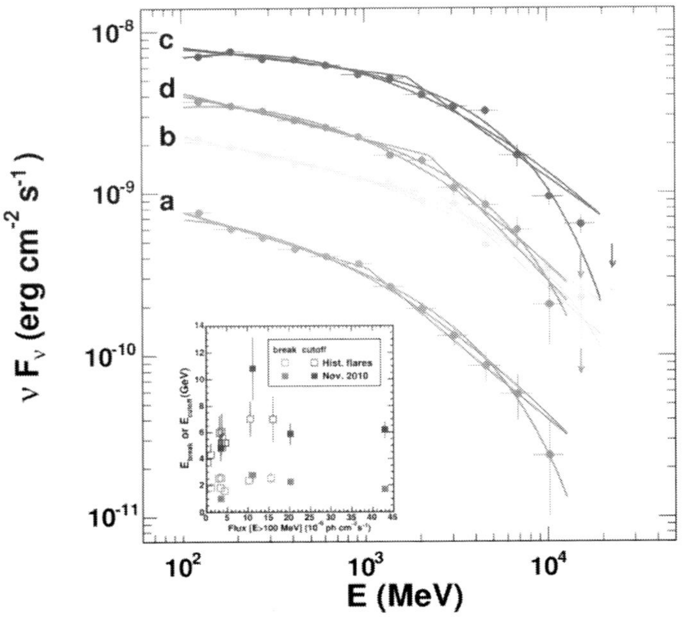

Figure 3. Spectral energy (νF_ν) distributions for four different time periods (pre-flare: a, plateau: b, flare: c, post-flare: d), along with the fitted BPL (solid), log-parabola (dashed) and power-law+exponential cutoff (PLEC) (dashed-3dotted) functions. The inset displays $E_{\rm break}$ (red) and $E_{\rm cutoff}$ (blue) as a function of flux for the different periods considered here (filled symbols) and for historical flares (open symbols).

3-hr and 6-hr light curves. The highest-energy photon collected during the MJD55516-55522 period within the energy and inclination angle-dependent 95% containment angle around the source position has $E_{max} = 31 \pm 3$ GeV and was detected at MJD55521.46. Its detection time is depicted with an arrow in the upper panel of Figure 2.

Significant differences between light curves for fluxes in the E=0.1-1 GeV ($F_{0.1-1\,{\rm GeV}}$) and E> 1 GeV ($F_{1{\rm GeV}}$) ranges are observed (see the second panel of Figure 2). The peak of the first subflare occurs approximately 15 hr after $F_{0.1-1{\rm GeV}}$ has leveled off, demonstrating clear spectral variability. This behavior is confirmed by a progressive decrease of Γ (spectral hardening) from $\Gamma \approx 2.35$ to $\Gamma \approx 2.1$ as the subflare develops beyond the 100 MeV peak (black points in the third panel of Figure 2). Overall, the $F_{1{\rm GeV}}$ light curve shows sharper structures than the $F_{0.1-1{\rm GeV}}$ light curve. A clear difference is also observed in the decaying stage, with the high-energy component starting to fade away *later* than the lower-energy component. In the lower two panels of Figure 2, the spectral indices obtained over the two restricted energy ranges are shown as well. As $F_{0.1-1\,{\rm GeV}}$ levels off in the first subflare while $F_{1\,{\rm GeV}}$ keeps rising, the $0.1 - 1$ GeV index is fairly constant, indicating that the hardening is limited to the range above 1 GeV. This hardening is confirmed by the evolution of Γ measured in the > 1 GeV range (blue points in the bottom panel in Figure 2).

Figure 3 shows the $\nu\ F_\nu$ distributions of the *Fermi*-LAT data for the four different time periods delineated in Figure 1. These distributions have been fitted with BPL (solid), log-parabola function (dashed), and PLEC (dashed-dotted) functions. For the pre-flare and plateau periods, both BPL and PLEC give fits of similar quality (as determined by a χ^2 of the $\nu\ F_\nu$

data points for the different functions) , while the log-parabola fit is notably worse. The PLEC function is preferred for the post-flare period. None of the tested functions provides a very good fit to the energy distribution in that period, which may be a result of the significant spectral evolution during the flare. The νF_ν spectra obtained over time intervals where the four subflares alternatively dominate are consistent for the first three subflares in terms of curvature, while a significantly harder spectrum is observed between MJD55520.0-55521.5. In that interval, the PLEC fit gives E_{cutoff} =8.3±1.7 GeV. A total of 10 photons with $E > 10$ GeV (out of 12 detected during the entire 5-day flare) were collected in that 1.5-day time lapse (second panel of Figure 2). The variation of E_{break} and E_{cutoff} with flux are displayed in the inset of Figure 3. As already found during the 2009 December and 2010 April flares, no strong evolution of either E_{break} or E_{cutoff} is found. E_{break} remains constant within a factor of ≈ 2 while the flux varies by a factor of ≈ 40.

4. Discussion

During its five-day outburst from 2010 November 17 to 21 (flare interval in Figure 1), 3C454.3 was the brightest GeV γ-ray source in the sky, with a flux $F_{100} = 66 \pm 2$ on 2010 November 18–19. Prior to the flaring phase, the *Fermi*-LAT light curve displays a 13-day long flux plateau preceding the major outburst. The onset of the plateau is marked by a rapid (< 1) day flux increase by a factor of ≈ 2. This feature appears to be a characteristic behavior indicating that 3C454.3 is about to flare, as noted in [6]. The spectrum hardens slightly from the pre-flare to the plateau preceding the giant flare. Spectral hardening and clustering of photons with $E > 10$ GeV is also seen in the decaying stage of the gamma-ray outburst at MJD55520.0-55521.5, which could point to the presence of radiating hadrons or $\gamma\gamma$-absorption effects. In the former case, protons require additional time to accelerate and cool while the < 1GeV flux, if due to rapidly cooling electrons, would decline more rapidly. In the latter case, the emergence of the hard component could occur if the radiating plasma becomes optically thin to $\gamma\gamma$-absorption, either due to a larger bulk Lorentz factor or increased size of the radiating plasma.

The minimum Doppler factor δ_{min} can be numerically evaluated from $\gamma\gamma$-opacity constraints. From Swift-XRT public data, the total energy flux in the 2–10 keV range is 0.8×10^{-10}ergcm^{-2}s^{-1} and the photon-number index is 1.70. Correlated X-ray and GeV variability supports the assumption that the γ-rays are made co-spatially with the X-rays. At the time that the 31-GeV photon was observed, t_{var}=0.4 d (taken as $\ln(2) \times T_f$ in the fading phase of the flare), giving $\delta_{min} = 16$, which is somewhat larger than the value of $\delta_{min} = 13$ found by [6]. The estimated comoving size of the emission region is $R' = c\, t_{var}\, \delta_{min}/(1+z) \approx 3 \times 10^{15}$ cm.

An upper limit on the optical depth $\tau_{\gamma\gamma}(E_{max}) \approx 2$ arising from the interaction of γ rays with broad-line-region (BLR) photons can be obtained by comparing the flux measured at E_{max}=31 GeV, with the flux extrapolated from lower energy. Assuming a BLR luminosity of 3×10^{45} erg s^{-1}[17] and adopting a BLR size $r_{BLR} \approx 6 \times 10^{17}$ cm ≈ 0.2 pc from reverberation mapping [14, 8], we calculate $\tau_{\gamma\gamma}(z_{em})$, where z_{em} is the distance of the emitting blob from the black hole following [18]. Assuming that the BLR clouds are distributed between 0.01 pc and r_{BLR}, the condition $\tau_{\gamma\gamma}(E_{max}) = 2$ is satisfied for $z_{em} \simeq 0.14$ pc. Therefore the emission region must have been located either close to the outer boundary of the BLR or beyond at the time of emission of the 31-GeV photon.

Approaches to the constant break-energy problem include $\gamma\gamma$ attenuation from He II line photons [16], intrinsic electron spectral breaks [1], Ly α scattering [6], and hybrid scattering [10] scenarios.

The 3-hr peak F_{100} is 85±5, corresponding to an apparent isotropic γ-ray luminosity $L_\gamma = 10^{50} L_{50}$ erg s^{-1} with $L_{50} = 2.1 \pm 0.2$. This is ≈ 4 times the luminosity of PKS 1622−297 ($L_{50} \simeq 0.5$ with the current cosmological model) during its 1995 flare [15], making this the largest γ-ray luminosity ever observed for a blazar. During the 5-day flare, $F_{100} = 43 \pm 1$

implies $L_{50} \simeq 1.0$. The black hole mass for 3C454.3 is estimated to be in the range of (0.5–4)$\times 10^9 M_\odot$[8, 11], so $L_{\rm Edd} \approx (0.6\text{–}5)\times 10^{47}$ erg s^{-1}. In order for the time-averaged flare luminosity to be lower than $L_{\rm Edd}$, $\theta_j < 2^\circ\text{–}6^\circ$, and a beaming factor $(1 - cos\theta_j)^{-1} > 200\text{–}1700$ is implied. For highly efficient γ-ray production, the absolute jet power is comparable to the disk luminosity of $L_d = 6.75\times 10^{46}$ erg s^{-1} estimated in [8].

This flaring episode marks a record among AGNs for its rate of change in apparent luminosity, $L_\gamma/\Delta t$. Using a 6-hour variability time scale, then $L_\gamma/\Delta t \simeq 10^{50} L_{50}$ erg s$^{-1}/(10^4\ t_4$ s) $\simeq 10^{46} L_{50}/t_4$ erg s^{-2}, noting that 6 hr/$(1+z) \simeq 10^4$ s. By comparison, the giant flares of PKS 2155−304 [7] reached only $< 10^{47}$ erg s^{-1}/300 s $\approx 3\times 10^{44}$ erg s^{-2} (in the TeV regime). This value now greatly exceeds $L_{\rm Edd}/(R_{\rm S}/c) \approx 1.3\times 10^{43}$ erg s^{-2} given by the ratio of the Eddington luminosity and the light crossing time across the Schwarzschild radius of a black hole, and strongly violates optically thin, Eddington-limited accretion scenarios [9], showing that such a condition is unlikely to apply to the highly asymmetric disk/jet system of a blazar.

Acknowledgments

The *Fermi*-LAT Collaboration acknowledges support from a number of agencies and institutes for both development and the operation of the LAT as well as scientific data analysis. These include NASA and DOE in the United States, CEA/Irfu and IN2P3/CNRS in France, ASI and INFN in Italy, MEXT, KEK, and JAXA in Japan, and the K. A. Wallenberg Foundation, the Swedish Research Council and the National Space Board in Sweden. Additional support from INAF in Italy and CNES in France for science analysis during the operations phase is also gratefully acknowledged.

References

[1] Abdo, A. A., et al. 2009, ApJ, 699, 817
[2] —. 2010b, ApJ, 722, 520
[3] —. 2010c, ApJ, 710, 810
[4] —. 2010d, ApJ, 710, 1271
[5] —. 2011, ApJL, 733, L26
[6] Ackermann, M., et al. 2010, ApJ, 721, 1383
[7] Aharonian, F., et al. 2007, ApJl, 664, L71
[8] Bonnoli, G., Ghisellini, G., Foschini, L., Tavecchio, F., & Ghirlanda, G. 2011, MNRAS, 410, 368
[9] Elliot, J. L., & Shapiro, S. L. 1974, ApJl, 192, L3+
[10] Finke, J. D. and Dermer, C. D. 2010, ApJl, 714, L303.
[11] Gu, M., Cao, X., & Jiang, D. R. 2001, MNRAS, 327, 1111
[12] Hartman, R. C., et al. 1999, ApJs, 123, 79
[13] Jorstad, S. G., et al. 2005, ApJ, 130, 1418
[14] Kaspi, S., Brandt, W. N., Maoz, D., Netzer, H., Schneider, D. P., & Shemmer, O. 2007, ApJ, 659, 997
[15] Mattox, J. R., Wagner et al. 1997, ApJ, 476, 692
[16] Poutanen, J. and Stern, B. 2010, ApJl, 717, L118 and this conference.
[17] Pian, E., Falomo, R., & Treves, A. 2005, MNRAS, 361, 919
[18] Reimer, A. 2007, ApJ, 665, 1023
[19] Striani, E., et al. 2010, ApJ, 718, 455
[20] Tavecchio, F., Ghisellini, G., Bonnoli, G., & Ghirlanda, G. 2010, MNRAS, 405, L94

Time lags of flaring AGN

B. Eichmann[1], **R. Schlickeiser**[2], **W. Rhode**[1]

[1] Experimentelle Physik V, Technische Universität Dortmund, 44221 Dortmund, Germany
[2] Institut für Theoretische Physik, Lehrstuhl IV: Weltraum- und Astrophysik, Ruhr-Universität Bochum, 44780 Bochum, Germany

E-mail: bjoern.eichmann@tu-dortmund.de

Abstract. In order to distinguish the leptonic from the hadronic origin of the non-thermal emission of a flaring AGN we present a semi-analytical model that describes the temporal development of the emergent particles (i.e. photons and neutrinos) based on their leptonic and hadronic origin respectively. The approach starts with the transport equation of the injected relativistic particles and takes spatial diffusion and continuous energy losses into account. On the one hand, a relativistic electron pick-up is considered, that leads to synchrotron, as well as external Compton emission and on the other hand, a relativistic proton pick-up, which results in high energy photons and neutrinos by inelastic proton-proton collisions. The paper ends up in the temporal development of the emergent photon and neutrino intensities of AGN flares in hadronic and leptonic interaction scenarios and gives useful predictions of flare durations and time lags between photons of different wavelength and high energy neutrinos.

1. Introduction

The primary hadronic or leptonic nature of the non-thermal radiation production in AGN is still an unsolved problem in high energy astroparticle physics and the detection of extragalactic high energy neutrinos could become the smoking gun evidence. Currently, several multifrequency campaigns on flaring blazars have shown a strong correlation between the optical, the X-ray and the TeV emission ([1], [2], [3]), that indicates the same origin of this non-thermal radiation. Consequently, there are four effects determining the temporal behaviour of the emission volume: (1) The injection condition of relativistic particles, its subsequent (2) spatial diffusion and (3) continuous cooling processes, as well as (4) propagation effects on the generated particles (i.e. photons and neutrinos respectively) in the emission knot.

Taking these four processes into account, we calculate the synchrotron and external Compton (EC) radiation by relativistic electrons, as well as the emission of high energy photons and neutrinos by inelastical p-p interactions. Hence, the calculations offers a systematic study of the temporal behaviour of a flaring AGN based on the leptonic respectively hadronic nature of the primary particles. Due to the mathematical complexity, this first approach neglects the radiation processes by synchrotron self-Compton (SSC) and photohadronic interactions. However, we show that these effects are also negligible at a certain range of parameters.

We consider a spherical emission knot of radius $R = 10^{15} R_{15}$ cm that is Doppler boosted towards the observer and consists of a non-relativistic plasma of electrons and protons of density $N_b = 10^{10} N_{10}\,\mathrm{cm}^{-3}$. Using such a high particle density in the emission volume, we obtain significant pion production by p-p interactions ([4]) and a diffusive buildup of synchrotron photons in the interior, as the optical depth $\tau = \sigma_T N_b R > 1$.

2. Kinetic equation of charged particles

All physical quantities are calculated in a coordinate system comoving with the radiation source.

2.1. Relativistic electrons and protons

The differential number density of relativistic electrons $n_e(r, \gamma, t)$ and protons $n_p(r, \gamma, t)$ respectively obeys the transport equation of [5] and takes spatial diffusion, continuous energy losses and a separable injection term $q_1(\gamma, t) q_2(r)$ into account:

$$\frac{\partial n_{e,p}(r,\gamma,t)}{\partial t} = \frac{1}{r^2}\frac{\partial}{\partial r}\left(r^2 D_1(\gamma)\frac{\partial n_{e,p}(r,\gamma,t)}{\partial r}\right) + \frac{\partial}{\partial \gamma}\left[|\dot{\gamma}| n_{e,p}(r,\gamma,t)\right] + q_1(\gamma,t) q_2(r) . \quad (1)$$

Here we idealized that the spatial diffusion is independent of its solide angle, as the influence of anisotropy decreases with increasing time and hence decreasing Lorentz factor.
The ultrarelativistic electrons and protons respectively are instantaneously injected at the time $t = t_0$ into the emission knot with a particle density $q_0 = 10^{-4}\, q_{-4}\ \mathrm{cm}^{-3}$ and an initial Lorentz factor $\gamma_0 = 10^6 \gamma_6$. This particle beam excites electrostatic turbulences, whereby the particles' energy is quickly changed until a plateau distribution in $p_\parallel$ is established ([6]), so that the injected particles can be considered as plateau distributed and we obtain

$$q_1(\gamma, t) = \frac{q_0}{\gamma_0 - 1}\, H[\gamma_0 - \gamma]\, \delta(t - t_0). \quad (2)$$

Additionally, electromagnetic turbulences (i.e. Alfvén waves) are excited, which result in pitch angle scattering of the injected charged particles with an energy dependent diffusion coefficient $D_1(\gamma)$. We assume an isospectral Kolmogorov turbulence with a spectral index $q = \{1, ..., 3\}$, so that

$$D_1(\gamma) = \frac{c\, l_{e,p}}{3} \gamma^\beta , \quad (3)$$

where $\beta = 2 - q$ and in the full diffusion limit the mean free path $l_{e,p}\gamma_0^\beta \ll R$. In addition the relativistic electrons cool by synchrotron and EC energy losses with the loss rate

$$|\dot{\gamma}| = D_s \gamma^2, \quad D_s = \frac{4}{3}\frac{c\sigma_T}{m_e c^2}\left(U_B + \frac{4}{3}\,\delta^2\, U^*_{Ph}\right) = 1.3 \cdot 10^{-9}\,(b^2 + l_{EC})\ \mathrm{s}^{-1} . \quad (4)$$

Here we assume a magnetic field of constant strength $B = 1\,b\,$Gauss, that is randomly distributed on scales larger than the Larmor radii of the injected electrons. Furthermore, $l_{EC} = 0.093\, L_{46}\, \delta_1^2\, \tau_{-2}\, z_{pc}^{-2}$ refers to the EC losses due to an isotropic external photon field ([7]) of energy density U^*_{Ph}, which depends on the luminosity $L_{ad} = 10^{46}\, L_{46}\,\mathrm{erg\ s}^{-1}$ of the accretion disk, the scattering optical depth $\tau_{sc} = 10^{-2}\,\tau_{-2}$ and the extension $z_{sc} = 1\, z_{pc}\,$pc of the scattering gas, as well as the Doppler factor $\delta = 10\,\delta_1$ of the emission volume.
Initial SSC energy losses are negligible compared to the synchrotron energy losses if the injection parameter is less than unity ([8]), i.e. $q_0 < 10^9\, R_{15}^{-1}\, \gamma_0^{-2}$.
The cross-section of electromagnetic interactions of protons is much smaller than that of electrons with the same Lorentz factor, as the Thomson cross-section for a proton $\sigma_{T,p} = 3 \times 10^{-7}\, \sigma_{T,e}$, where $\sigma_{T,e}$ is the Thomson cross-section for an electron. Thus, the relevant interaction processes are on the one hand interactions with cosmic photons and on the other hand interactions with cosmic matter fields. We focus on the latter and especially on the energy loss by pion production:

$$|\dot{\gamma}|_\pi^{pp} \simeq D_c\, H[\gamma - 1.30]\,\gamma , \quad D_c = 7 \times 10^{-16}\,(N_b/1\ \mathrm{cm}^{-3})\ \mathrm{s}^{-1} . \quad (5)$$

Due to the thermal radiation field (of the accretion disk or the Broad Line Region) with a photon energy $\theta = 10\theta_1$ eV the production of N pions by photohadronic interactions occure only for ultrarelativistic protons with a minimum Lorentz factor of ([9])

$$\gamma_{min}^{p\gamma} \simeq \frac{N\, m_\pi\, c^2}{2\theta}\left(1 + \frac{N\, m_\pi}{2m_p}\right) \simeq 7.3 \cdot 10^6\, N\, \theta_1^{-1} \,. \tag{6}$$

Therefore, the photohadronic energy losses are negligible for protons with a Lorentz factor $\gamma_6 < 7.3\, N\, \theta_1^{-1}$. Using the spatial boundary conditions for the particle propagation in the spherical emission volume ([5]):

(1.) a finite density at the centre of the knot; and

(2.) an exclusively outward directed particle flux at the boundary surface of the knot;

we finally obtain the differential number density of relativistic electrons

$$\begin{aligned} n_e(\gamma,t,r) =& \frac{q_0}{\gamma_0 - 1}\, H\left[\gamma_0 - \frac{\gamma}{1 - D_s\gamma(t-t_0)}\right] H\left[\frac{\gamma}{1 - D_s\gamma(t-t_0)} - 1\right] \\ &\times \frac{H[t-t_0]}{(1 - D_s\gamma(t-t_0))^2} \sum_{k=1}^{\infty} b_k\, \frac{\sin(\lambda_k r)\, E_{k,\beta}^e(\gamma,t)}{r} \,, \end{aligned} \tag{7}$$

with

$$E_{k,\beta}^e(\gamma,t) = \begin{cases} \exp\left(\frac{\lambda_k^2\, c\, l_e \gamma^{\beta-1}}{3(\beta-1)D_s}\left(1 - \frac{1}{(1-D_s\gamma(t-t_0))^{\beta-1}}\right)\right); & \text{for } \beta \neq 1, \\ (1 - D_s\,\gamma(t-t_0))^{\frac{c\, l_e \lambda_k^2}{3\, D_s}}; & \text{for } \beta = 1\,, \end{cases} \tag{8}$$

as well as the differential number density of relativistic protons

$$\begin{aligned} n_p(\gamma,t,r) =& \frac{q_0}{\gamma_0 - 1}\, H[t-t_0]\, H[\gamma_0 - \gamma\, \exp(D_c(t-t_0))] \\ &\times \exp(D_c(t-t_0)) \sum_{k=1}^{\infty} b_k\, \frac{\sin(\lambda_k r)\, E_{k,\beta}^p(\gamma,t)}{r} \,, \end{aligned} \tag{9}$$

where

$$E_{k,\beta}^p(\gamma,t) = \begin{cases} \exp\left(-\frac{\lambda_k^2\, c\, l_p}{3\,\beta D_c}\gamma^\beta\left(\exp(\beta D_c(t-t_0)) - 1\right)\right); & \text{for } \beta \neq 0, \\ \exp\left(-\lambda_k^2\, c\, l_p (t-t_0)/3\right); & \text{for } \beta = 0\,. \end{cases} \tag{10}$$

Here the eigenvalues are

$$\lambda_k \simeq \frac{k\pi}{R}, \quad k = 1, 2, 3, ..., k_{max}, \tag{11}$$

and the expansion coefficient is calculated by

$$b_k \simeq \frac{2}{R}\int_0^R dr\, q_2(r) r \sin(\lambda_k r)\,. \tag{12}$$

Based on mathematical convenience, we consider a homogeneous particle injection into the whole emission volume ($q_2(r) = 1$), since we have no a priori knowledge about the true location of the relativistic particle source. However, the solution for more complex spatial injection assumptions are simply obtained by changing the expansion coefficient (12).

3. Leptonic emission

In this section we consider the radiation resulting from the relativistic electrons of Eq. (7) and neglect any further relativistic electrons in the emission region.

3.1. Synchrotron radiation

First, we calculate the synchrotron photon emission generated by the relativistic electron density distribution (7). The broad continuum of the synchrotron power, that is described by an integral over the modified Bessel function of the second kind $K_{5/3}(x)$ can be approximated in vacuum ([10], [11]) by

$$P(\nu,\gamma) \simeq P_0 \left(\frac{\nu}{\nu_s \gamma^2}\right)^{\frac{1}{3}} \exp(-\frac{\nu}{\nu_s \gamma^2}), \qquad (13)$$

with $P_0 = 2.647 \times 10^{-10}\,\mathrm{eV\,s^{-1}\,Hz^{-1}}$ and the characteristic frequency $\nu_s \gamma^2 = 3eB\gamma^2/(4\pi m_e c)$. Hence, the spontaneous synchrotron emission coefficient $j_s(r,\nu,t)$ that results from an isotropic relativistic electron distribution $n_e(r,\gamma,t)$ yields

$$j_s(r,\nu,t) = \frac{P_0}{4\pi}\left(\frac{\nu}{\nu_s}\right)^{\frac{1}{3}} \int_1^\infty d\gamma\, \gamma^{-\frac{2}{3}} \exp(-\frac{\nu}{\nu_s \gamma^2})\, n_e(r,\gamma,t)\,. \qquad (14)$$

For an optically thin source, the synchrotron intensity on the surface ($r = R$) at a time t and an energy $\epsilon = h\nu$ is given by ([12])

$$I_{Sy}(R,\epsilon,t) = \frac{1}{2R}\int_0^R dr'\, r' \int_{R-r'}^{R+r'} ds\, \frac{\rho_{Sy}(r',\epsilon,t-\frac{s}{c})}{s}, \qquad (15)$$

with the omnidirectional photon production rate

$$\rho_{Sy}(r',\epsilon,t-\frac{s}{c}) = 4\pi \exp\left(-\frac{g(x)\,s}{R}\right) j_s\left(r',\epsilon,t-\frac{s}{c}\right), \qquad (16)$$

where the exponential term expresses the survival propability of traveling the distance $s = \sqrt{R^2 + r'^2 - 2Rr'\mu'}$ from the spherical coordinates r', $\mu' = \cos\phi'$ to the boundary surface of the knot ([5]). Due to the synchrotron cooling of the relativistic electrons the amplitude of the emergent intensity, as well as the total flare duration and the time of maximal synchrotron flare emission increase with decreasing photon energy ϵ (see Fig 1). In the full diffusion limit ($R \gg \gamma_0 l_e$) the spatial diffusion of the relativistic electrons has no influence on the temporal behaviour of the emergent synchrotron intensity.

3.2. External Compton radiation

Secondly, we consider the inverse Compton scattering of low energy photons by the relativistic electron density distribution (7). Since we assume a low number density of injected particles ($q_0 < 10^9\, R_{15}^{-1}\, \gamma_0^{-2}$) to keep the SSC losses negligible, the seed photons for the EC scattering result from an external photon field like the accretion disk of the AGN. Thus we adopt an isotropic and temporal constant number density of thermal photons

$$n_{Ph}(\epsilon) = n_0 \left(\exp\left(\frac{\epsilon}{\theta}\right) - 1\right)^{-1} \qquad (17)$$

with a disk temperature $\theta = 10\,\theta_1\,\mathrm{eV}$ and $n_0 = 34\, L_{46}\, \tau_{-2}\, z_{pc}^{-2}\,\mathrm{cm^{-3}}$. Using the δ-function approximation for inverse Compton scattering ([13]) in the limit $\gamma \gg 1 \gg \epsilon/(m_e c^2)$ the

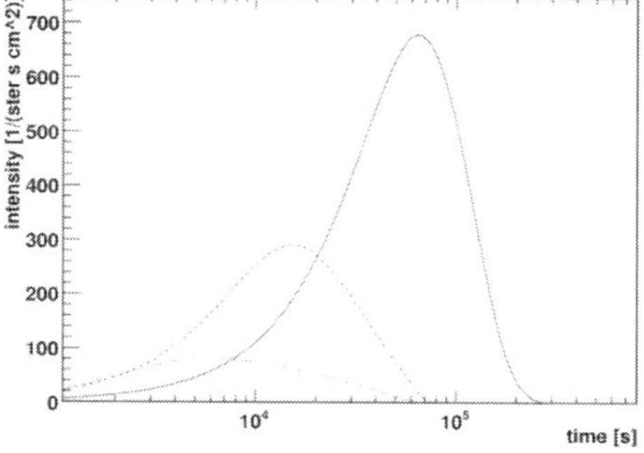

Figure 1. Temporal development of the emergent synchrotron intensity at an energy $\epsilon = 1\,\mathrm{eV}$ (solid line), $\epsilon = 100\,\mathrm{eV}$ (dashed line) and $\epsilon = 1\,\mathrm{keV}$ (dotted line). Furthermore, we used $\beta = 0$, $l_e = 10^{10}\,\mathrm{cm}$, $R_{15} = 1$, $\gamma_6 = 1$, $q_{-4} = 1$ and $b^2 + l_{EC} = 1$.

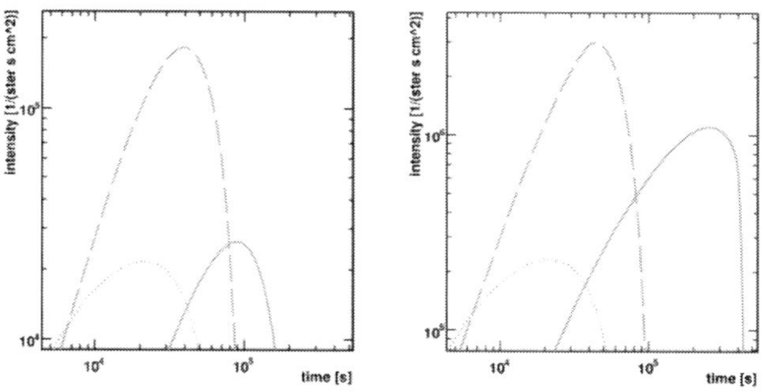

Figure 2. Temporal development of the emergent EC intensity at an energy $E_\gamma = 1\,\mathrm{GeV}$ (solid line), $E_\gamma = 10\,\mathrm{GeV}$ (dashed line) and $E_\gamma = 100\,\mathrm{GeV}$ (dotted line). Furthermore, we used an accretion disk temperature of $\theta_1 = 1$(left panel) and $\theta_1 = 10$ respectively, as well as $n_0 = 1\,\mathrm{cm}^{-3}$, $\beta = 0$, $l_e = 10^{10}\,\mathrm{cm}$, $R_{15} = 1$, $\gamma_6 = 1$, $q_{-4} = 1$ and $b^2 + l_{EC} = 1$.

spontaneous EC emission coefficient yields

$$j_c(r, E_\gamma, t) = \frac{E_\gamma c \sigma_T}{12\pi^2\, m_e c^2} \int_0^\infty d\epsilon\, n_{ph}(\epsilon) \int_0^{m_e c^2/\epsilon} d\gamma\, n_e(r, \gamma, t) \delta\left(E_\gamma - \gamma^2 \epsilon\right) . \quad (18)$$

Analogous to the previous section the emergent intensity of EC photons at a time t and an energy E_γ is calculated by

$$I_{EC}(R, E_\gamma, t) = \frac{1}{2\,R} \int_0^R dr'\, r' \int_{R-r'}^{R+r'} ds\, \frac{\rho_{EC}(r', E_\gamma, t - \frac{s}{c})}{s} , \quad (19)$$

with the omnidirectional photon production rate

$$\rho_{EC}(r', E_\gamma, t - \frac{s}{c}) = 4\pi \exp\left(-s/R\right)\, j_c\left(r', E_\gamma, t - \frac{s}{c}\right) , \quad (20)$$

where the exponential term expresses the propability of escaping the emission volume ([5]). Fig. 2 shows that the temporal development of the emergent EC intensity is similar to the case

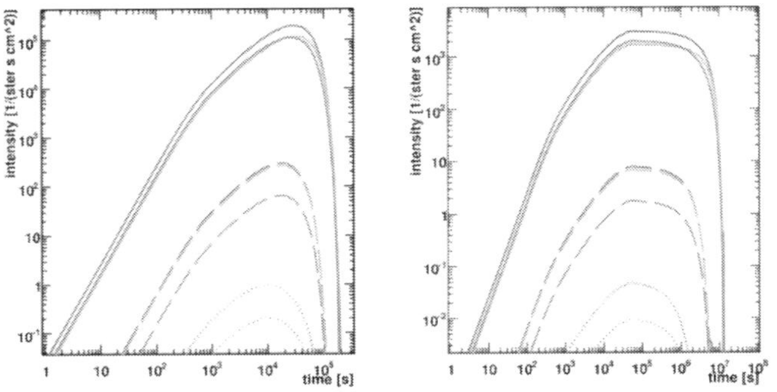

Figure 3. Temporal development of the emergent intensity of gammas (red), muon-neutrinos (magenta) and electron-neutrinos (orange) by inelastic p-p interactions with a background particle density of $N_{10} = 10$ (left panel) and $N_{10} = 0.1$ (right panel) respectively. We considered the emission at an energy $E_l = 0.1\,\mathrm{TeV}$ (solid line), $E_l = 1\,\mathrm{TeV}$ (dashed line) and $E_l = 100\,\mathrm{TeV}$ (dotted line) and used $\beta = 0$, $l_p = 10^{10}\,\mathrm{cm}$, $R_{15} = 1$, $\gamma_6 = 1$ and $q_{-4} = 1$.

of synchrotron radiation and the time of maximal flare emission, as well as the flare duration increase with decreasing gamma energy E_γ. The spatial diffusion has no influence on the light-curves in the full diffusion limit. With increasing disk temperature the amplitude of the light curves increase and the time of maximal flare emission shifts to later times (especially at low gamma energies E_γ).

4. Hadronic emission

In this section we neglect the electrons, as well as the photon field and consider only the relativistic protons of Eq. (9) and their interactions with the non-relativistic protons of the background plasma. According to Eq. (6) the pion production by photohadronic interactions is negligible. However, using the minimal Lorentz factor $\gamma_{min}^{e^+e^-} = 5.1 \cdot 10^4\, \theta_1^{-1}$ for photo-pair production ([9]), the emission knot has to be considered at a distance

$$z_{pc} > 1.6 \cdot 10^{-4}\, \delta_1\, L_{46}^{1/2}\, \tau_{-2}^{1/2}\, \theta_1^{-1/2}\, N_{10}^{-1/2}$$

in order to neglect (in comparison to the inelastic proton-proton losses) the photo-pair production by the scattered radiation field n_{ph} of the accretion disk.

4.1. Gamma and neutrino emission by inelastic p-p interactions

The relativistic protons with an energy $E_p = \gamma\, m_p c^2$ collide inelastically with the non-relativistic protons of the background plasma of constant density N_b and generate neutral and charged ultrarelativistic π-mesons which decay quasi instantaneously into gamma-rays ($l = \gamma$) and neutrinos ($l = \{\nu_e, \nu_\mu\}$) respectively with an energy E_l. Using the energy spectra $F(E_l/E_p, E_p)$ of secondary particles ([14]) its omnidirectional production rate is calculated by

$$\rho_l^{pp}(r, E_l, t) = 4\pi\, \exp(-s/R)\, c\, N_b \int_{E_l}^{\infty} \frac{dE_p}{E_p}\, \sigma_\pi^{pp}(E_p)\, n(r, E_p, t)\, F(E_l/E_p, E_p)\,, \tag{21}$$

where high energy photons ($x \geq 1$) and neutrinos have the same escape propability $\exp(-s/R)$.

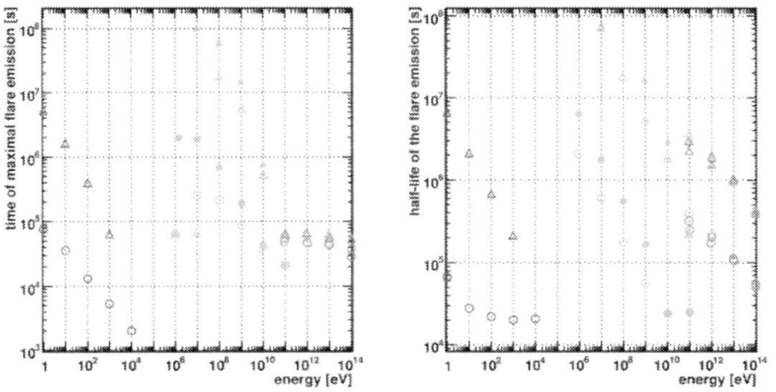

Figure 4. Characteristical temporal features of the emission of synchrotron photons (blue symbols), EC gammas (green symbols), as well as gammas (red symbols), muon-neutrinos (magenta symbols) and electron-neutrinos (orange symbols) by inelastic proton-proton interactions. In case of the leptonic scenarios we used $b^2 + l_{EC} = 25$ (circles) and $b^2 + l_{EC} = 0.25$ (triangles) respectively, as well as the disk temperature $\theta_1 = 5$ (non-filled symbols) and $\theta_1 = 50$ (filled symbols) respectively for the EC process. For the hadronic scenario the cooling parameter is $N_{10} = 1$ (circles) and $N_{10} = 0.1$ (triangles) respectively. Furthermore, we used $\beta = 0$, $l_{p,e} = 10^{10}$ cm, $R_{15} = 1$, $\gamma_6 = 1$, $q_{-4} = 1$.

Consequently, the emergent intensity of gamma-rays and neutrinos by inelastic p-p collisions yields

$$I_l^{pp}(R, E_l, t) = \frac{1}{2\,R} \int_0^R dr'\, r' \int_{R-r'}^{R+r'} ds\, \frac{\rho_l^{pp}(r', E_l, t - \frac{s}{c})}{s}\,. \tag{22}$$

Due to the slower cooling mechanism (5) of the relativistic protons, the time of maximal emission of gammas and neutrinos hardly increases with decreasing energy E_l (see Fig. 3). However, the flare duration considerably increases with decreasing E_l, as well as with a decreasing number density N_b of background protons. Except for the amplitudes of the emergent intensities there is only a marginal difference in the emission of gammas and neutrinos. Like in the case of leptonic radiation processes, the spatial diffusion of the relativistic protons has in the full diffusion limit ($R \gg \gamma_0 l_p$) no influence on the temporal flare development.

5. Time lags and conclusions

The amplitude of the emergent intensity by the leptonic respectively hadronic scenarios strongly depends on several unknown parameters, like the spatial distribution of seed electrons and protons respectively, as well as the properties of the emission volume (R, N_b, B) or the external photon field (n_0, θ). However, the temporal development of the emergent intensity is mainly defined by the parameters B, θ and N_b of the particular cooling and emission process. The temporal flare behaviour is not significantly changed by the injection assumption, but a finite injection duration τ_{inj} enlarges the flare duration by the time τ_{inj}. Based on the leptonic respectively hadronic scenario Fig. 4 shows the half-life of the emergent intensity and the time of maximal flare emission at an energy range from 1 eV up to 100 TeV. The flare duration, as well as the time of maximal emission for each radiation scenario increases with decreasing photon energy and/ or decreasing cooling parameters B and N_b respectively. The EC emission gets increasingly delayed (especially at low E_γ) with increasing disk temperature θ, however at the smallest E_γ the time of maximal flare emission stays at the beginning of the flare. Furthermore, there is a plain difference between the hadronic and the leptonic scenario, since the time of maximal emission of gammas and neutrinos by pion production hardly differs by changing the cooling parameter N_b or the energy E_l. The minimal total flare duration is limited by the radius of the emission knot due to the light crossing time R/c. Another feature of Fig 4 is the small time lag between the neutrino and the gamma emission by p-p interactions. The time of maximal flare emission, as well as the half-life of the flare decrease with increasing energy E_l, however, the decay is steeper in case of neutrinos, which results from the bigger difference in the maximal value of its energy spectra F_l at different proton energies (see Fig. 7-9 in [14]). Consequently, the model yields significant time lags between the time of leptonic and hadronic emission maxima, as well as different flare durations of gammas and neutrinos resulting from a hadronic emission scenario.

References

[1] Aharonian, F. A. et al., 2009, A&A, **502**, 749A
[2] Donnarumma, I. et al., 2009, ApJ, **707**, 1115D
[3] Vercellone, S. et al., 2010, ApJ, **712**, 405
[4] Pohl, M., Schlickeiser, R., 2000, A&A, **355**, 829P
[5] Eichmann, B., Schlickeiser, R. & Rhode, W., 2010, A&A, **511**, A26
[6] Pohl, M., Lerche, I. & Schlickeiser, R., 2001, AIPC, **587**, 474P
[7] Dermer, C. D., Schlickeiser, R., 1994, ApJS, **90**, 945D
[8] Schlickeiser, R., Böttcher, M. & Menzler, U., 2010, A&A, **519**, A9
[9] Mannheim, K., Schlickeiser, R., 1994, A&A, **286**, 983M
[10] Crusius, A., Schlickeiser, R., 1986, A&A, **164**, L16
[11] Crusius, A., Schlickeiser, R., 1988, A&A, **196**, 327C
[12] Gould, R. J., 1979, A&A, **76**, 306
[13] Dermer, C. D., Schlickeiser, R., 1993, ApJ, **416**, 458D
[14] Kelner, S. R., Aharonian, F. A. & Bugayov, V. V., 2006, PhRvD, **74**, 3

International Workshop on Beamed and Unbeamed Gamma-Rays from Galaxies IOP Publishing
Journal of Physics: Conference Series **355** (2012) 012032 doi:10.1088/1742-6596/355/1/012032

γ-ray emission region located in the parsec scale jet of OJ287

I. Agudo[1,2], S. G. Jorstad[2,3], A. P. Marscher[2], V. M. Larionov[3,4], J. L. Gómez[1], A. Lähteenmäki[5], M. Gurwell[6], P. S. Smith[7], H. Wiesemeyer[8,9], C. Thum[10] and J. Heidt[11]

[1]Instituto de Astrofísica de Andalucía, CSIC, Granada, Spain
[2]Institute for Astrophysical Research, Boston University,Boston, USA
[3]Astronomical Institute, St. Petersburg State University, St. Petersburg, Russia
[4]Isaac Newton Institute of Chile, St. Petersburg Branch, St. Petersburg, Russia
[5]Aalto University Metsähovi Radio Observatory, Kylmälä, Finland
[6]Harvard–Smithsonian Center for Astrophysics, Cambridge, USA
[7]Steward Observatory, University of Arizona, Tucson, USA
[8]Instituto de Radio Astronomía Milimétrica, Granada, Spain
[9]Max-Plack-Institut für Radioastronomie, Bonn, Germany
[10]Institut de Radio Astronomie Millimétrique, St. Martin d'Hères, France
[11]ZAH, Landessternwarte Heidelberg, Königstuhl, Heidelberg, Germany

E-mail: iagudo@iaa.es

Abstract. We report on the location of the γ-ray emission region in flares of the BL Lacertae object OJ287 at > 14 pc from the central supermassive black hole. We employ data from multi-spectral range (total flux and linear polarization) monitoring programs combined with sequences of ultra-high-resolution 7 mm VLBA images. The correlation between the brightest γ-ray and mm flares is found to be statistically significant. The two γ-ray peaks, detected by *F*ermi-LAT, that we report here happened at the rising phase of two exceptionally bright mm flares accompanied by sharp linear polarization peaks. The VLBA images show that these mm flares in total flux and polarization degree occurred in a jet region at > 14 pc from the innermost jet region. The time coincidence of the brighter γ-ray flare and its corresponding mm linear polarization peak evidences that both the γ-ray and mm outbursts occur > 14 pc from the central black hole. We find two sharp optical flares occurring at the peak times of the two reported γ-ray flares. This is interpreted as the γ-ray flares being produced by synchrotron self-Compton scattering of optical photons from the flares triggered by the interaction of moving knots with a stationary conical shock in the jet.

1. Introduction

The hypothesis of the location of the γ–ray flare emission in blazars close to the mm jet cores (>> 1 pc downstream the supermassive black hole) started to have support from the observations made by the Energetic Gamma Ray Experiment Telescope (EGRET), aboard the Compton Gamma Ray Observatory. The EGRET data suggested the connection of γ–ray flares with their mm wavelength counterpart [13], the connection of γ–ray flares with superluminal ejection of blobs in the jets of some blazars [10], and the relation of the flux density of the mm wavelength VLBI core with the γ–ray flux [9].

The unprecedented sensitivity of the Large Area Telescope (LAT) aboard the *Fermi* Gamma Ray Space Telescope provides detailed γ–ray light curves of tens of bright blazars that, together with those from multi–spectral–range observing programs, has allowed to identify abrupt γ–ray flares with those at other spectral ranges and to provide robust evidence of the location of the γ–ray flare emitting region $>>$ 1 pc from the central black hole; e.g. the cases of AO 0235+164 [6], PKS 1510−089 [12], 3C 279 [1], and 3C 454.3 [16]. However, the location of the γ–ray emission –which is a critical issue to identify the γ-ray emission mechanism– is still under debate for some blazars, as was raised in this conference, see Poutanen et al. (these proceedings) and Becerra et al. (these proceedings).

The technique used by [16], [12] and [6] is based on the use of ultra–high–resolution VLBI monitoring to resolve the innermost jet regions and to look for the correlation of their polarized emission with shorter wavelength observations along the spectrum.

In this paper, where we extend on previous work [5], we use this technique to investigate the location of the flaring γ–ray emission in the BL Lacertae (BL Lac) object OJ287 (z = 0.306), by applying, for the first time in this kind of studies, a Monte–Carlo–based method to evaluate the significance of the correlation with the *Fermi*–LAT γ–ray light curve.

2. Observations

The photo-polarimetric observations presented here include 7 mm VLBA maps at 0.15 mas resolution (Fig. 1), 3 mm observations with the IRAM 30 m Telescope, and optical observations (Figs. 1 and 2) that include data from the Calar Alto (2.2 m Telescope), Steward (2.3 and 1.54 m telescopes), Lowell (1.83 m Perkins Telescope), St. Petersburg State University (0.4 m Telescope), and Crimean Astrophysical (0.7 m Telescope) observatories, as well as photo-polarimetric data from [18]. Our total flux light curves (Fig. 1) come from the *Fermi*-LAT γ-ray (0.1–200 GeV) and *Swift* X-ray (0.3–10 keV) archives, and from the one by the Yale University SMARTS program. To these, we add data from the SMA at 1.3 mm and 850 μm, from the IRAM 30 m Telescope at 1.3 mm, and from the Metsähovi Radio Observatory 14 m Telescope at 8 mm.

The reduction of our data followed the procedures explained in [5]. In particular we followed [11] for the VLBA data, [12] for the optical polarimetric data, [3, 4] for the IRAM data, [12] for the *Swift* data, and [16] for the *Fermi* LAT data. However, for the γ-ray light curve in Fig. 1, which covers a longer time range than the one presented in our previous work on OJ287 [5], release v9r18p6 of the *F*ermi Science Tools was used, instead of previous release (v9r15p2). Release v9r18p6 produced significantly higher SNR results, and hence larger number of detections. This allowed us to use here a time binning of 5 days during quiescent periods, and even 1 day or better around the two γ-ray peaks discussed below.

3. Results

3.1. 1 mm and 7 mm flares located at the core

We modeled the brightness distribution of the 7mm VLBA images of the source with a small number of circular Gaussian components. The two brightest ones are Q0, that we identify here as the *core* (i.e., the innermost jet region observable in our 7 mm images.) and Q1, see Fig. 1. The identification of C0 as the innermost jet feature is justified by the decrease of the emission to the west of C1, and by the detection of superluminal motion of features M1, M2 (to the west-southwest of C1), and M3. The latter crossed C1 in 2010 Oct. (preliminary speed $\sim 10\,c$[1]) and represents the clearer case of motion outwards C0 (Fig. 1). This is also evident from this movie: `https://w3.iaa.es/~iagudo/research/OJ287/OJ287End2010.mov` , that starts just before the time of ejection of M3.

[1] We assume along this paper the standard cosmology with H_0=71 km s^{-1} Mpc^{-1}, $\Omega_M = 0.27$, and $\Omega_\Lambda = 0.73$. Under this assumption the angular scale is 4.48 pc/mas, and 1 mas/yr proper motion translates into a superluminal apparent speed of 19 c.

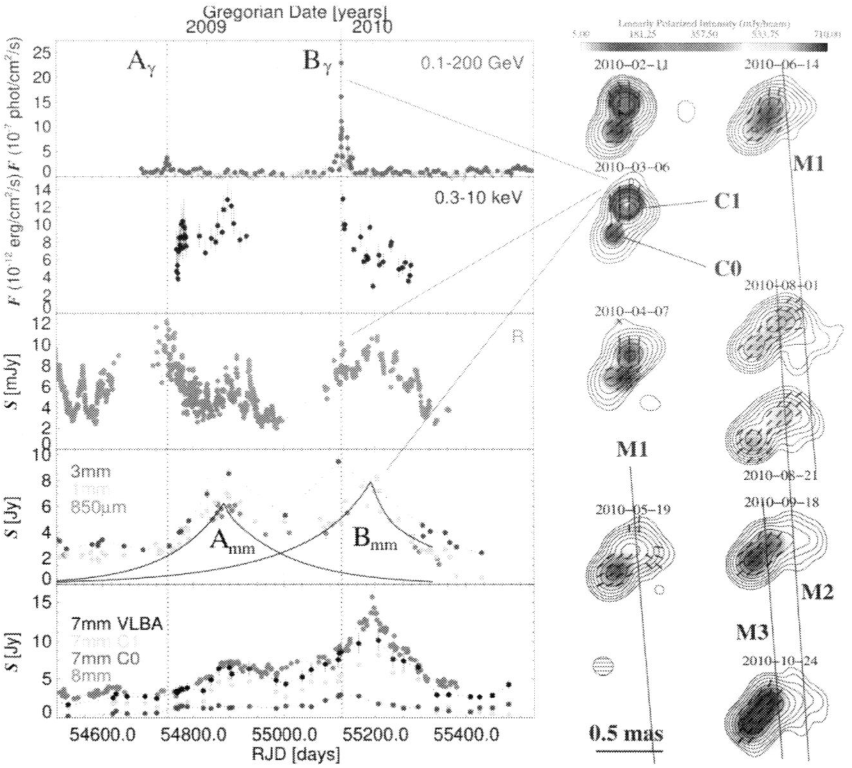

Figure 1. *Left*: Light curves of OJ287 from γ-rays (top), to mm wavelengths (bottom). RJD = Julian Date -2400000.0. *Right*: Sequence of 7 mm VLBA images from February to October 2010. Images were convolved with a FWHM = 0.15 mas circular Gaussian beam. Contour levels start at 0.2 of the peak total intensity (6.32 Jy/beam) and increment by factors of two. The color scale represents linear polarization intensity, and black segments symbolize the distribution of position angles of the linear polarization. A movie of the total intensity time evolution corresponding to the same time span can be found here: `https://w3.iaa.es/~iagudo/research/OJ287/OJ287FebOct2010.mov`.

The most prominent 1 mm flares ever reported in OJ287 (A_{mm} and B_{mm}, see Fig. 1) took place in C1, as shown by the match of the 7 mm light curve of C1 and those at other mm wavelengths. From the half opening angle of the inner 7 mm jet in OJ287 – 1°.9–4°.1 [11, 17] –, and the mean distance of C1 and C0 (0.23 ± 0.01 mas), we estimate that C1 is located > 14 pc downstream from C0 (i.e. the core).

3.2. γ–ray flaring activity at the 7 mm core

The two A_{mm} and B_{mm} flares in OJ287 start to rise at the time of two sharper γ–ray flares, as interpreted by [13] for a number of sources from the analysis of previous EGRET and Metsähovi data. These two flares, on 2008 October 4 and 2009 October 24, display 0.1–200 GeV photon fluxes a factor ~ 2 and ~ 5 larger than the quiescent γ–ray level of the source, respectively.

The discrete correlation function (DCF) between the γ–ray to 1 mm light curves (Fig. 3), shows a prominent peak at a time lag ~ -80 days (γ–rays leading). Our Monte-Carlo simulations to test the statistical significance of such correlation peak show that it is significant at 99.7 % confidence, hence confirming the correlation between B_γ and B_{mm} –the two most luminous γ–ray

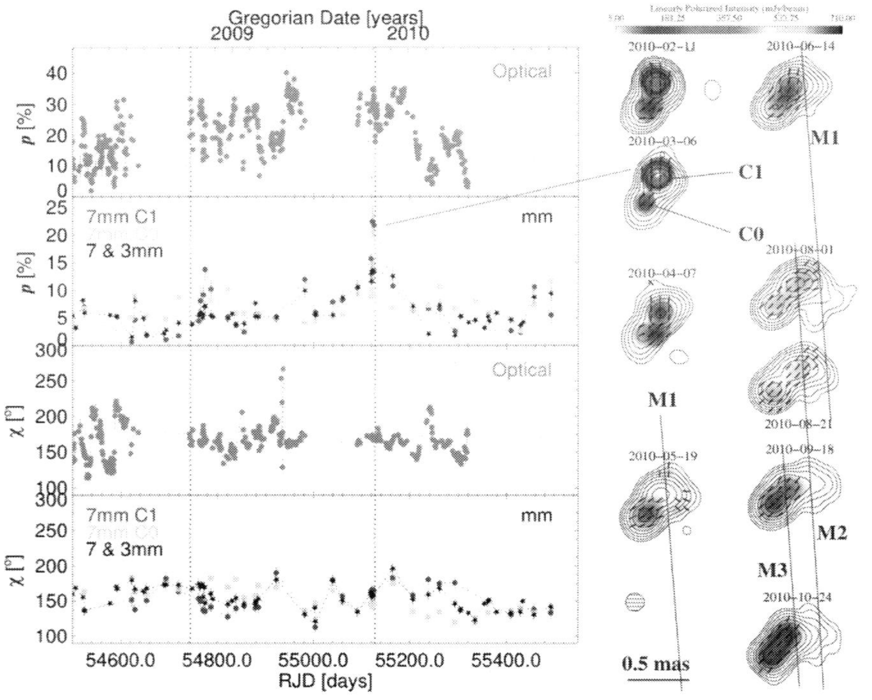

Figure 2. *L*eft: Optical and mm-wave linear polarization of OJ287 as a function of time. *R*ight: 7 mm VLBA sequence of images as in Fig. 1.

and 1 mm flares in our data. Finally, the match between the C1 7 mm light curve and the 1 mm light curve locates the B_γ flare at the position of the C1 jet feature.

The optical light curve shows two sharp peaks at the peak time of A_γ and B_γ. However, the poorer X-ray time sampling does not allow us to claim for connection between the γ-ray and X-ray variability.

3.3. Related linear polarization variability

Despite the similarity of C0 and C1 in their χ evolution, these two emission regions differ in their p behavior (Fig. 2). Whereas C0 never reaches $p \gtrsim 10\,\%$, C1 exhibits the two largest p peaks (of 14 % around 2008–11–04, and 22 % on ∼2009–10–16) ever observed in OJ287 at 7 mm. These two violent p increases of the core are contemporaneous with A_γ and B_γ. The mm p peak in the end of 2008 seems to be delayed by ~ 15 days with regard to A_γ, while the second ($p \gtrsim 22\,\%$) peak is consistent with being simultaneous with the extreme B_γ γ–ray peak. We take this coincidence of these two exceptional events as evidence that both of them occurred at the same place: the C1 jet feature at > 14 pc from the central black hole.

The optical polarization fraction, which varies faster and stronger than that at mm wavelengths –consistent with the frequency dependence turbulence model of [15]–, also evidences sharp p peaks at essentially the same time of the peak times of A_γ and B_γ. However, similar optical polarization fractions also happen at other times.

In contrast, the optical and mm linear polarization angle is surprisingly stable at $\chi \approx 160°$–$170°$ (i.e. similar to the position angle of the inner jet) both during the time around A_γ and B_γ and in the overall χ evolution curves. This stability is only greatly altered in the optical by sporadic short-term rotations of χ by up to 180° [18], consistent with turbulent plasma behavior.

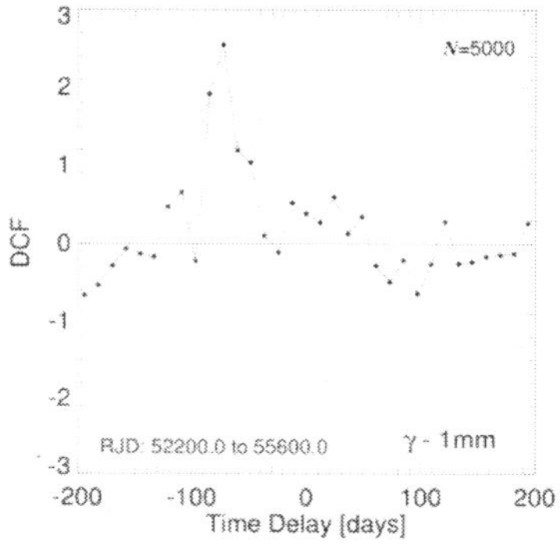

Figure 3. DCF (as defined in [5]) between the γ-ray and 1 mm light curves of OJ287. Grey dotted curves of DCF values denote 99.7% confidence limits for correlation (or anti-correlation at negative DCF) against stochastic variability as derived from Monte-Carlo simulations (see [5] for details about the method).

4. Discussion and Conclusions

We report on a number of time coincidences in the multi–spectral–range and polarimetric behavior of OJ287. These include: *i)* two sharp 0.1–200 GeV flares in October 2008 and 2009 (A_γ and B_γ, respectively), *ii)* the rising of the two most luminous 1 mm flares detected in the source (A_{mm} and B_{mm}), that our 7 mm VLBA monitoring locates unambiguously at the position of C1, *iii)* two sharp optical flares, and *iv)* two sharp p increases at the location of C1 –that are not accompanied by significant χ variability around a mean value $\bar{\chi}_{mm} \approx 160°$–$170°$. Also, the extremely high linear polarization fraction of C1 during B_γ supports even stronger a location of the γ–ray events in C1.

The γ-ray inverse Compton (IC) emission is produced from either the synchrotron self-Compton (SSC) mechanism or IC scattering of infrared radiation from the dusty torus (IC/dust). An SSC model is certainly possible given the relatively small γ-ray to synchrotron luminosity ratio; ≈ 2 in OJ287, see [5]. From the IC/dust mechanism it is expected that the optical and γ-ray emission should vary together [16]. This is the case during the γ-ray flares, however IR emission from the dusty torus has not been detected so far in BL Lacs like OJ287, which favors the SSC mechanism.

We propose a scenario where optical and γ-ray flares are produced by particle acceleration in C1 (interpreted as a conical standing shock [8, 2]) when a moving jet feature interacts with C1, which is located well downstream the jet's acceleration and collimation zone (ACZ [14]), and at > 14 pc from C0 (i.e. the core). The moving feature should be a turbulent perturbation with higher relativistic electron density and magnetic field than in the jet flow. The turbulent character is required to explain the weakness of the observed millimeter polarization outside the high polarization peaks. This scenario was treated by [7], whose model reproduces the polarization properties observed in OJ287 in the reported flares, i.e. large polarization peaks occurring at the time of interaction of turbulent blobs with a conical shock, and χ parallel to the jet axis (see [5]).

The mm and optical flares start with the increase of the magnetic field and electron energies of the moving feature after the interaction with the standing conical shock. The mm outburst continues as the lower-energy electrons fill the shocked region. The optical and γ-ray flares, produced by higher-energy electrons, are dumped near the conical shock. However, if synchrotron losses were the only mechanism producing the optical emission dumping in flares, the optical and γ-ray emission would be constant after the shocked jet feature fills the entire layer after the shock (at least before the entire jet feature crosses the conical shock). If the SSC mechanism dominates, the radiative losses are enhanced until the γ–ray losses start to be efficient, when electrons emitting optical synchrotron loose most of their energy and hence the optical flare is dumped. The time-scale for this dumping is $(1+z)\delta^{-1} \sim 20(\delta/19)^{-1}(a/0.054\,\mathrm{mas})$ days (where a is the angular cross section of the jet and the Doppler factor (δ) is the one reported

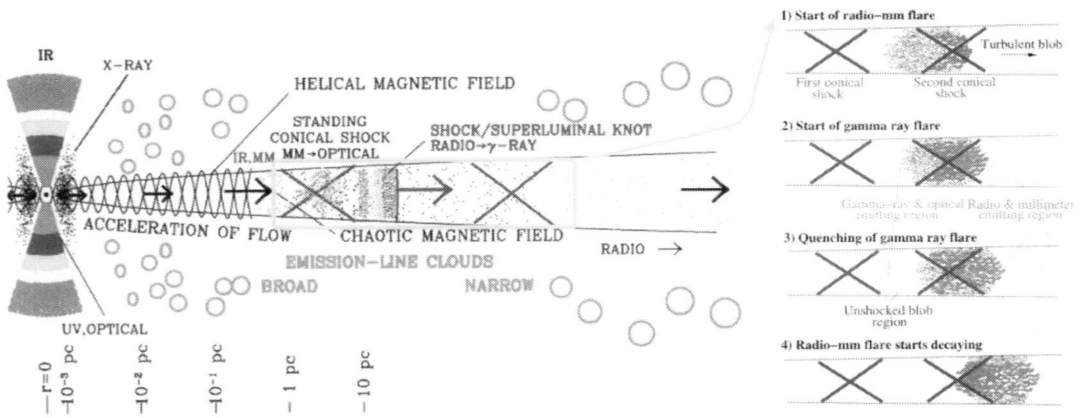

Figure 4. Illustration of the proposed scenario for the multi-spectral-range behavior of OJ287.

by [11]), which is similar to the B_γ γ–ray time-scale; $\sim$ 3-4 weeks).

Acknowledgments

This research was partly funded by NASA grants NNX08AJ64G, NNX08AU02G, NNX08AV61G, NNX08AV65G, NNX08AW56S and NNX09AU10G, NSF grant AST-0907893, NRAO award GSSP07-0009; RFBR grant 09-02-00092; MICIIN grants AYA2007-67627-C03-03 and AYA2010-14844, and CEIC grant P09-FQM-4784. The VLBA is an instrument of the NRAO, a facility of the NSF operated under cooperative agreement by AUI. The PRISM camera at Lowell Observatory was developed by Janes et al. The Calar Alto Observatory is jointly operated by MPIA and IAA-CSIC. The IRAM 30 m Telescope is supported by INSU/CNRS, MPG, and IGN. The SMA is a joint project between the SAO and the Academia Sinica.

References

[1] Abdo, A. A., et al. 2010, *Nature*, **463**, 919
[2] Agudo, I., et al. 2001, *ApJ*, **549**, L183
[3] Agudo, I., et al. 2006, *A&A*, **456**, 117
[4] Agudo, I., Thum, C., Wiesemeyer, H., & Krichbaum, T. P. 2010, *ApJS*, **189**, 1
[5] Agudo, I., et al. 2011a, *ApJ*, **726**, L13
[6] Agudo, I., et al. 2011b, *ApJ*, **735**, L10
[7] Cawthorne, T. V. 2006, *MNRAS*, **367**, 851
[8] Gómez et al. 1997, *ApJ*, **482**, L33
[9] Jorstad, S. G., et al. 2001a, *ApJS*, **134**, 181
[10] Jorstad, S. G., et al. 2001b, *ApJ*, **556**, 738
[11] Jorstad, S. G., et al. 2005, *AJ*, **130**, 1418
[12] Jorstad, S. G., et al. 2010, *ApJ*, **715**, 362
[13] Lähteenmäki, A., & Valtaoja, E. 2003, *ApJ*, **590**, 95
[14] Marscher, A. P., et al, *Nature*, **452**, 966
[15] Marscher, A. P., & Jorstad, S. G. 2010, *Proc. Fermi meets Jansky - AGN at Radio and Gamma-Rays*, p 171
[16] Marscher, A. P., et al. 2010, *ApJ*, **710**, L126
[17] Pushkarev, A. B., Kovalev, Y. Y., Lister, M. L. & Savolainen, T. 2009, *A&A*, **507**, L33
[18] Villforth, C., et al. 2010, *MNRAS*, **402**, 2087

International Workshop on Beamed and Unbeamed Gamma-Rays from Galaxies IOP Publishing
Journal of Physics: Conference Series **355** (2012) 012033 doi:10.1088/1742-6596/355/1/012033

Multiple synchrotron self-Compton modeling of gamma-ray flares in 3C 279

Marc Türler[1] **and Claes-Ingvar Björnsson**[2]

[1] ISDC, Geneva Observatory, University of Geneva, ch. d'Ecogia 16, 1290 Versoix, Switzerland
[2] Department of Astronomy, Alba Nova University Center, Stockholm University, SE–106 91 Stockholm, Sweden

E-mail: `marc.turler@unige.ch`

Abstract. The correlation often observed in blazars between optical-to-radio outbursts and gamma-ray flares suggests that the high-energy emission region shall be co-spatial with the radio knots, several parsecs away from the central engine. This would prevent the important contribution at high-energies from the Compton scattering of seed photons from the accretion disk and the broad-line region that is generally used to model the spectral energy distribution of low-frequency peaking blazars. While a pure synchrotron self-Compton model has so far failed to explain the observed gamma-ray emission of a flat spectrum radio quasar like 3C 279, the inclusion of the effect of multiple inverse-Compton scattering might solve the apparent paradox. Here, we present for the first time a physical, self-consistent SSC modeling of a series of shock-waves in the jet of 3C 279. We show that the analytic description of the high-energy emission from multiple inverse-Compton scatterings in the Klein-Nishina limit can fairly well account for the observed gamma-ray spectrum of 3C 279 in flaring states.

1. Introduction

The radio-to-infrared emission of blazars is well understood as being synchrotron emission associated to bright structures observed to propagate — often with apparent superluminal motion — in a relativistic jet at parsec-scale distances [1]. The preferred model for the emission of these knots are shock waves resulting from disturbances at the base of the jet that move outward, become supersonic, compress the gas locally, and accelerate particles at the shock front [2]. In order to test the validity of this shock-in-jet model, a tool was developed to fit the multi-frequency light-curves of the bright quasar 3C 273 with a series of model outbursts [3; 4]. The same approach was then also used to describe the flaring behavior of other sources: the blazar 3C 279 [5], and the micro-quasars GRS 1915+105 [6] and Cyg X-3 [7; 8]. While these studies still relied on model parameters that are observables, we now developed a fully physical parametrization of the synchrotron emission of shock waves [9].

The correlation often observed in blazars between optical-to-radio outbursts and gamma-ray flares suggests that the high-energy emission is closely linked to the synchrotron component (e.g. [10]). If the emission arises parsecs away from the central engine, the seed photons for inverse-Compton scattering cannot be of external origin. However, for a low-energy peaking blazar like 3C 279, it has been shown that such a synchrotron self-Compton (SSC) scenario is not able to account for the observed gamma-ray emission [11; 12]. While this is true for single scatterings, the effect of multiple-inverse Compton (MIC) orders has been reconsidered recently to solve this apparent paradox [13]. We present here a first attempt to apply an analytical model for multiple synchrotron self-Compton (MSSC) emission of shock waves to a broad dataset of 3C 279.

2. Theory

This section aims at linking the theoretical model for MIC scattering [13; 14] with the parameters and the formalism used for the synchrotron emission of shock waves in a relativistic jet [9]. As a starting point we link the optical depth τ_0 of [13] with the normalization K of the electron energy distribution, assumed to be of the form $n_e(\gamma) = K\,\gamma^{-p}$ for $\gamma_{min} \leq \gamma \leq \gamma_{max}$, where n_e is the electron number density and γ is the Lorentz factor of the relativistic electrons. For $p > 2$ and $\gamma_{max} \gg \gamma_{min}$, integrating $n_e(\gamma)$ leads to $K = (p-1)\gamma_{min}^{p-1}\,n_e$ such that Eq. (7) of [13] becomes:

$$\tau_0 = \frac{8\,\sigma_T}{3(p-2)}\,x_R\,K\,, \tag{1}$$

where σ_T is the Thomson cross-section and $x_R = f_R\,R$ is the thickness of the emitting slab containing the shocked plasma behind the shock front, which is assumed to be a constant fraction f_R of the jet radius R [9].

For a given electron energy distribution $n_e(\gamma)$, the MIC spectrum is determined by two main parameters α and γ_{cool}. The spectral index α ($F_\nu \propto \nu^{-\alpha}$) connects the $\nu\,F_\nu$ maxima of the synchrotron and the inverse-Compton components. The Lorentz factor γ_{cool} corresponds to the energy of electrons that have just enough time to cool down while crossing the emission region, i.e. $x(\gamma_{cool}) = \beta_{rel}ct_{cool}(\gamma_{cool}) \equiv x_R$, where $\beta_{rel}c$ is the average speed of electrons along the jet relative to the shock front. In the MIC scenario, there are three distinct cases [13]:

$$\text{A)}\quad \gamma_{cool} > \gamma_{min} \text{ and } \alpha > (p-1)/2 \implies \tau_0 = \gamma_{cool}^{p-1-2\alpha} < 1 \tag{2}$$

$$\text{B)}\quad \gamma_{cool} > \gamma_{min} \text{ and } \alpha < (p-1)/2 \implies \tau_0 = \gamma_{min}^{p-1-2\alpha} > 1 \tag{3}$$

$$\text{C)}\quad \gamma_{cool} < \gamma_{min} \implies \tau_0 = \gamma_{min}^{p-1-2\alpha}\left(\frac{\gamma_{min}}{\gamma_{cool}}\right)^{\alpha}. \tag{4}$$

In cases A and B, $\alpha = (p-1)/2$ coincides with $\tau_0 = 1$ and we note the corresponding value of γ_{cool} as $\gamma_{cool,0}$. One can derive a simple equation linking γ_{cool} to α as:

$$\left(\frac{\gamma_{cool}}{y}\right)^{\alpha} = \left(\frac{\gamma_{cool,0}}{y}\right)^{(p-1)/2}, \tag{5}$$

where $y \equiv mc^2/(h\nu_B)$ and $\nu_B \equiv eB/(2\pi mc)$ is the cyclotron frequency for an electron mass, m, and charge, e, in a magnetic field, B. Eq. (5) is valid in case A and also in case C, if in the latter the label "0" denotes $\alpha = (p-1)/2$ rather than $\tau_0 = 1$. For case B, there is an additional term $(\gamma_{min}/\gamma_{abs})^{p-1-2\alpha}$ on the left-hand side (LHS). This is due to the fact that in this case the Compton-scattered flux is dominated by the lowest energy synchrotron photons, i.e. those radiating at the synchrotron self-absorption frequency $\nu_{abs} = \nu_B\gamma_{abs}^2$.

In Eq. (5) we implicitly assumed that $\gamma_{abs} < \gamma_{cool}$. When $\gamma_{abs} > \gamma_{cool}$, the corresponding equation is:

$$\left(\frac{\gamma_{cool}}{\gamma_{abs}}\right)^{p-1-2\alpha}\left(\frac{\gamma_{cool}^2}{y\,\gamma_{abs}}\right)^{\alpha} = \left(\frac{\gamma_{cool,0}^2}{y\,\gamma_{abs,0}}\right)^{(p-1)/2}, \tag{6}$$

where $\gamma_{abs,0}$ is the value of γ_{abs} when $\tau_0 = 1$ and the equation reduces to Eq. (5) when $\gamma_{abs} \to \gamma_{cool}$. For case B, a modification is again needed to the LHS of Eq. (6): the ratio

$\gamma_{\rm cool}/\gamma_{\rm abs}$ has to be replaced by $\gamma_{\rm min}/\gamma_{\rm abs}$. From Eqs. (13) and (18) of [13], one notes that the right-hand side (RHS) of both Eqs. (5) and (6) is given by:

$$\mathrm{RHS} = \frac{\gamma_{\rm min}^{p-2}}{y}\,\frac{2\beta_{\rm rel}}{\tau_0}\,\frac{U_{\rm e}}{U_B} = \frac{6\pi\, mc^2\,\beta_{\rm rel}}{\sigma_{\rm T}\, y\, x_R\, B^2} = \frac{3\,h\,e\,\beta_{\rm rel}}{mc\,\sigma_{\rm T}\, x_R\, B}\,, \tag{7}$$

where $U_{\rm e} = K\,mc^2\,\gamma_{\rm min}^{2-p}/(p-2)$ and $U_B = B^2/(8\pi)$ are the energy densities of the electrons and the magnetic field, respectively, and where we used Eq. (1) for τ_0. This equation is to be calculated for a given jet model. We now have to obtain equations for α that are independent of $\gamma_{\rm cool}$ in each case A, B and C, and both for 1) $\gamma_{\rm abs} < \gamma_{\rm cool}$ with Eq. (5) and for 2) $\gamma_{\rm abs} > \gamma_{\rm cool}$ with Eq. (6). We get the following equations for α:

A1) $$[2\ln y]\,\alpha^2 + [\ln\tau_0 + 2\ln(\mathrm{RHS}) - (p-1)\ln y]\,\alpha - (p-1)\ln(\mathrm{RHS}) = 0 \tag{8}$$

A2) $$[2\ln(y/\gamma_{\rm abs})]\,\alpha^2 + [(p-1)\ln(\gamma_{\rm abs}^3/y) + 2\ln(\mathrm{RHS})]\,\alpha - (p-1)\ln(\mathrm{RHS}\,\gamma_{\rm abs}^{p-1}/\tau_0) = 0 \tag{9}$$

B) $$\alpha = (p-1)/2 - \ln\tau_0/(2\ln\gamma_{\rm min}) \tag{10}$$

C1) $$\alpha = \ln(\gamma_{\rm min}^{p-1}/(\mathrm{RHS}\,\tau_0))/\ln(y\,\gamma_{\rm min}) \tag{11}$$

C2) $$[\ln(y/\gamma_{\rm abs})]\,\alpha^2 + [(p-1)\ln(\gamma_{\rm min}\gamma_{\rm abs}) + \ln(\mathrm{RHS})]\,\alpha - (p-1)\ln(\gamma_{\rm min}^{p-1}/\tau_0) = 0 \tag{12}$$

Pratically, as $\gamma_{\rm abs}$ is not know a priori, we first derive α for the cases A1, B and C1, which are independent of $\gamma_{\rm abs}$. We then calculate $\gamma_{\rm cool}$ for each of the three cases A1, B1 and C1 using Eq. 5 and then $\gamma_{\rm abs}$ given by:

$$\gamma_{\rm abs}^{p+4+\alpha} = \frac{e^2 g_\kappa(p)}{16\,mc}\,\frac{x_R\,K}{\nu_B}\,\gamma_{\rm cool}^{\alpha}\,, \tag{13}$$

where we used Eq. (7) of [9] and Eq. (20) of [13] and where $g_\kappa(p)$ is defined in Eq. (4) of [9]. In the special case B1, we actually first combine Eq. (5) and Eq. (13) to get an expression of $\gamma_{\rm cool}$ independent of $\gamma_{\rm abs}$. We can now discriminate between the various cases. If $\gamma_{\rm abs} > \gamma_{\rm cool}$ occurs we calculate α for the relevant case (A2, B2 or C2) and then update the values of $\gamma_{\rm cool}$ with Eq. (6) and of $\gamma_{\rm abs}$ with Eq. (13).

We note that the three different cases A, B and C concern the inverse-Compton spectrum and are distinct from the three-stage evolution of the synchrotron self-absorption turnover in the shock model of Marscher & Gear [2]. The latter is governed by the width x of the emission region that can be smaller than x_R if synchrotron or inverse-Compton cooling timescales, $t_{\rm cool}$, limit the effective emission region to $x_{\rm abs} = \beta_{\rm rel} c t_{\rm cool}(\gamma_{\rm abs}) < x_R$. In the MSSC scenario, $x_{\rm abs} = x_R(\gamma_{\rm cool}/\gamma_{\rm abs})^{\alpha}$ and this inverse-Compton cooling dominates over first-order Compton or synchrotron cooling [14]. When $\alpha > 1$, $x_{\rm abs} = x_R(\gamma_{\rm cool}/\gamma_{\rm abs})$, where $\gamma_{\rm cool}$ and $\gamma_{\rm abs}$ are now controlled by U_B alone (i.e. with $U_S = 0$ in Eq. (8) of [9]). This synchrotron-cooling stage only sets in if $\gamma_{\rm cool} < \gamma_{\rm abs}$ is still satisfied when $\alpha = 1$. Otherwise, we have a direct transition from the inverse-Compton stage to the adiabatic expansion cooling stage, where radiative cooling is negligible (i.e. $x_{\rm abs} > x_R$) and thus the emission region is limited by x_R.

3. Method

In order to test the MSSC scenario for a shock wave propagating in a relativistic jet flow we had to fully describe the expected evolution with time of both the synchrotron and the inverse-Compton spectra. For this we basically took the same assumptions as described in Sect. 3 of [9] to define the parameters relevant for the shock-in-jet modeling. We ended-up with an evolution of several key parameters with distance, X, of the shock-wave from the apex of the jet, which is itself related to observed time. Those parameters were the normalization, K, and the two limits, $\gamma_{\rm min}$ and $\gamma_{\rm max}$, of the electron energy distribution, as well as the strength of the magnetic field,

B, and the maximal thickness of the emission region, x_R, which is directly proportional to the jet radius R (cf. Sect. 2). We then calculated the evolution of τ_0 with Eq. (1) and derived also that of α, $\gamma_{\rm cool}$ and $\gamma_{\rm abs}$, which are all smoothly changing with time across the transitions from one case to the other.

We then calculated the evolution of the self-absorption flux density $F_{\rm abs} \equiv F_\nu^{\rm thin}(\nu_{\rm abs})$ with Eq. (6) of [9], which normalizes both the synchrotron and inverse-Compton spectra. The exact shape of these components is controlled by p and α for what concerns the spectral slopes and by $\gamma_{\rm min}$, $\gamma_{\rm cool}$, $\gamma_{\rm abs}$ for the position of the various breaks. Depending on the relative order of these characteristic Lorentz factors, the synchrotron spectrum can take five different shapes [15], whereas the inverse-Compton spectrum has basically three different shapes for cases A, B and C [13]. The analytical formulation used for the breaks has been simplified as described in [9]. A high-energy cut-off is applied to both spectra at the frequency corresponding to $\gamma_{\rm max}$. The inverse-Compton spectrum is also limited at low-frequency by a break towards a spectral index ($F_\nu \propto \nu^{-s}$) of $s=(p-1)/2$ below the rest-frame frequency $\nu_B\gamma_{\rm cool}^4$ and another one to an index of $s=-1$ below $\nu_B\gamma_{\rm min}^4$, or $\nu_B\gamma_{\rm abs}^4$ if $\gamma_{\rm abs} > \gamma_{\rm min}$. Finally, we attempted for a smooth transition from a multiple-Compton to a first-order Compton spectrum.

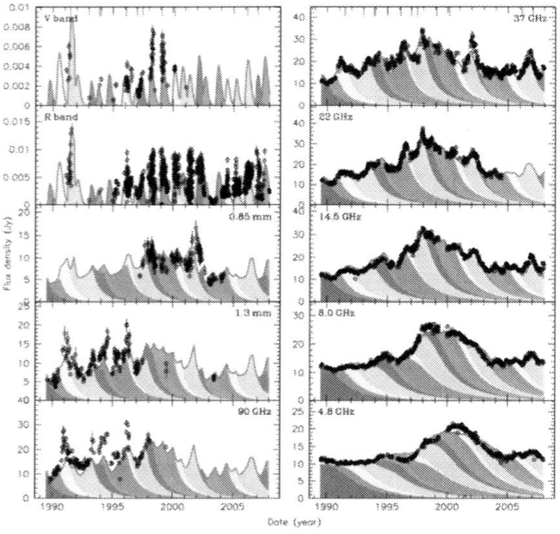

Figure 1. Fit of ten lightcurves of 3C 279 in the optical-to-radio domain with a series of self-similar synchrotron model outbursts following the spectral evolution shown in Fig. 2. The contribution of individual outbursts is shown with different colors and vertical lines at the top show the times of the available SEDs.

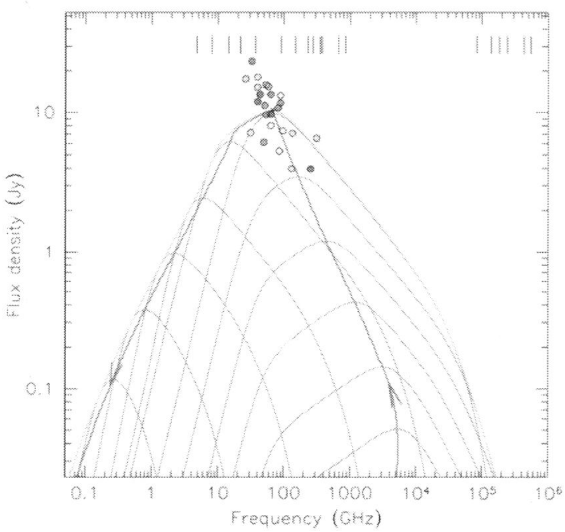

Figure 2. Spectral evolution of the average synchrotron outburst in 3C 279 as derived from the fit of 19 lightcurves (vertical lines) as shown in Fig. 1. Synchrotron spectra at different times (grey curves) rise and decay with a maximum following the red line. The peak of individual outbursts is shown by colored dots.

4. Results

To test whether the model outlined above for the high-energy emission of blazars is valid for a flat spectrum radio quasar (FSRQ) like 3C 279, we took all the spectral energy distributions (SEDs) of this object obtained during the operations of the Compton Gamma-Ray Observatory [11].

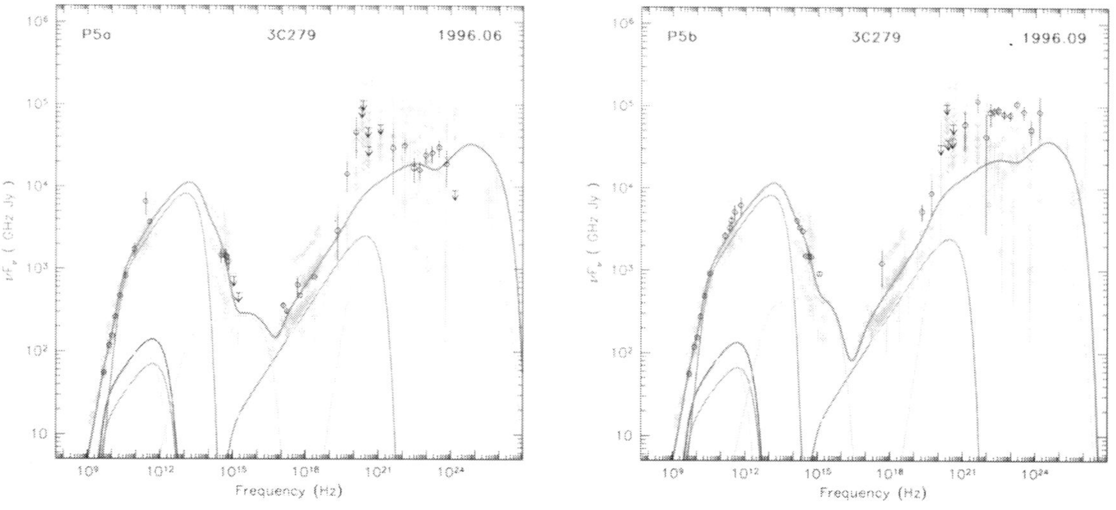

Figure 3. Spectral decomposition of two SEDs of 3C 279 derived by fitting them self-consistently together with all the optical-to-radio lightcurves (see Fig. 1). The emission of different model outbursts – corresponding to a succession of different shock waves in the jet – are shown with the same colors as in Figs. 1 and 2. The important GeV variability between epoch P5b (right panel) and epoch P5a (left panel) in only ~10 days is not fully accounted for by the early emission of a rising outburst (green curves), where MIC scattering dominates strongly over first-order Compton (dashed line). The points from the ten other SEDs and the MAGIC data of 2006 and 2007 [16] are shown in grey for comparison.

We then fitted these eleven SEDs together with 19 lightcurves in the radio-to-infrared domain[1] with a set of 24 model outbursts corresponding to a succession of shock waves propagating in the jet. Individual outbursts only differ in shock compression factor, η, and distance from the apex of the jet, $X_{\rm p}$, needed to build-up the shock [9]. The obtained fit to the lightcurves is shown in Fig. 1, while the corresponding synchrotron spectral evolution is shown in Fig. 2. We note that the latter has a similar shape to what was obtained previously with only about half of the current dataset [5], but here the rising and slowly decaying peak of the evolution is controlled by a relatively high value of $\gamma_{\rm min}$ and with no synchrotron stage. The overall spectral decomposition is illustrated in Fig. 3 with the most extreme high-energy spectrum in the dataset (epoch P5b) and the one taken just ~10 days before. In contrary to an early attempt [12], a pure SSC model seems now to be able to account both for the X-ray and high-energy flaring emission of 3C 279. We note, however, that the presented model is not a true one-zone model. Indeed, to have a deep enough gap between the synchrotron and the inverse-Compton humps, the latter had to be scaled down by a factor 0.1 that corresponds to assuming that the high-energy emission site is 10 times smaller than the synchrotron emission region [13].

The model we used here assumes a conical jet with a constant Doppler factor, an opening radius of 2° pointing 3° away from the line of sight and with a bulk Lorentz factor for the shocks of $\Gamma = 6.8$. The main parameters for an average outburst at synchrotron peak flux are: a distance $X_{\rm p} = 16.7\,{\rm pc}$ from the apex of the jet; a magnetic field of $B = 48\,{\rm mG}$; an electron energy distribution characterized by an index $p = 2.10$, a normalization $K_{\rm p} = 6460\,{\rm cm}^{-3}$, and cuts at $\gamma_{\rm min} = 254$ and $\gamma_{\rm max} = 7450$. This leads to an excess by a factor of $\sim$ 100 of the

[1] The description of the dataset is postponed to another paper in preparation.

electron energy density U_e compared to U_B. Although these values are rapidly changing further upstream or downstream, they seem reasonable compared to other studies (e.g. [11]). The important difference is that our emission site is far more distant from the central engine, where the contribution from photons external to the jet becomes negligible. This is at least true for the long-term X- and gamma-ray emission. The very early evolution of the outburst in our model (see the green spectrum in Fig. 3) seems to produce flaring emission on time-scales of days at GeV energies and up to to the TeV domain as detected by MAGIC on a few occasions [16].

5. Conclusion

We presented a new approach to model the variable high-energy emission of blazars and demonstrated its viability by applying it to a very rich dataset of 3C 279 spanning almost 20 years and the full SED from the radio to VHE gamma-rays. These preliminary results suggest that a shock-in-jet model with a proper treatment of MIC scattering can account for the high-energy emission observed in the source at some epochs, while simultaneously reproducing the frequency-dependent variability monitored in the radio-to-optical range. Such a pure-SSC scenario for FSRQs shall be further tested in a forthcoming paper with the inclusion of the underlying, steady jet emission that shall not exceed the observed quiescent state of the source and by including addition observational constraints in the X-ray to gamma-ray range, in particular with lightcurves of 3C 279 from the satellites *Fermi*, *Swift*, and the *Rossi X-ray Timing Explorer* (e.g. [17]).

Acknowledgments

This work was done in the frame of the International Team collaboration number 160 supported by the International Space Science Institute in Bern, Switzerland.

References

[1] Savolainen T, Wiik K, Valtaoja E, Jorstad S G and Marscher A P 2002 *A&A* **394** 851
[2] Marscher A P and Gear W K 1985 *ApJ* **298** 114
[3] Türler M, Courvoisier T J L and Paltani S 1999 *A&A* **349** 45
[4] Türler M, Courvoisier T J L and Paltani S 2000 *A&A* **361** 850
[5] Lindfors E J, Türler M, Valtaoja E, Aller H, Aller M *et al* 2006 *A&A* **456** 895
[6] Türler M, Courvoisier T J L, Chaty S and Fuchs Y 2004 *A&A* **415** L35
[7] Lindfors E J, Türler M, Hannikainen D C, Pooley G, Tammi J *et al* 2007 *A&A* **473** 923
[8] Miller-Jones J C A, Rupen M P, Türler M, Lindfors E J *et al* 2009 *MNRAS* **394** 309
[9] Türler M 2011 *MmSAI* **82** 104 (*Preprint* `arXiv:1010.0907`)
[10] León-Tavares J, Valtaoja E, Tornikoski M, Lähteenmäki A and Nieppola E 2011 *A&A* **532** A146
[11] Hartman R C, Böttcher M and Aldering G *et al* 2001 *ApJ* **553** 683
[12] Lindfors E J, Valtaoja E and Türler M 2005 *A&A* **440** 845
[13] Björnsson C I 2010 *ApJ* **723** 417
[14] Björnsson C I and Aslaksen T 2000 *ApJ* **533** 787
[15] Granot J and Sari R 2002 *ApJ* **568** 820
[16] Aleksić J, Antonelli L A, Antoranz P, Backes M, Barrio J A *et al* 2011 *A&A* **530** A4
[17] Abdo A A, Ackermann M, Ajello M *et al* 2010 *Nature* **463** 919

Probing the origin of VHE emission from M 87 with MWL observations in 2010

M. Raue[1], L. Stawarz[2], D. Mazin[3], P. Colin[4], C. M. Hui[5], M. Beilicke[6], R.C. Walker[7]
for the H.E.S.S., MAGIC, VERITAS, and Fermi/LAT Collaborations and the M 87 MWL Monitoring Teams

[1] Institut für Experimentalphysik, Universität Hamburg, Hamburg, Germany
[2] Obserwatorium Astronomiczne, Uniwersytet Jagielloński, ul. Orla 171, 30-244 Kraków, Poland
[3] IFAE, Edifici Cn., Campus UAB, E-08193 Bellaterra, Spain
[4] Max-Planck-Institut für Physik, D-80805 München, Germany
[5] Department of Physics and Astronomy, University of Utah, Salt Lake City, UT 84112, USA
[6] Department of Physics, Washington University, St. Louis, MO 63130, USA
[7] National Radio Astronomy Observatory (NRAO), Socorro, NM 87801, USA

E-mail: `martin.raue@desy.de`

Abstract. The large majority of extragalactic very high energy (VHE; E >100 GeV) sources belongs to the class of active galactic nuclei (AGN), in particular the BL Lac sub-class. AGNs are characterized by an extremely bright and compact emission region, powered by a super-massive black hole (SMBH) and an accretion disk, and relativistic outflows (jets) detected all across the electro-magnetic spectrum. In BL Lac sources the jet axis is oriented close to the line of sight, giving rise to a relativistic boosting of the emission. In radio galaxies, on the other hand, the jet makes a larger angle to the line of sight allowing to resolve the central core and the jet in great details. The giant radio galaxy M 87 with its proximity (16 Mpc) and its very massive black hole ($(3-6)\times 10^9 M_\odot$) provides a unique laboratory to investigate VHE emission in such objects and thereby probe particle acceleration to relativistic energies near SMBH and in jets. M 87 has been established as a VHE emitter since 2005. The VHE emission displays strong variability on time-scales as short as a day. It has been subject of a large joint VHE and multi-wavelength (MWL) monitoring campaign in 2008, where a rise in the 43 GHz VLBA radio emission of the innermost region (core) was found to coincide with a flaring activity at VHE. This had been interpreted as a strong indication that the VHE emission is produced in the direct vicinity of the SMBH black hole. In 2010 again a flare at VHE was detected triggering further MWL observations with the VLBA, Chandra, and other instruments. At the same time M 87 was also observed with the Fermi/LAT telescope at GeV energies and the European VLBI Network (EVN). In this contribution preliminary results from the campaign will be presented.

1. Introduction

The giant radio galaxy M 87 provides a unique environment to study relativistic plasma outflows and the surrounding of super-massive black holes (SMBH). Its prominent jet [2] is resolved from radio to X-rays displaying complex structures (knots, diffuse emission; [3, 4]), strong variability [5, 6], and super-luminal motion [7, 8] (Fig. 1). With its proximity (16.7 ± 0.2 Mpc; [9]) and its very massive black hole of $M_{\rm BH} \simeq (3-6)\times 10^9\ {\rm M}_\odot$ [10, 11] high resolution radio observations enables one to directly probe structures with sizes down to < 200 Schwarzschild radii.

M 87 has been established as a very high energy (VHE; E > 100 GeV) emitter since 2005 [12, 13].

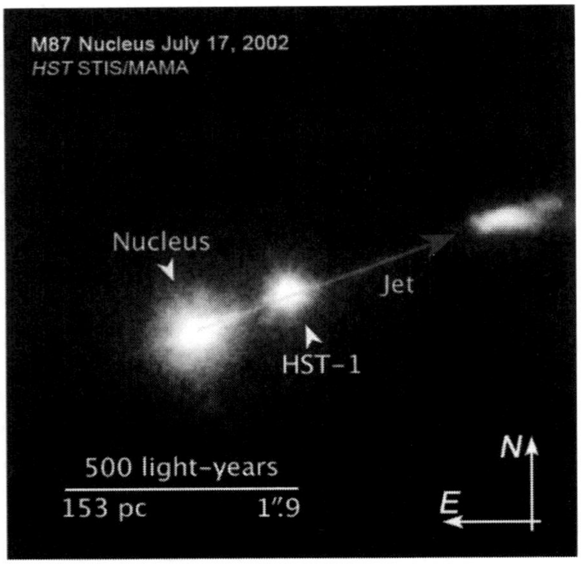

Figure 1. Hubble space telescope (HST) image of M 87 [1]. *(Illustration Credit: NASA, ESA, and Z. Levay (STScI); Credit: NASA, ESA, and J. Madrid (McMaster University))*

It is one of only 4 extragalactic VHE sources belonging to the class of radio galaxies, for which only weak or moderate beaming of the emission is expected. M 87 shows strong variability at VHE with time-scales of the order of days [13, 14, 15]. This indicates a compact emission region $< 5 \times 10^{15}\delta\,\mathrm{cm}$ (δ: bulk Doppler factor of the emitting plasma), corresponding to only a few tens of Schwarzschild radii. At GeV energies M 87 has recently been detected by *Fermi*/LAT [16].

The angular resolution of ground based VHE instruments[1] does not allow for a direct determination of the origin of the VHE emission in the inner kpc structures. To further investigate the location of the VHE emission site and the associated production mechanisms variability studies and the search for correlations with other wavelengths need to be utilized (e.g. [15]).

Up to now, three episodes of enhanced VHE activity have been detected from M 87. The first one, detected in 2005 [13], occurred during an extreme radio/optical/X-ray outburst of the jet feature HST-1 [5, 6], which has been discussed as possible emission site for the VHE emission (e.g. [17, 8, 18]). During the second flaring episode, detected in 2008, HST-1 was in a low flux state, but radio measurements showed a flux increase of the core region within a few hundred Schwarzschild radii of the SMBH, suggesting the direct vicinity of the SMBH as the origin of the VHE emission [15]. This conclusion was further supported by the detection of an enhanced X-ray flux from the core region by *Chandra*.

The third episode of increased VHE activity occurred in 2010 during a joint VHE monitoring campaign by MAGIC and VERITAS. The detection of the high state [19, 20] triggered further MWL observations by the VLBA, *Chandra*, and other instruments. Preliminary results from the campaign are presented in this paper, while the final campaign results will be reported in an upcoming publication.

2. The 2010 VHE flare

The preliminary combined VHE lightcurve of the 2010 monitoring campaign is shown in Fig. 2. During the campaign, two VHE flares have been detected [19, 20]: The first episode took place in Feb. 2010 where a single night of increased activity was detected by MAGIC. Follow up observations did not reveal further activity. The second episode took place in Apr. 2010 and showed a pronounced VHE

[1] Typical ~0.1 deg per event, corresponding to ~ 30 kpc projected size.

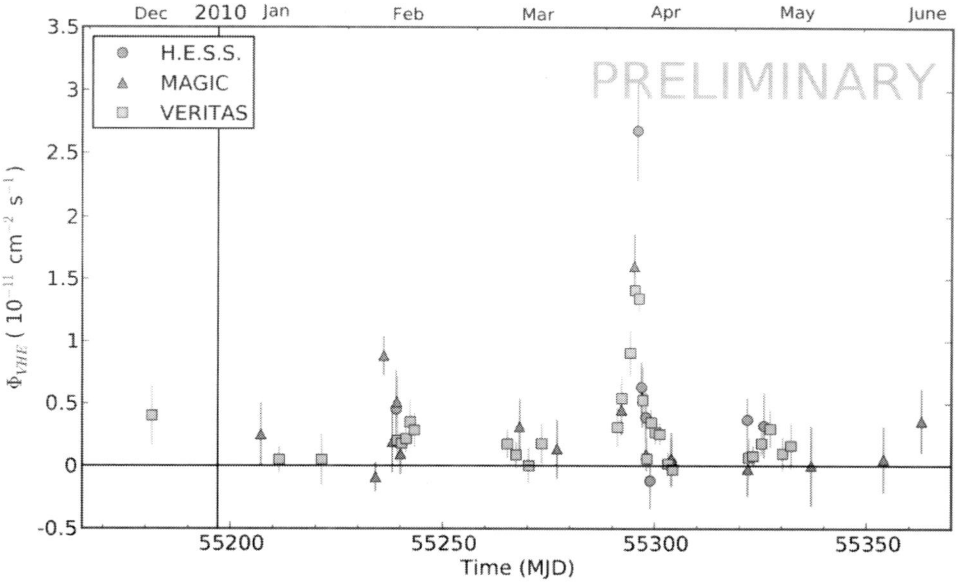

Figure 2. Preliminary VHE lightcurve from the 2010 monitoring campaign on M 87 as recorded by H.E.S.S., MAGIC, and VERITAS. The integral flux above an energy of 300 GeV is shown.

flare detected by several instruments triggering further MWL observations. The VHE activity of this second flaring episode is concentrated in a single observation period between MJD 55290 and MJD 55305 ($\sim$15 days). This time-period is exceptionally well covered by observations with 21 pointings by different VHE instruments resulting in an observation almost every night. The detected flare displays a smooth rise and decay in flux with a peak around MJD 55296 (April 9-10, 2010; see Fig. 2). In general, during nights with (quasi) simultaneous observations by different instruments, the measured fluxes are found to be in excellent agreement.

In comparison to previous VHE flares detected in 2005 and 2008, the 2010 flare shows similar time-scales and peak flux levels, but the overall variability pattern is somewhat different to what has been observed before (more extended periods of flaring activities with several flux maxima), though the statistics and the sampling of the previous VHE flares limit a definitive conclusion. From the VHE long-term lightcurve spanning from 2004 to 2010 the duty cycle for VHE flares is estimated to be $\sim 14 - 4\%$ (depending on the assumed threshold flux defining a VHE high state).

3. 2010 MWL observations

The discovery of a VHE high state in April 2010 triggered further MWL observations. *Chandra* started observing $\sim$ 2 days after the peak VHE flux had been detected performing five pointings spaced in intervals between 1.5 and 3 days (5ks each). A second set of four observations was taken starting about two weeks later. HST-1 was found in a low flux state while the core showed an increase in X-ray flux in the first observation that followed the VHE flare.

Five VLBA 43 GHz observations where taken in 2010, three monitoring observations and two additional observations following the detection of the VHE high state. No enhanced radio flux from the core region was detected, contrary to what has been observed during the 2008 VHE outburst.

During the 2010 campaign, further MWL data were taken by the Liverpool Telescope, the EVN, and the VLBA, and M 87 was continually monitored at MeV/GeV energies with the Fermi-LAT. Results from these observations will be presented in the upcoming publication.

Acknowledgments

The author acknowledges support by the LEXI program of the state of Hamburg, Germany. — The H.E.S.S. Collaboration acknowledges support of the Namibian authorities and of the University of Namibia in facilitating the construction and operation of H.E.S.S., as is the support by the German Ministry for Education and Research (BMBF), the Max Planck Society, the French Ministry for Research, the CNRS-IN2P3 and the Astroparticle Interdisciplinary Programme of the CNRS, the U.K. Science and Technology Facilities Council (STFC), the IPNP of the Charles University, the Polish Ministry of Science and Higher Education, the South African Department of Science and Technology and National Research Foundation, and by the University of Namibia. We appreciate the excellent work of the technical support staff in Berlin, Durham, Hamburg, Heidelberg, Palaiseau, Paris, Saclay, and in Namibia in the construction and operation of the equipment. — The MAGIC Collaboration thanks the Instituto de Astrofsica de Canarias for the excellent working conditions at the Observatorio del Roque de los Muchachos in La Palma. The support of the German BMBF and MPG, the Italian INFN and Spanish MICINN is gratefully acknowledged. — The VERITAS Collaboration acknowledges support from the U.S. Department of Energy, the U.S. National Science Foundation and the Smithsonian Institution, by NSERC in Canada, by Science Foundation Ireland, and by STFC in the UK. — The *Fermi* LAT Collaboration acknowledges support from a number of agencies and institutes for both development and the operation of the LAT as well as scientific data analysis. These include NASA and DOE in the United States, CEA/Irfu and IN2P3/CNRS in France, ASI and INFN in Italy, MEXT, KEK, and JAXA in Japan, and the K. A. Wallenberg Foundation, the Swedish Research Council and the National Space Board in Sweden. Additional support from INAF in Italy and CNES in France for science analysis during the operations phase is also gratefully acknowledged. — The Very Long Baseline Array is operated by the National Radio Astronomy Observatory, a facility of the NSF, operated under cooperative agreement by Associated Universities, Inc. — The European VLBI Network is a joint facility of European, Chinese, South African and other radio astronomy institutes funded by their national research councils. This effort is supported by the European Community Framework Programme 7, Advanced Radio Astronomy in Europe, grant agreement No. 227290. — This research has made use of data from the MOJAVE database that is maintained by the MOJAVE team [21]. — This research has made use of NASA's Astrophysics Data System.

References

[1] Madrid J P 2009 *AJ* **137** 3864–3868 (*Preprint* `0904.3546`)
[2] Curtis H D 1918 *Publications of Lick Observatory* **13** 31
[3] Perlman E S, Biretta J A, Zhou F, Sparks W B and Macchetto F D 1999 *AJ* **117** 2185–2198 (*Preprint* `arXiv:astro-ph/9901176`)
[4] Perlman E S, Biretta J A, Sparks W B, Macchetto F D and Leahy J P 2001 *ApJ* **551** 206–222 (*Preprint* `arXiv:astro-ph/0012044`)
[5] Harris D E, Biretta J A, Junor W, Perlman E S, Sparks W B and Wilson A S 2003 *ApJ* **586** L41–L44 (*Preprint* `arXiv:astro-ph/0302270`)
[6] Harris D E, Cheung C C, Biretta J A, Sparks W B, Junor W, Perlman E S and Wilson A S 2006 *ApJ* **640** 211–218 (*Preprint* `arXiv:astro-ph/0511755`)
[7] Biretta J A, Sparks W B and Macchetto F 1999 *ApJ* **520** 621–626
[8] Cheung C C, Harris D E and Stawarz L 2007 *ApJ* **663** L65–L68 (*Preprint* `arXiv:0705.2448`)
[9] Mei S, Blakeslee J P, Côté P, Tonry J L, West M J, Ferrarese L, Jordán A, Peng E W, Anthony A and Merritt D 2007 *ApJ* **655** 144–162 (*Preprint* `arXiv:astro-ph/0702510`)
[10] Macchetto F, Marconi A, Axon D J, Capetti A, Sparks W and Crane P 1997 *ApJ* **489** 579–+ (*Preprint* `arXiv:astro-ph/9706252`)
[11] Gebhardt K and Thomas J 2009 *ApJ* **700** 1690–1701 (*Preprint* `0906.1492`)
[12] Aharonian F, Akhperjanian A, Beilicke M *et al.* 2003 *A&A* **403** L1–L5 (*Preprint* `arXiv:astro-ph/0302155`)
[13] Aharonian F, Akhperjanian A G, Bazer-Bachi A R *et al.* 2006 *Science* **314** 1424 (*Preprint* `astro-ph/0612016`)
[14] Albert J, Aliu E, Anderhub H, Antonelli L A, Antoranz P *et al.* 2008 *ApJ* **685** L23–L26 (*Preprint* `0806.0988`)
[15] Acciari V A, Aliu E, Arlen T, Bautista M, Beilicke M *et al.* 2009 *Science* **325** 444–448 (*Preprint* `http://www.sciencemag.org/cgi/reprint/325/5939/444.pdf`) URL `http://www.sciencemag.org/cgi/content/abstract/325/5939/444`

[16] Abdo A A, Ackermann M, Ajello M, Atwood W B, Axelsson M *et al.* 2009 *ApJ* **707** 55–60 (*Preprint* 0910.3565)
[17] Stawarz Ł, Aharonian F, Kataoka J, Ostrowski M, Siemiginowska A and Sikora M 2006 *MNRAS* **370** 981–992 (*Preprint* arXiv:astro-ph/0602220)
[18] Harris D E, Cheung C C, Stawarz Ł, Biretta J A and Perlman E S 2009 *ApJ* **699** 305–314 (*Preprint* 0904.3925)
[19] Mariotti M and MAGIC Collaboration 2010 *The Astronomer's Telegram* **2431**
[20] Ong R A and Mariotti M 2010 *The Astronomer's Telegram* **2542**
[21] Lister M L, Aller H D, Aller M F, Cohen M H, Homan D C, Kadler M, Kellermann K I, Kovalev Y Y, Ros E, Savolainen T, Zensus J A and Vermeulen R C 2009 *AJ* **137** 3718–3729 (*Preprint* 0812.3947)

Unbeamed γ-rays from low luminosity AGNs

Andrzej Niedźwiecki,[1] **Fu-Guo Xie**[2,3] **and Volker Beckmann**[4]

[1]Department of Astrophysics, University of Łódź, Pomorska 149/153, 90-236 Łódź, Poland
[2]Key Laboratory for Research in Galaxies and Cosmology, Shanghai Astronomical Observatory, Chinese Academy of Sciences, 80 Nandan Road, Shanghai 200030, China
[3]Kavli Institute for Astronomy and Astrophysics, Peking University, Beijing 100871, China
[4]François Arago Centre, APC, Université Paris Diderot, CNRS/IN2P3, CEA/DSM, Observatoire de Paris, 13 rue Watt, 75205 Paris Cedex 13, France

E-mail: niedzwiecki@uni.lodz.pl, fgxie@shao.ac.cn, beckmann@apc.univ-paris7.fr

Abstract. Proton-proton collisions in two-temperature accretion flows may lead to substantial γ-ray emission through neutral pion production and decay. We present preliminary results of our study of such a hadronic γ-ray emission. We find that radiatively inefficient flows around rapidly rotating black holes can produce the MeV/GeV luminosities up to $\sim 10^{44}$ erg s^{-1}. The hot-flow model predictions for the hadronic and the Comptonization components can put interesting constraints on the properties of the source of high-energy radiation in AGNs. As an example, we compare our model with X/γ-ray observations of Centaurus A and we find that the model with a maximally rotating black hole strongly overpredicts the γ-ray flux.

1. Introduction

Optically thin, two-temperature accretion flows with low radiative efficiency have been considered as the explanation of a variety of black hole systems and substantial work has been done to investigate their dynamical and spectral properties (see, e.g., a review in [1]). However, the work done so far focused on the radiation produced by electrons, while little attention has been paid to hadronic processes which could produce a considerable amount of high energy radiation since the proton temperature is very large in such flows. Already Shapiro, Lightman & Eardley realized, in their classical paper [2], that two-temperature flows can develop a significant γ-ray luminosity above 10 MeV, L_γ, through π^0 production; their rough estimate indicated a strong dependence of L_γ on the spin of the black hole and, then, they suggested that this effect may serve as a means to measure the spin value. A few attempts of detailed prediction of the γ-ray flux from π^0 decay in hot flows followed this initial finding and only two such studies [3,4], embodying the recent progress in the hot accretion theory, have been done during the last fifteen years. As we note in Section 2, both models [3,4] are subject to certain shortcomings; their results have not been compared with γ-ray measurements from black holes, apart from the *CGRO*/EGRET detection from the direction of the Galactic Center.

A more careful investigation of the γ-ray emission from hot accretion flows appears to be a very timely effort, given the significant progress in our exploration of the γ-ray activity of AGNs thanks to the *Fermi* mission. Motivated by this, we extend our model of radiative processes in hot flows [5] by including the description of the hadronic γ-ray production. The basic features of our model and predictions for the γ-ray emission are presented in Section 2. The detections of radio galaxies by *Fermi* [6] establish a particularly interesting area for application of our results.

In Section 3 we briefly compare our model with FR I radio galaxies, focusing on the best studied object from this class, Centaurus A.

2. High-energy emission from hot accretion flows

We have recently developed a model of optically thin, two-temperature accretion flows involving an exact Monte Carlo treatment of global Comptonization as well as a fully general relativistic (GR) description of both the radiative and hydrodynamic processes, see [5]. In our model we consider a black hole, characterised by its mass, M, and angular momentum, J, surrounded by a geometrically thick accretion flow with an accretion rate $\dot{M}$. The inclination angle of the line of sight to the symmetry axis is given by $\theta_{\rm obs}$. Below we use the dimensionless accretion rate, $\dot{m} = \dot{M}/\dot{M}_{\rm Edd}$, and the spin parameter, $a = J/(cR_{\rm g}M)$, where $\dot{M}_{\rm Edd} = L_{\rm Edd}/c^2$, $L_{\rm Edd} \equiv 4\pi GMm_{\rm p}c/\sigma_{\rm T}$ and $R_{\rm g} = GM/c^2$.

In [5] we considered only the cooling of electrons and in our next work, in preparation, we extend the model to study the γ-ray production through proton-proton interactions. As pointed out in [7], Coulomb collisions are too inefficient to thermalize protons in optically thin flows; indeed, we find that in our solutions the accretion time-scale, t_a, is much shorter than the proton relaxation time-scale, t_p; e.g. $t_a \lesssim 1$ hour and $t_p > 1$ year for $M = 2 \times 10^8 M_\odot$ and $\dot{m} = 0.1$ within a few innermost $R_{\rm g}$. Then, the protons are unable to redistribute their energy through Coulomb collisions and the proton distribution function is determined by the poorly understood viscous heating mechanism. This introduces a significant uncertainty in modeling the hadronic component, as the γ-ray flux and spectrum strongly depend on the distribution of proton energies, see [3]. We consider two cases:
(i) a Maxwellian distribution of protons, with temperature, $T_{\rm p}(R)$, determined by the hydrodynamic solution; this case is most likely unrealistic (as $t_a \ll t_p$) but it allows to estimate the minimum level of γ-ray luminosity for a given set of model parameters,
(ii) a power-law distribution, $n_{\rm p}(E) \propto E^{-s}$, with the radial-dependent power-law index, $s(R)$, determined by the requirement that the average energy of protons is the same as in case (i) at each radius R. Note that [3] and [4] considered a different non-thermal model, namely, the total dissipated power was assumed to heat only a small part of the proton population, leading to even more efficient γ-ray production than in our model.
For a given distribution of protons (thermal or non-thermal) we determine the γ-ray spectrum in the plasma rest frame, strictly following [8], and then we compute the transfer of γ-ray photons in curved space-time.

Figure 1 shows the radial distributions of parameters crucial for radiative processes and Figure 2 shows the broadband spectra produced in our models with $M = 2 \times 10^8 M_\odot$, $\dot{m} = 0.1$ and two extreme values of the spin parameter, $a = 0$ and $a = 0.998$. Furthermore, these models assume the viscosity parameter, $\alpha = 0.3$, and a rather weak magnetic field, with the magnetic pressure equal 1/10th of the total pressure. Note a strong difference between the dependence of the Comptonization and the hadronic components on the black-hole spin parameter.

Properties of the Comptonized radiation in our models are discussed in detail in [5], here we give a brief summary. We have studied in detail the hot flow solutions only for $\dot{m} = 0.1$–0.5, for which the luminosities of the flows are between $\sim 10^{-3}$ and $10^{-2} L_{\rm Edd}$ and the X-ray spectra are rather hard, with the photon spectral index $\Gamma \sim 1.6$. These parameters, just below the turning point in the Γ–$L/L_{\rm Edd}$ relation observed in both AGNs and black-hole binaries (see discussion and references in [5]), probably correspond to maximum luminosities of flows cooled by the Comptonization of synchrotron emission; we currently investigate solutions at lower $\dot{m}$. Rotation of the black hole stabilizes the circular motion of the flow which yields a higher density (through the continuity equation) within the innermost $\simeq 10R_{\rm g}$, see Figure 1b. The resulting large difference between the Coulomb heating in models with a non-rotating and a rapidly rotating black hole is, however, outweighed by a strong contribution of the compressive heating

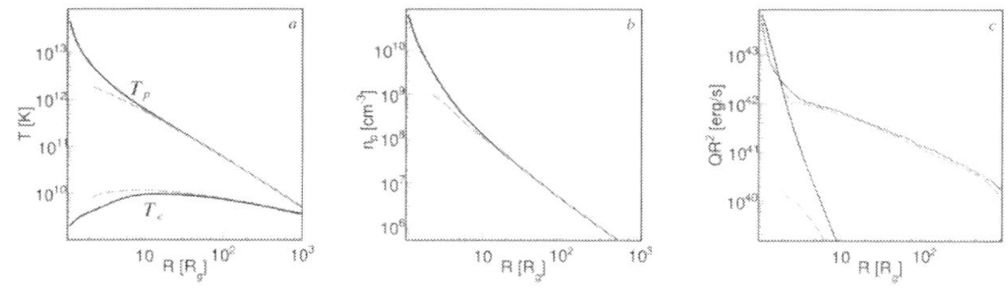

Figure 1. The radial profiles of the electron and proton temperature (a); the proton number density (b); and the proton (black, red) and electron (blue, magenta) cooling rates (c), in hot flow models with $M = 2 \times 10^8 M_\odot$ and $\dot{m} = 0.1$. In all panels, the solid (black, blue) lines are for $a = 0.998$ and the dashed (red, magenta) lines are for $a = 0$. In panel (c), Q denotes a vertically integrated rate, so QR^2 gives cooling rates (per unit volume) times volume.

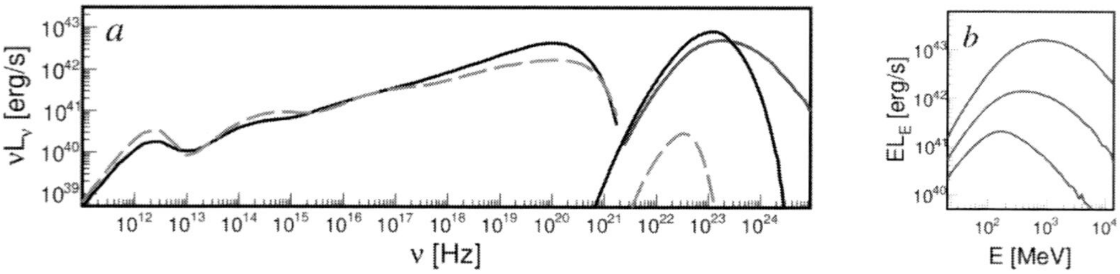

Figure 2. (a) The angle-averaged spectra produced in optically thin, two-temperature flows; the synchrotron+Compton (radio–hard-X-rays) and the hadronic (γ-rays) components are shown separately. The solid lines are for $a = 0.998$ and the dashed lines are for $a = 0$. The blue line is for a power-law distribution of protons, the remaining γ-ray spectra are for a thermal distribution. (b) Angular distribution of the γ-ray spectrum for $a = 0.998$ and the power-law distribution of protons; $\theta_{\rm obs} = 15$, 40 and 80° from bottom to top.

of electrons, much less dependent on spin. The flows have radiative efficiencies of $\eta \simeq 0.004$–0.01 in the considered range of luminosities.

In turn, the hadronic radiation is much more sensitive to the spin parameter. The stabilized rotation of the flow results in a much stronger dissipative heating in the inner region for $a = 0.998$, yielding a much larger proton temperature, see Figure 1a. As a result, the γ-ray luminosity is comparable to the X-ray luminosity for $a = 0.998$, while for $a = 0$ it is weaker by over 3 orders of magnitude, see Figure 2a. For the power-law distribution of protons, the energy index approaches $s \simeq 2.1$ in the innermost region in the model with $a = 0.998$ and the resulting γ-ray spectrum extends to much higher energies than the spectrum produced in the thermal proton model.

Our model of hadronic emission involves crucial improvements with respect to the previous studies. The self-similar solution of the flow structure, underlying the calculation of the γ-ray spectrum in [3], gives a very inaccurate approximation of the innermost region even for flows around a non-rotating black hole, whereas detailed predictions of the hadronic emission are very sensitive to the details of the flow structure in that region. [4] refined the model by using a GR model of the flow, however, they neglected the Doppler and gravitational shifts of energy as well as gravitational focusing and capturing by the black hole. As we can see in Figure 1c, the emissivity of the γ-rays is strongly concentrated within the innermost few R_g, then, neglecting the effects of gravitational redshift and the capturing of photons results in a significant overestimation of L_γ detected far away from the flow.

Journal of Physics: Conference Series **355** (2012) 012035 doi:10.1088/1742-6596/355/1/012035

Figure 2b illustrates another GR effect, related to the properties of the Kerr metric. Trajectories of photons emitted close to a rapidly rotating black hole are strongly bent toward the equatorial plane and the produced radiation may have a significant intrinsic anisotropy (see e.g. [9]). The magnitude of this anisotropy depends on the radial emissivity of a spectral component. For the Comptonized radiation the emissivity profile is rather flat and the anisotropy is reduced due to mixing of radiation produced in the innermost part with that from more distant regions. For the hadronic component the dependence on $\theta_{\rm obs}$ is very pronounced, with the difference by two orders of magnitude between the fluxes detected by face-on and edge-on observers, and a significant difference between the spectral shapes.

We have not included the effect of $\gamma\gamma$ absorption in our computations yet, which is currently the major shortcoming of our model. Very rough estimations, with an isotropic approximation of the photon field, indicate that for $\dot{m} = 0.1$ the flow is transparent for photons with energies $\lesssim 1$ GeV but the innermost part is opaque for photons with higher energies. Exact calculations, taking into account the angular distribution of photons in the flow, are required for accurate assessment of this effect. We can expect that the probability of absorption strongly depends on the direction of photon escape. The density of the photon field increases with increasing $\dot{m}$ and we expect that the γ-rays cannot escape from the central region for $\dot{m} > 0.1$. Then, the models shown in Figures 1 and 2 correspond roughly to the largest L_γ (of $\sim 10^{44}$ erg s^{-1} for large values of M, a and $\theta_{\rm obs}$) which can be expected from accretion flows in the specific accretion scenario considered in our model.

Finally, we note that proton-photon interactions are much less efficient than proton-proton interactions in the central region of the flow. Namely, the number density of γ-ray photons, which can effectively interact with protons in photomeson production, is by a factor of $\sim 10^3$ smaller than the number density of protons, $n_{\rm p}$, within the innermost few $R_{\rm g}$; the number density of hard X-ray photons, which can interact with protons in photopair production, is similar to $n_{\rm p}$. The cross-section for both channels of proton-photon interactions is by over two orders of magnitude smaller than the cross-section for proton-proton interaction, making the photo-hadronic production of secondary particles negligible.

3. Comparison with observations

It was suggested already by [10] that tenuous, two-temperature flows may be responsible for the low nuclear luminosities in radio galaxies with large radio lobes. Currently, a substantial observational evidence indicates that accretion flow typically proceeds through the radiatively-inefficient mode below a characteristic luminosity of $\simeq 0.01 L_{\rm Edd}$ in AGNs (e.g., [11]). FR I radio galaxies make a class of objects observed at such luminosities which are particularly interesting for our study, as the key property required for large L_γ in our model - i.e. rapid black-hole rotation - can be expected primarily in jet sources. Remarkably, an average radiative efficiency, $\eta \lesssim 0.005$, estimated for FR Is by [12] is consistent with the values of η obtained in our hot flow solutions. Furthermore, our model, in which the synchrotron emission is the main source of seed photons, predicts a hardening of the X-ray spectrum with increasing luminosity, which is indeed observed in low-luminosity AGNs; interestingly, the values of Γ and the 2–10 keV luminosity, L_{2-10}, measured in two well known, γ-ray emitting FR I galaxies, Cen A (e.g. [13]) and M87 [14], agree with the Γ–$L_{2-10}/L_{\rm Edd}$ relation determined for low luminosity AGNs, e.g., in [15]. [16] investigate the X-ray emission, using the coupled jet/accretion-flow model, in a sample of FR I radio galaxies and they conclude that the jet emission may dominate only in the least luminous sources, with $L_{2-10} \lesssim 5 \times 10^{-6} L_{\rm Edd}$.

The MeV/GeV luminosities measured in FR Is are between $\simeq 10^{41}$ and 10^{44} erg s^{-1} [6], and, as we note above, such L_γ can be produced by hot flows with $\dot{m} \lesssim 0.1$ around rotating black holes. The steep slopes of the γ-ray spectra determined in most FR Is, with $\Gamma_\gamma \gtrsim 2.2$, are consistent with the softness of spectra from π^0 decay; a simple power-law gives a good representation of the observed data, but their quality does not allow to rule out spectral curvature which characterizes the π^0 decay spectra. NGC 1275, with a flat γ-ray spectrum, cannot be explained by our hadronic model; this source may contain a significant contribution from the jet, as suggested also by its location in the region occupied by BL Lacs in the Γ_γ–L_γ plane (see fig. 6 in [6]) as well as by the very small inclination of the jet axis to the line of sight, of $< 3^\circ$ close to the core [17].

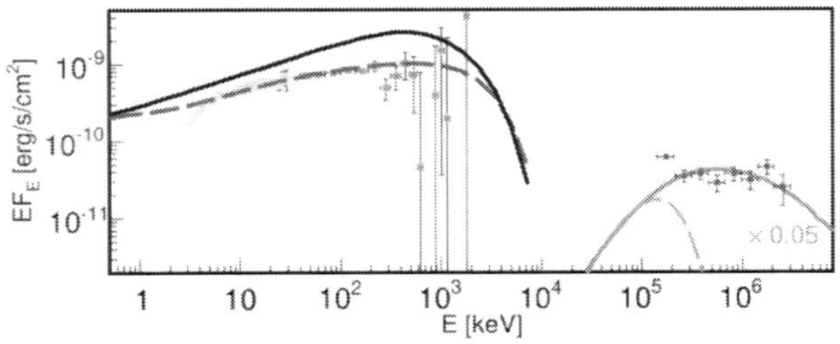

Figure 3. Nonsimultaneous observational data of Cen A by *INTEGRAL* and *Fermi*, see [18], compared with our model spectra. The solid-black and dashed-blue lines show the Comptonization spectra for $a = 0.998$ and $a = 0$, respectively. The solid-red line shows the hadronic spectrum, rescaled by a factor of 0.05, for $a = 0.998$ in the model assuming a power-law distribution of protons. The dashed-magenta line shows the thermal hadronic component for $a = 0$. The spectra for $a = 0.998$ correspond to $\theta_{\rm obs} = 50^\circ$. The green squares, red squares and blue circles show the *INTEGRAL*/JEM-X, *INTEGRAL*/SPI and *Fermi* data, respectively. The observed spectrum is strongly absorbed below 10 keV, which effect is not taken into account in the model spectra. Normalization of the model spectra corresponds to the distance of 3.8 Mpc.

Thanks to its proximity, Centaurus A is the best studied FR I radio galaxy, suitable for detailed testing of various accretion models through modeling of its broadband spectrum. Its X-ray luminosity, $L_{2-10}/L_{\rm Edd} \simeq 10^{-4}$, satisfies the criterion, suggested in [16], for a FR I galaxy to be dominated by a hot flow rather than a jet emission. In Figure 3 we compare the Comptonization and hadronic spectra from our model with the X/γ-ray observations of Cen A, described in [18].

As we discuss in [5], our model approximately reproduces the luminosity and the slope of the X-ray emission in Cen A; furthermore, it predicts very small changes of Γ corresponding to the change of L_{2-10} by a factor of 3 (as typically observed in Cen A) which appears to be compatible with the reported spectral variability within $\Gamma \simeq 1.7$–1.8 only. On the other hand, the spectral break energy of $\sim$200 keV, reported in [18], is inconsistent with breaks at $\sim$500 keV predicted in our models. Furthermore, the $a = 0.998$ model predicts a slightly harder spectrum than that observed, as can be seen in Figure 3. However, an additional process, which would lead to a softening of the spectrum and a decrease of the cut-off energy, has to be included in the model for large values of a. Namely, an efficient non-thermal synchrotron emission from relativistic $e^\pm$, coming from the decay of charged pions copiously produced, together with π^0, through proton-proton collisions, would provide a strong seed photon input in addition to the thermal synchrotron emission. We have performed initial calculations confirming that this effect should be crucial in approaching the agreement between the model with a rotating black hole and the hard X-ray data.

As illustrated in Figure 3, the prediction for the γ-ray flux may be strongly constraining for the black hole spin if $\dot{m}$ is determined by the X-ray flux. For Cen A, the detected MeV/GeV flux cannot be explained by emission from the flow around a slowly rotating black hole. For a maximally rotating black hole, the predicted γ-ray flux exceeds the measurement by a factor of $\sim$ 10 (the exact value depends on $\theta_{\rm obs}$) and we conclude that the black hole in Cen A cannot rotate at the maximum velocity, if the X-ray emission is produced by thermal Comptonization in an optically thin flow.

The (rescaled) hadronic spectrum for the power-law distribution of protons is in agreement with the *Fermi* data points except for the lowest energy bin. An excess at these energies could be possibly explained for the plasma containing two populations of protons: a subrelativistic thermal-like population, giving a contribution similar to that shown by the dashed curve for $a = 0$, and a non-thermal one.

4. Discussion

FR I radio galaxies are supposed to be the parent population of BL Lacs and their radiation is often interpreted as jet emission observed at large inclination angle to the jet axis and thus subject to less efficient Doppler boosting. Spectral energy distributions of FR Is are indeed consistent with a jet model, however, the Lorentz and Doppler factors required by such models are much lower than typical values found in models of BL Lac objects, see [6]. Therefore, the radiation observed in FR Is and BL Lacs must have a different origin; otherwise the debeamed radiation seen in the latter would be much weaker than that observed in FR Is. This has led to the development of complex models explaining a significant spread of the bulk velocity of the emitting plasma in the jet. However, this deprives the jet model of its major attracting feature, i.e. the simplicity by explaining differences between FR I and BL Lac objects through orientation effects only.

Addressing the problem of the origin of the high-energy radiation from FR Is, we note that γ-ray detections, usually regarded as the evidence that the observed radiation originates from the jet, can be alternatively explained by the hadronic emission from a hot flow. Our preliminary comparison of the X/γ-ray spectrum observed from Cen A with the hot flow model spectra suggests that this object can be explained by the model with a submaximally rotating black hole. A more detailed comparison will be done in our future work, in which we will take into account additional effects from hadronic processes in the X-ray spectra as well as the absorption of γ-rays.

ACKNOWLEDGMENTS

This research has been supported in part by the NCN grant N N203 582240, the NSFC grants 10973003 and 10843007 and the NBRPC grants 2009CB824800 and 2009CB24901.

References

[1] Narayan R., McClintock J. E., 2008, *New Astronomy Review* **51** 733
[2] Shapiro S. L., Lightman A. P., Eardley D. M., 1976 *ApJ* **204** 187S
[3] Mahadevan R., Narayan R., Krolik J., 1997 *ApJ* **486** 268
[4] Oka K., Manmoto T., 2003 *MNRAS* **340** 543
[5] Niedźwiecki A., Xie F.-G., Zdziarski A. A., 2011 *MNRAS* submitted [arXiv:1107.0860]
[6] Abdo A. A., et al., 2010 *ApJ* **720** 912
[7] Mahadevan R., Quataert E., 1997 *ApJ* **490** 605
[8] Dermer C., 1986 *ApJ* **307** 47
[9] Niedźwiecki A., 2005 *MNRAS* **356** 913
[10] Rees M. J., Begelman M. C., Blandford R. D., Phinney E. S., 1982 *Nature* **295** 17
[11] Ho L. C., 2008 *ARA&A* **46** 475
[12] Balmaverde B., Baldi R. D., Capetti A., 2008 *A&A* **486** 119
[13] Evans D., et al., 2004, *ApJ* **612** 786
[14] Di Matteo T., Allen S. W., Fabian A., Wilson A., Young A., 2003 *ApJ* **582** 133D
[15] Younes G., Porquet D., Sabra B., Reeves J. N., 2011 *A&A* **530** 149
[16] Wu Q., Yuan F., Cao X., 2007 *ApJ* **669** 96
[17] Krichbaum T. P., et al., 1992 *A&A* **260** 33
[18] Beckmann V., Jean P., Lubiński P., Soldi S., Terrier R., 2011 *A&A* **531** 70

Contribution from star-forming galaxies to the cosmic gamma-ray background radiation

Ryu Makiya

Department of Astronomy, Kyoto University, Kitashirakawa-Oiwake-cho, Sakyo-ku, Kyoto 606-8502, Japan

E-mail: makiya@kusastro.kyoto-u.ac.jp

Abstract. The origin of extragalactic gamma-ray background (EGRB) has been discussed for a long time and various sources have been discussed as possible contributors to EGRB, such as active galactic nuclei, star-forming galaxies, galaxy clusters, structure formation, and dark matter annihilation. We present a new theoretical calculation of the contribution to the EGRB from star-forming galaxies based on a state-of-the-art model of hierarchical galaxy formation that is in quantitative agreement with a variety of observations of local galaxies as well as high redshifts. Gamma-ray luminosity and spectrum of galaxies are related to star formation rate, gas mass, and star formation mode (quiescent or starburst) of model galaxies using latest observed data of nearby galaxies. Our standard model predicts $\sim$ 14% contribution from star-forming galaxies to the total EGRB flux recently reported by the Fermi Gamma-Ray Space Telescope. The predicted number of nearby galaxies detectable by Fermi is consistent with the observation. The sum of the known contributions from AGNs and star-forming galaxies can explain a large part of EGRB, with a remarkable agreement between the predicted model spectrum and observation.

1. Introduction

Diffuse γ-rays coming from all directions of sky was first observed by SAS-2 satellite (EGRB[1][2]). The origin of EGRB has been thought that superposition of faint discrete sources, such as active galactic nuclei (AGNs), galaxy clusters, structure formation, and dark matter annihilation. In this work, we calculated the contribution from star-forming galaxies to the EGRB, using the state-of-art model of hierarchical galaxy formation and latest observational data of nearby galaxies. Cosmic-rays accelerated by supernova remnants (SNR) are interacted with interstellar gas and radiation field, and emit γ- rays via bremsstrahlung, inverse Compton (IC), and π_0 decay process. Diffuse gamma-rays from our galaxy and large magellanic cloud (LMC) has been observed, and recently, small magellanic cloud (SMC), M82 and NGC253 are detected in γ-rays for the first time by H.E.S.S. [3], VERITAS [4], and *Fermi* [5].

2. Model

To calculate the contribution from star-forming galaxies to the EGRB, we need a γ-ray luminosity function (LF) of galaxies (comoving number density of galaxies per unit γ- ray luminosity). We use a mock galaxy catalog produced by one of the latest semi- analytic models of hierarchical galaxy formation (the Mitaka model [6][7]) to construct the LF. The Mitaka model can quantitatively reproduce a wide variety of observed characteristics of galaxies.

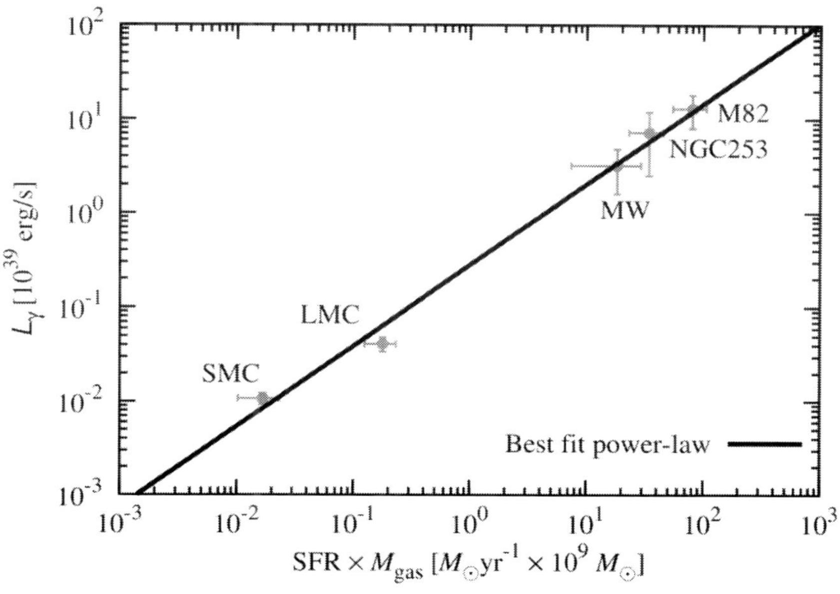

Figure 1. SFR × gas mass against the γ-ray luminosity.

We estimated the γ-ray luminosity of galaxies using star formation rate (SFR) and gas mass. SFR is an indicator of the cosmic-ray luminosity, while gas mass account for the amount of target atom. In Fig.1 it is clearly seen that there is a good correlation between the γ-ray luminosity and SFR × gas mass. We fitted these data to power-law function, and the result is

$$L_\gamma = 0.28 \times \left(\frac{\text{SFR}}{M_\odot \text{yr}^{-1}} \times \frac{M_{\text{gas}}}{10^9 M_\odot}\right)^{0.86} [10^{39}\text{erg/s}], \quad (1)$$

where L_γ is γ-ray luminosity in 100 MeV - 5 GeV. For the γ-ray spectrum of galaxies, we simply assumed that all quiescent galaxies have the same spectral shape as the MW. For starburst galaxies, spectrum could become harder than quiescent galaxies, therefore we consider another model that MW-like spectrum plus hard power-law component (see fig.2).

Using the mock numerical galaxy catalog including the SFR and gas mass of each galaxies and eq.(1), we constructed the γ-ray LF of galaxies. The EGRB flux and spectrum can be calculated by integrating the γ-ray LF. We also calculated the attenuation of γ-rays by cosmic infrared background (CIB) photons, and reprocessed cascade emission.

3. Results

In Fig.3 the EGRB flux from all star-forming galaxies (quiescent+starburst), as well as the total flux including AGNs, are plotted. Model 1 and 2 correspond to two cases of spectrum of starburst galaxies. For the model of the AGN contribution we used Inoue & Totani (2009)[8]. The data points are the *Fermi* measurements of the EGRB including the contribution from resolved sources [9]. To show the effect of absorption, Model 1 with no- absorption and the flux of cascade emission are also plotted. As a result, the EGRB photon flux from all star-forming galaxies become 7% of the total EGRB, and the sum of the contributions from AGNs and star-forming galaxies account for the 50% of the total EGRB. It should also be noted that the combined spectrum by AGNs and star-forming galaxies is remarkably similar to the observed

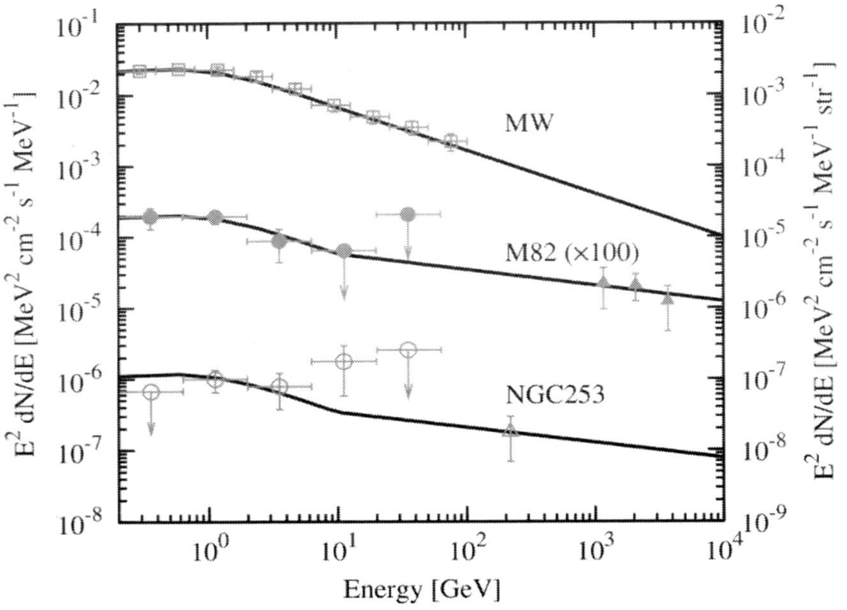

Figure 2. γ-ray spectrum of galaxies. MW is in units shown on the right ordinate, while the others are in left ordinate.

EGRB spectrum. Further examination is required to see whether the 50% residual of the EGRB is mainly a result of modeling uncertainty, experimental uncertainty, or significant contributions from other sources, e.g., dark matter annihilation.

4. Summary

We have presented a new calculation of EGRB from cosmic-ray interactions in star-forming galaxies, based on a state-of-the-art galaxy formation model in the framework of hierarchical structure formation. We found that star-forming galaxies make $\sim 7\%$ contribution to the total EGRB flux reported by *Fermi* in our standard model. Combined with the contribution from blazars as estimated by the *Fermi* data, more than 50% of EGRB can be accounted for. The combined spectrum by AGNs and star-forming galaxies is remarkably similar to the observed EGRB spectrum.

More detailed description of our model and precise discussion of model uncertainties can be found in Makiya et al. (2011) [10].

Acknowledgments

This work was supported by the Grant-Aid for the Global COE Program "The Next Generation of Physics, Spun from University and Emergence " from the Ministry of Education, Culture, Sports, Science and Technology (MEXT) of Japan. The numerical calculations were in part carried out on SGI Altix3700 BX2 at Yukawa Institute for Theoretical Physics of Kyoto University. TT was supported by the Grant-in-Aid for Scientific Research (19047003, 19740099, 2004005) from MEXT. MARK was supported by the Research Fellowship for Young Scientists from the Japan Society for the Promotion of Science (JSPS).

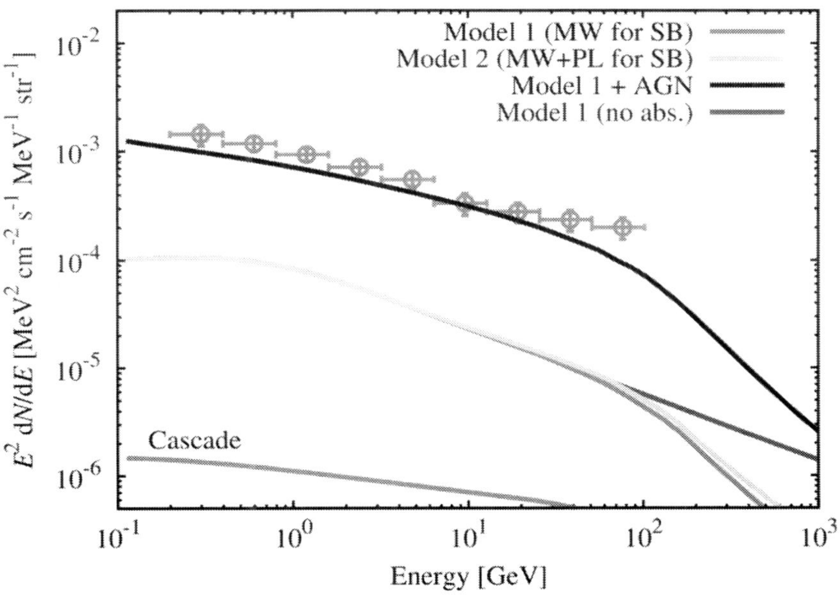

Figure 3. The EGRB spectrum.

References

[1] Fichtel, C. E., Hartman, R. C., Kniffen, D. A., Thompson, D. J., Ogelman, H. B., Ozel, M. E., & Tumer, T. 1977, ApJ, 217, L9
[2] Fichtel, C. E., Simpson, G. A., & Thompson, D. J. 1978, ApJ, 222, 833
[3] Acero, F., et al. 2009, Science, 326, 1080
[4] VERITAS Collaboration 2009, Nature, 462, 770
[5] Abdo, A. A., et al. 2010, ApJ, 709, L152
[6] Nagashima, M., & Yoshii, Y. 2004, ApJ, 610, 23
[7] Nagashima, M., Yahagi, H., Enoki, M., Yoshii, Y., & Gouda, N. 2005, ApJ, 634, 26
[8] Inoue, Y., & Totani, T. 2009, ApJ, 702, 523
[9] Abdo, A. A., et al. 2010b, Phys. Rev. Lett., 104, 101101
[10] Makiya, R., Totani, T., & Kobayashi, M. A. R. 2011, ApJ, 728, 158

Journal of Physics: Conference Series **355** (2012) 012037 doi:10.1088/1742-6596/355/1/012037

Extragalactic Gamma-ray Background Radiation from Beamed and Unbeamed Active Galactic Nuclei

Yoshiyuki Inoue

Department of Astronomy, Kyoto University, Kitashirakawa, Sakyo-ku, Kyoto 606-8502, Japan

E-mail: yinoue@kusastro.kyoto-u.ac.jp

Abstract. The origin of the extragalactic gamma-ray background (EGRB) radiation has been a mystery in astrophysics for a long time. Recently the Fermi gamma-ray satellite (*Fermi*) has revealed that ∼22% of the unresolved EGRB would be explained by blazars, which are one population of beamed active galactic nuclei (AGNs). The remaining ∼78% of the unresolved EGRB is still unknown. We estimate the contribution of gamma-ray loud radio galaxies, which are misaligned radio loud AGNs recently detected by *Fermi*, to EGRB using the radio luminosity function of radio-loud AGNs with the correlation between the radio and gamma-ray luminosities. We find that ∼25% of the unresolved EGRB will be explained by gamma-ray loud radio galaxy population. We also discuss further about the origin of EGRB by comparing the *Fermi* EGRB data with our studies on various AGN populations' contribution to EGRB, which are radio quiet AGNs, blazars, and gamma-ray loud radio galaxies.

1. Introduction

The origin of the extragalactic gamma-ray background (EGRB) radiation has been argued for a long time in astrophysics. Although several sources have been suggested to explain the MeV and GeV background, the origin of the MeV and GeV background is still a debated issue also due to the uncertainties and difficulties connected to the measurements of their intensities. In the case of the GeV background, since blazars which are a type of active galactic nuclei (AGNs) are dominant extragalactic gamma-ray sources [1, 2], it is expected that unresolved population of blazars would explain the GeV EGRB (see e.g. [3]). Recently, [4], analyzing data from the 11months catalog, showed that unresolved blazars can explain $\sim 22\%$ of the EGRB above 0.1 GeV. Other gamma-ray emitting extragalactic sources have also been discussed as the origin of the GeV EGRB (see [5] for details).

Fermi has recently detected GeV gamma-ray emissions from 11 misaligned AGNs (i.e. radio galaxies), which are one type of AGNs with the direction of a relativistic jet *not* coinciding with our line of sight [6]. Their average photon index at 0.1-10 GeV is ~ 2.4 which is same as that of GeV EGRB [7] and blazars [4]. Although they are fainter than blazars, the expected number in the entire sky is larger than blazars. Then, it is naturally expected that they will make a significant contribution to EGRB.

In this proceeding paper, we discuss the contribution of gamma-ray loud radio galaxies (not blazars) to EGRB based on [5]. Furthermore, we also discuss the EGRB contribution from various AGN populations such as radio-quiet AGNs and blazars based on our previous studies [19, 3]. Throughout this paper, we adopt the standard cosmological parameters of $(h, \Omega_M, \Omega_\Lambda) = (0.7, 0.3, 0.7)$.

2. Gamma-ray Luminosity Function of Gamma-ray Loud Radio Galaxies

To estimate the EGRB contribution of gamma-ray loud radio galaxies, we need to construct a gamma-ray luminosity function (GLF). However, because of the small statistics, it is difficult to construct a GLF using the current gamma-ray data only. Here, the radio luminosity function (RLF) of radio galaxies is extensively studied by previous works (see e.g. [10, 11]). If there is a correlation between the radio and gamma-ray luminosities, we are able to convert the RLF to the GLF using that correlation.

Figure 1 shows the 5 GHz and 0.1-10 GeV luminosity relation of the *Fermi* gamma-ray loud radio galaxies. Square and triangle data represents FRI and FRII radio galaxies, respectively. The solid line shows the fitting line to all the data. The function is given by

$$\log_{10}(L_\gamma) = (-3.90 \pm 0.61) + (1.16 \pm 0.02)\log_{10}(L_{5\rm GHz}), \tag{1}$$

where errors show 1-σ uncertainties. In the case of blazars, the slope of the correlation between $L_\gamma(>100{\rm MeV})$, luminosity above 100 MeV, and radio luminosity at 20 GHz is 1.07 ± 0.05 [8]. The correlation slopes of gamma-ray loud radio galaxies are similar to that of blazars. This may indicate that emission mechanism is similar in gamma-ray loud radio galaxies and blazars. The partial correlation coefficient [9] for this data set becomes 0.866 with chance probability 1.65×10^{-6}. Therefore, we conclude that there is a correlation between the radio and gamma-ray luminosities of gamma-ray loud radio galaxies.

We derive the GLF of gamma-ray loud radio galaxies, $\rho_\gamma(L_\gamma, z)$, using the correlation between the radio and gamma-ray luminosities as Equation 1 and the RLF of radio galaxies, $\rho_r(L_r, z)$, with radio luminosity, L_r. The GLF is given as

$$\rho_\gamma(L_\gamma, z) = \kappa \frac{dL_r}{dL_\gamma}\rho_r(L_r, z), \tag{2}$$

where κ is a normalization factor. We use the RLF by [11].

The normalization factor κ, which corresponds to the fraction of gamma-ray loud radio galaxies against all radio galaxies, is determined by normalizing our GLF to the source count distribution of the *Fermi* radio galaxies, which is sometimes called logN–logS plot or cumulative flux distribution. Source count distribution is calculated by

$$N(>F_\gamma) = 4\pi \int_0^{z_{\rm max}} dz \frac{d^2V}{d\Omega dz} \int_{L_\gamma(z,F_\gamma)}^{L_{\gamma,{\rm max}}} dL_\gamma \rho_\gamma(L_\gamma, z), \tag{3}$$

where $L_\gamma(z, F_\gamma)$ is the gamma-ray luminosity of a blazar at redshift z whose photon flux at >100 MeV is F_γ. Hereinafter, we assume $z_{\rm max} = 5$ and $L_{\gamma,{\rm max}} = 10^{48}$erg/s in this study. These assumptions hardly affect the results in this study.

Since the completeness of the *Fermi* sky survey depends on the photon flux and photon index of a source, we need to take into account this effect (so called detection efficiency) to compare GLF with the cumulative source count distribution of gamma-ray loud radio galaxies. The detection efficiency of *Fermi* is shown in Figure 7 of [4].

Figure 2 shows the source count distribution of gamma-ray loud radio galaxies. The data is after the conversion of the detection efficiency. Solid red curve shows in the case of $\kappa = 1$ which corresponds to the case that all radio galaxies emit gamma-rays. Dashed blue curve corresponds to the GLF fitted to the *Fermi* data with $\kappa = 0.081 \pm 0.011$. ~ 1000 gamma-ray loud radio galaxies are expected with 100% complete entire sky survey above the flux threshold $F_\gamma(>100{\rm MeV}) = 1.0 \times 10^{-9}$ photons cm^{-2} s^{-1} above 100 MeV. We note that the current detection efficiency of *Fermi* at $F_\gamma(>100{\rm MeV}) = 1.0 \times 10^{-9}$ photons cm^{-2} s^{-1} is $\sim 10^{-3}$.

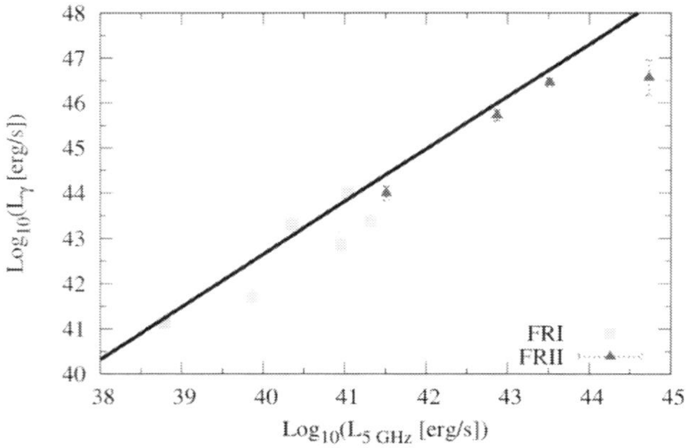

Figure 1. Gamma-ray luminosity at 0.1–10 GeV versus radio luminosity at 5 GHz. The square and triangle data represents FRI and FRII galaxies, respectively. The solid line is the fit to all sources.

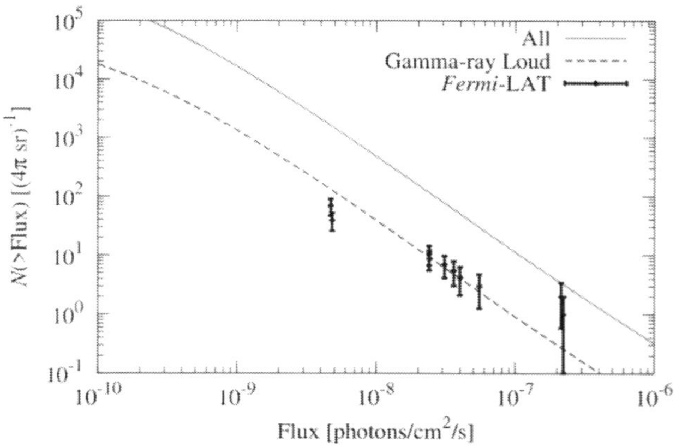

Figure 2. The source count distribution of gamma-ray loud radio galaxies in the entire sky. Solid and dashed curve corresponds to the all radio galaxies and gamma-ray loud radio galaxies, respectively. The *Fermi* data are also shown.

3. EGRB from Gamma-ray Loud Radio Galaxies

We calculate the EGRB spectrum by integrating our GLF in the redshift and luminosity space, using the a simple broken power law SED model (see [5] for details). The EGRB spectrum is calculated as

$$\frac{d^2F(\epsilon)}{d\epsilon d\Omega} = \frac{c}{4\pi}\int_0^{z_{\max}} dz \left|\frac{dt}{dz}\right| \int_{L_{\gamma,\min}}^{L_{\gamma,\max}} dL_\gamma \rho_\gamma(L_\gamma, z) \frac{dL[L_\gamma,(1+z)\epsilon]}{d\epsilon} \exp[-\tau_{\gamma,\gamma}(\epsilon,z)] \times \{1.0 - \omega(F_\gamma[L_\gamma, z])\}, \tag{4}$$

where t is the cosmic time and dt/dz can be calculated by the Friedmann equation in the standard cosmology. The minimum gamma-ray luminosity is set at $L_{\gamma,\min} = 10^{39}$ erg/s. $\omega(F_\gamma[L_\gamma, z])$ is the detection efficiency of *Fermi* at the photon flux $F_\gamma(L_\gamma, z)$.

Here, high energy γ-rays propagating the universe are absorbed by the interaction with the extragalactic background light (EBL), also referred to as cosmic optical and infrared background [12, 13]. $\tau_{\gamma,\gamma}(\epsilon, z)$ is the optical depth of this background radiation. In this study, we adopt the EBL model by [13]. We also include the contribution of the cascade emission. The gamma-ray absorption creates electron–positron pairs. These pairs scatter the cosmic microwave background radiation to make the secondary emission component (so called cascade emission) to the absorbed primary emission [14].

Fig. 3 shows the unresolved νI_ν EGRB spectrum of gamma-ray loud radio galaxies in the unit of MeV^{2}cm^{-2}s^{-1}MeV^{-1}sr^{-1} predicted by our GLF. The intrinsic (the spectrum without the EBL absorption effect), absorbed, and cascade components of the EGRB spectrum and the total EGRB spectrum (absorbed+cascade) are shown. As in Fig. 3, the cascade emission does not contribute to the EGRB spectrum significantly.

The expected EGRB photon flux above 100 MeV from gamma-ray loud radio galaxy populations is 0.26×10^{-5} photons cm^{-2} s^{-1} sr^{-1}. As the unresolved *Fermi* EGRB flux above 100 MeV is 1.03×10^{-5} photons cm^{-2} s^{-1} sr^{-1} [7], the gamma-ray loud radio galaxies explains $\sim$ 25% of the unresolved EGRB flux. For the comparison, recent analysis of *Fermi* blazars showed that blazars explains $\sim$ 22% of the unresolved EGRB [4]. Therefore, we are able to

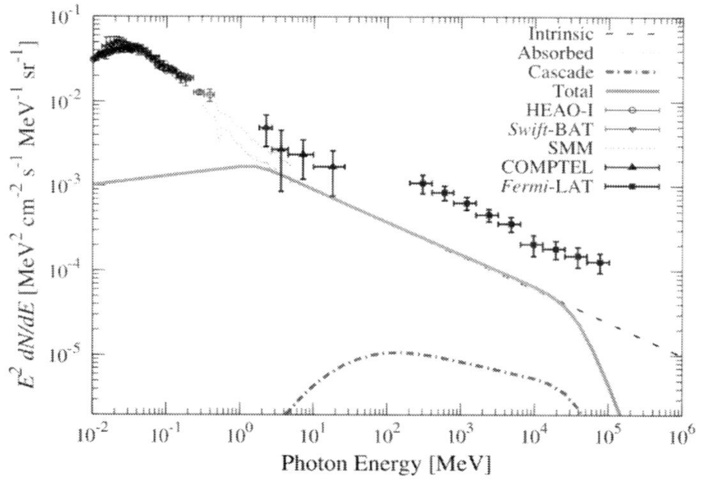

Figure 3. The unresolved EGRB spectrum from gamma-ray loud radio galaxies. Dashed, dotted, dot-dashed, and solid curves show the intrinsic spectrum (no absorption), absorbed, cascade, and total (absorbed+cascade) EGRB spectrum, respectively. The observed data of *HEAO-I* [15], *Swift*-BAT [16], *SMM* [17], COMPTEL [18], and *Fermi*-LAT [7] are also shown by the symbols indicated in the figure.

explain $\sim$ 47% of EGRB by radio-loud AGN populations. We note that, because of small samples and the uncertainty in the correlation between radio and gamma-ray luminosities, the fraction of gamma-ray loud radio galaxies in unresolved EGRB varies from $\sim$ 10% to $\sim$ 63%.

4. EGRB from Beamed and Unbeamed AGNs

We have also conducted works on EGRB from radio-quiet AGNs [19] and blazars [3]. Radio-quiet AGNs would significantly contribute to the MeV background by including non-thermal electron population in the AGN hot corona above the accretion disk [19]. Our blazar GLF model [3] is constructed based on the EGRET data taking into account blazar sequence [20, 21]. This blazar GLF is consistent with various *Fermi* data (source counts and EGRB contribution) [22, 23]. Fig. 4 shows the EGRB contribution of radio-quiet AGNs, blazars, and gamma-ray loud radio galaxies (i.e. beamed and unbeamed AGN populations). As in the figure, AGN populations are dominant in the cosmic background radiation from X-ray to gamma-ray. In this figure, we use the total (resolved + unresolved) EGRB data to avoid the instrument dependence, since the model by [3] is based on the EGRET data.

5. Origin of the GeV EGRB

Although AGN populations dominate EGRB, it is difficult to explain 100% of EGRB. Fig. 5 shows the unresolved gamma-ray loud radio galaxy EGRB spectra at each redshift bins. Because of EBL, the spectrum above > 30 GeV shows the absorbed signature. Here, cosmological sources have their evolution peaks at $z = 1 \sim 2$ such as cosmic star formation history and AGN activity (see e.g. [24, 25]). This means that the gamma-rays from extragalactic sources (e.g. galaxies and AGNs) will experience EBL absorption. However, as shown in Fig. 5, the *Fermi* EGRB spectrum does not clearly show such an absorbed signature. This might suggest that nearby gamma-ray emitting sources or sources with very hard gamma-ray spectrum would be the dominant population of EGRB above 10 GeV such as high-frequency-peaked BL Lacs. To address this issue, we should await the EGRB information above 100 GeV by future observations such as those provided by *Fermi*. CTA would also be possible to see the EGRB at much higher energy band. We also need to examine the EBL models at high redshift. It is expected that CTA will see blazars up to $z \sim 1.2$ at very high energy > 30 GeV [22]. Therefore, *Fermi* and CTA will be a key to understanding the origin of EGRB.

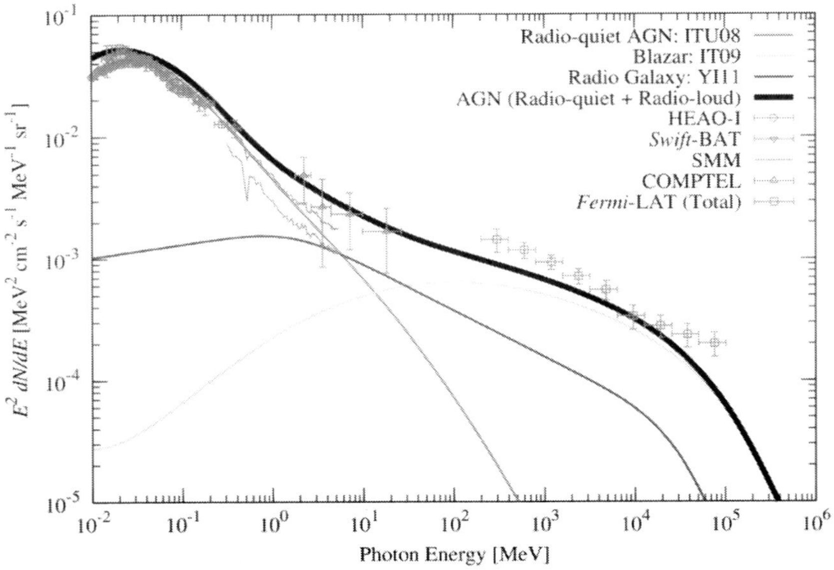

Figure 4. EGRB spectrum from various AGN populations. Magenta, green, blue, and black curves show the EGRB contribution from radio-quiet AGN [19], blazars [3], gamma-ray loud radio galaxies [5], and total AGN populations (radio-quiet AGNs, blazars, and gamma-ray loud radio galaxies), respectively. The observed data are same as Fig. 3, but total (resolved + unresolved) EGRB data are shown here.

6. Conclusion

In this paper, we studied the contribution of gamma-ray loud radio galaxies to the EGRB by constructing their GLF. First, we explored the correlation between the radio and gamma-ray luminosities of gamma-ray loud radio galaxies which are recently reported by *Fermi*. We found that there is a correlation $L_\gamma \propto L_{5\rm GHz}^{1.16}$ by a partial correlation analysis.

Based on this correlation, we defined the GLF of gamma-ray loud radio galaxies using the RLF of radio galaxies. We normalized the GLF to fit to the cumulative flux distribution of *Fermi* samples. Then, we predicted the contribution of gamma-ray loud radio galaxies to the MeV and GeV EGRB. We found that gamma-ray loud radio galaxies will explain $\sim 25\%$ of the EGRB flux above 100 MeV and also make a significant contribution to the 1–30 MeV EGRB. Since blazars explain $\sim$ 22% of EGRB, $\sim$ 47% of EGRB would be explained by blazars and gamma-ray loud radio galaxies.

Based on our previous studies of radio-quiet AGNs and blazars, we also showed the contribution of beamed and unbeamed AGN populations to EGRB. AGN populations dominate the cosmic X-ray and gamma-ray background in about 7 orders magnitude of photon energy.

We also make an interpretation on the origin of the EGRB above 10 GeV from the point of view of the EBL absorption effect. If the EBL absorption signature is still not appeared in the EGRB spectrum, the origin would be nearby sources or sources with hard gamma-ray spectrum. We should await the EGRB data at higher energy band for this issue.

Acknowledgments

The author thanks the organizing committees of the conference *Beamed and Unbeamed Gamma-rays from Galaxies* at Lapland Hotel Olos, Muonio, Finland for providing great opportunities of

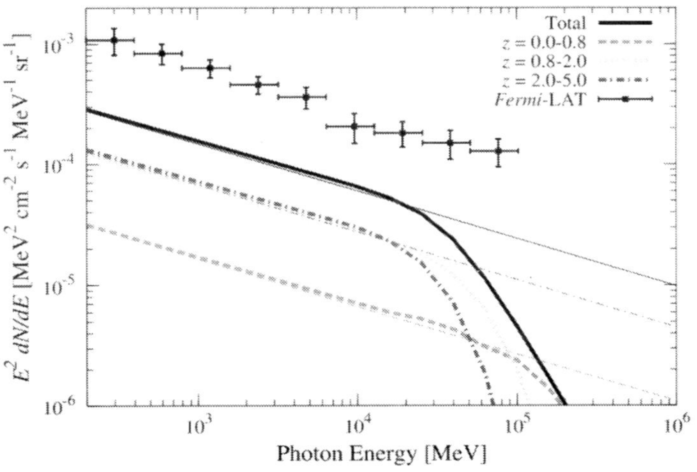

Figure 5. Gamma-ray loud radio galaxy EGRB spectra at each redshift bins. Solid, dashed, dotted, and dot-dashed curves shows the EGRB spectrum at $z = 0.0 - 5.0$, $z = 0.0 - 0.8$, $z = 0.8 - 2.0$, and $z = 2.0 - 5.0$, respectively. The thin and thick curve corresponds to the intrinsic spectrum (not absorbed) and the total spectrum (absorbed + cascade), respectively. The unresolved EGRB data of *Fermi*-LAT [7] is also shown.

lots of discussions. The author acknowledges support by the Research Fellowship of the Japan Society for the Promotion of Science (JSPS). This work was supported in part by the Global COE Program "The Next Generation of Physics, Spun from Universality and Emergence" at Kyoto University from the Ministry of Education, Culture, Sports, Science and Technology (MEXT) of Japan.

[1] Hartman, R. C. et al. 1999, *Astrophysical Journal, Supplement*, 123, 79
[2] Abdo, A. A., et al. 2010a, *Astrophysical Journal, Supplement*, 188, 405
[3] Inoue, Y. & Totani, T. 2009, *Astrophysical Journal*, 702, 523
[4] Abdo, A. A., et al. 2010d, *Astrophysical Journal*, 720, 435
[5] Inoue, Y. 2011, *Astrophysical Journal*, 733, 66
[6] Abdo, A. A., et al. 2010b, *Astrophysical Journal*, 720, 912
[7] Abdo, A. A., et al. 2010c, Physical Review Letters, 104, 101101
[8] Ghirlanda, G., Ghisellini, G., Tavecchio, F., Foschini, L., & Bonnoli, G. 2010, arXiv:1007.2751
[9] Padovani, P. 1992, *A&A*, 256, 399
[10] Dunlop, J. S. & Peacock, J. A. 1990, *Monthly Notices of the Royal Astronomical Society*, 247, 19
[11] Willott, C. J., Rawlings, S., Blundell, K. M., Lacy, M., & Eales, S. A. 2001, *Monthly Notices of the Royal Astronomical Society*, 322, 536
[12] Kneiske, T. M., Bretz, T., Mannheim, K., & Hartmann, D. H. 2004, *A&A*, 413, 807
[13] Finke, J. D., Razzaque, S., & Dermer, C. D. 2010, *Astrophysical Journal*, 712, 238
[14] Aharonian, F. A., Coppi, P. S., & Voelk, H. J. 1994, *Astrophysical Journal*, 423, L5
[15] Gruber, D. E., Matteson, J. L., Peterson, L. E., & Jung, G. V. 1999, *Astrophysical Journal*, 520, 124
[16] Ajello, M. et al. 2008, *Astrophysical Journal*, 689, 666
[17] Watanabe, K., Hartmann, D. H., Leising, M. D., The, L., Share, G. H., & Kinzer, R. L. 1997, in American Institute of Physics Conference Series, Vol. 410, Proceedings of the Fourth Compton Symposium, ed. C. D. Dermer, M. S. Strickman, & J. D. Kurfess, 1223–1227
[18] Kappadath, S. C. et al. 1996, A&AS, 120, C619+
[19] Inoue, Y., Totani, T., & Ueda, Y. 2008, *Astrophysical Journal*, 672, L5
[20] Fossati, G., Maraschi, L., Celotti, A., Comastri, A., & Ghisellini, G. 1998, *Monthly Notices of the Royal Astronomical Society*, 299, 433
[21] Kubo, H., Takahashi, T., Madejski, G., Tashiro, M., Makino, F., Inoue, S., & Takahara, F. 1998, *Astrophysical Journal*, 504, 693
[22] Inoue, Y., Totani, T., & Mori, M. 2010, *PASJ*, 62, 1005
[23] Inoue, Y., Inoue, S., Kobayashi, M. A. R., Totani, T., Kataoka, J., & Sato, R. 2011, *Monthly Notices of the Royal Astronomical Society*, 411, 464
[24] Hopkins, A. M. & Beacom, J. F. 2006, *Astrophysical Journal*, 651, 142
[25] Ueda, Y., Akiyama, M., Ohta, K., & Miyaji, T. 2003, *Astrophysical Journal*, 598, 886

Cosmic rays in star-forming galaxies

Massimo Persic[1], **Yoel Rephaeli**[2]

[1] INAF, Osservatorio Astronomico di Trieste, via G B Tiepolo 11, I-34143 Trieste, Italy; and INFN, Trieste

[2] School of Physics and Astronomy, Tel Aviv University, Tel Aviv 69978, Israel; and University of California, San Diego

E-mail: persic@oats.inaf.it, yoelr@wise.tau.ac.il

Abstract.
The energy density of cosmic ray protons (CRp) in star-forming environments can be *(i)* measured from γ-ray π^o-decay emission, *(ii)* inferred from the measured radio non-thermal synchrotron emission (once a theoretical p/e ratio and particle-field equipartition have been assumed), and *(iii)* estimated from the observed supernova rate and the deduced CRp residency time. For most of the currently available galaxies where these methods can be simultaneously applied, the results of the various methods agree and suggest that CRp energy densities range from $\mathcal{O}(10^{-1})\,\mathrm{eV\,cm^{-3}}$ in very quiet environments up to $\mathcal{O}(10^{2})\,\mathrm{eV\,cm^{-3}}$ in very active ones. The only case for which the methods do not agree is the Small Magellanic Cloud, where the discrepancy between measured and estimated CRp energy density may be due to a smaller characteristic CR consfinement volume.

1. Introduction

Active star formation in galaxies leads to production of CR protons and electrons via the Fermi-I diffusive shock acceleration mechanism in supernova (SN) remnants. Basic considerations suggest that the timescales required for CRp to be Fermi-I accelerated (τ_+) and to lose energy via pion decay into photons and e^+e^- pairs or via advection by bulk outflows (τ_-) are shorter than the timescale of starburst activity in galaxies ($\tau_{\rm SB}$), which is roughly comparable to a dynamical timescale. Consequently, in a starburst region a rough balance can be achieved between energy gains and losses for galactic CRs during a typical burst of star formation. Under equilibrium conditions in a galaxy, a minimum-energy configuration of the magnetic field and the CRs may be attained. This implies that energy densities of particles and magnetic fields may be in approximate equipartition.

The equipartition assumption enables deduction of the CRp energy density, $U_{\rm p}$ (which essentially measures the particle energy content), from the relatively easily observable synchrotron radio emission. In this *radio-based* approach, $U_{\rm p}$ can be estimated if the source's size, distance, radio flux, and radio spectral index are known, and particle-field equipartition and a p/e ratio are assumed.

In the γ*-based* approach, $U_{\rm p}$ can be obtained from numerically fitting the GeV-TeV spectral flux, which is mostly due – via π^0 decay – to CRp interactions with ambient gas protons, based on the solution of the diffusion-loss equation for accelerated CR particles. Only recently have such measurements become possible – and only for ten sources to date.

Journal of Physics: Conference Series **355** (2012) 012038 doi:10.1088/1742-6596/355/1/012038

Table 1. Star-forming galaxies: the data.

Object	$D_L^{[1]}$ (Mpc)	$r_s^{[2]}$ (kpc)	$f_{1\,\mathrm{GHz}}^{[3]}$ (Jy)	$\alpha_{\mathrm{NT}}^{[4]}$	$\nu_{\mathrm{SN}}^{[5]}$ (yr^{-1})	$M_{\mathrm{gas}}^{[6]}$ ($M_\odot$)	$L_\gamma^{[7]}$ ($\mathrm{erg\,s^{-1}}$)	$\tau_{\mathrm{res}}^{[8]}$ (yr)	Notes
Arp 220	74.7	0.25	0.3	0.65	3.5	9.24	<42.25	9.0E+3	SB
M 82	3.4	0.26	10.0	0.71	0.25	9.37	40.21	2.6E+3	SB
NGC 253	2.5	0.20	5.6	0.75	0.12	9.20	39.76	2.0E+4	SB
Milky Way	–	4.4	–	–	2E-2	9.81	38.91	2.7E+7	quiescent
M 31	0.78	5.17	4.8	0.88	1E-2	9.88	38.66	4.0E+7	quiescent
M 33	0.85	2.79	3.30	0.95	3E-3	9.35	<38.54	2.6E+7	quiescent
LMC	0.049	3.0	285.0	0.84	2E-3	8.86	37.67	1.0E+7	quiescent
SMC	0.061	1.53	45.3	0.85	1E-3	8.66	37.04	4.0E+7	quiescent

[1]Distance (from Abdo et al. 2011). [2]Effective radius of star-forming region. See text. Data are from Persic & Rephaeli 2010 and refs. therein (Arp 220, M 82, NGC 253), Beck & Gräve 1982 (M 31), Tabatabaei et al. 2007 (M 33), Weinberg & Nikolaev 2001 (LMC), and Wilke et al. 2003 (SMC). [3]1 GHz flux density. Data are from Persic & Rephaeli 2010 and refs. therein (Arp 220, M 82, NGC 253)), Beck & Gräve 1982 (M 31), Tabatabaei et al. 2007 (M 33), Klein et al. 1989 (LMC), and Haynes et al. 1991 (SMC). [4]Non-thermal spectral radio index. Data are from Persic & Rephaeli 2010 and refs. therein (Arp 220, M 82, NGC 253), Beck & Gräve 1982 (M 31), Tabatabaei et al. 2007 (M 33), Klein et al. 1989 (LMC), and Haynes et al. 1991 (SMC). [5]SN rate. Data are from Persic & Rephaeli 2010 and references therein (Arp 220, M 82, NGC 253), Diehl et al. 2006 (Milky Way), and van den Bergh & Tammann 1991 (M 31, M 33, SMC, LMC; see also Pavlidou & Fields 2001). [6]Gas mass (neutral plus molecular hydrogen: $M_{\mathrm{HI}} + M_{\mathrm{H_2}}$), in log. Data are from: Torres 2004 for Arp 220; Abdo et al. 2010a for M 82, NGC 253, and the Milky Way; Abdo et al. 2010b for M 31 and M 33; Abdo et al. 2010c for the LMC; and Abdo et al. 2010d for the SMC. [7]High-energy (>100 MeV) γ-ray luminosity, in log (from Abdo et al. 2011). [8]CRp residency time.

In the *SN method*, once a fraction of SN kinetic energy that is channeled into particle acceleration is assumed, U_{p} can be estimated if the size of the star-forming region, and the SN rate there, are known – and if the CRp residency timescale can be evaluated from the presence (or lack) of a galactic wind emanating from the star-forming region and from the gas density there. The residency timescale is dominated by the advection timescale ($\sim 10^5$yr) in high-SFR (starburst) galaxies and by the *pp*-generated π^0-decay timescale ($\sim 10^7$yr) in low-SFR (quiescent) galaxies. (Both CRp energy loss timescales are lower than the typical starburst timescale, $\tau_{\mathrm{SB}} \sim 10^8$ yr.)

Expanding on a previous paper (Persic & Rephaeli 2010), here we show that the three methods give consistent results on U_p for a sample of 8 galaxies with widely varying levels of star formation activity from very quiescent to extreme starbursts. These are the only galaxies of their kind for which γ-ray data (plus radio data and SN rates) are available (see Table 1). After reviewing the radio, γ-ray, and SN methods (sect. 2,3,4), the corresponding values of U_{p} are inferred for our sample galaxies in sect. 5. The results are discussed in sect. 6 and summarized in sect. 7.

2. Particles and magnetic field.

CRe populations consist of primary (directly accelerated) and secondary (produced via charged pion decays) electrons. Their combined spectral density distribution is assumed to be a single power-law of the Lorentz factor, γ, in some interval $\gamma_1 \leq \gamma \leq \gamma_2$

$$N_e(\gamma) = N_{e,0}\,(1+\chi)\,\gamma^{-q} \tag{1}$$

where $N_{e,0}$ is a normalization factor of the primary electrons, and χ the secondary-to-primary electron number density ratio.

Electron synchrotron emission from a region with effective radius r_s (i.e., the radius of a same-volume sphere) and mean magnetic field B, located at a distance d, is

$$f_\nu = 5.67\times 10^{-22}\,\frac{r_s^3}{d^2}\,N_{e,0}\,(1+\chi)\,a(q)\,B^{\frac{q+1}{2}}\,(\frac{\nu}{4\times 10^6})^{-\frac{q-1}{2}}\quad \mathrm{erg/(s\ cm^2 Hz)}\,, \tag{2}$$

where $a(q)$ is defined and tabulated in, e.g., Tucker (1975). Setting

$$\psi \equiv (\frac{r_s}{0.1\,\mathrm{kpc}})^{-3}\,(\frac{d}{\mathrm{Mpc}})^2\,(\frac{f_{1\,\mathrm{GHz}}}{\mathrm{Jy}}) \tag{3}$$

and $\nu = 1\,\mathrm{GHz}$, the normalization of the electron spectrum is

$$N_{e,0} = \frac{5.72}{1+\chi}\times 10^{-15}\,\psi\,a(q)^{-1}\,B^{-\frac{q+1}{2}}\,250^{\frac{q-1}{2}} \tag{4}$$

A second relation is required to separately estimate N_e and B. This is provided by the condition of equipartition between particles (electrons and protons) and the magnetic field, $U_p + U_e \simeq B^2/8\pi$, which could possibly be attained due to a high degree of coupling between all the relevant degrees of freedom in the starburst region. In terms of the proton-to-electron energy density ratio, $\kappa(q)/(1+\chi) = U_p/U_e$, equipartition translates to

$$U_p\,[1+\frac{1+\chi}{\kappa(q)}] \simeq \frac{B^2}{8\,\pi}\,. \tag{5}$$

The electron energy density is $U_e = N_{e,0}\,(1+\chi)\,m_e c^2\,\int_{\gamma_1}^{\gamma_2}\gamma^{1-q}\,d\gamma$. An approximate expression for U_e is obtained by ignoring the contribution of low energy electrons with $\gamma < \gamma_1$: while the change of spectral slope at $\gamma < \gamma_1$ is important as it marks the transition from synchrotron to Coulomb losses (due to Coulomb interactions of the CRe with gas particles; see Rephaeli 1979 and Sarazin 1999), for our purposes here we consider only the spectrum above γ_1 which is directly deduced from radio measurements. In our approximate treatment here we assume $\gamma_1 = 10^3$ in all regions of interest, a possible simplification given that our numerical estimates depend only relatively weakly on the exact value of γ_1. Under this assumption, the electron energy density is $U_e = N_{e,0}\,(1+\chi)\,m_e c^2\,\int_{\gamma_1}^{\gamma_2}\gamma^{1-q}d\gamma$, where γ_2 is an upper cutoff whose exact value is irrelevant in the applicable limit $\gamma_2 >> \gamma_1$. For $q > 2$ and $\gamma_2 >> \gamma_1$, $U_e \simeq N_{e,0}(1+\chi)m_e c^2 \gamma_1^{2-q}/(q-2)$, which upon substitution of the expression for $N_{e,0}(1+\chi)$ from Eq.(4) yields

$$U_e = \frac{2.96}{(1+\chi)}\times 10^{-22}\,250^{\frac{q}{2}}\,\psi\,\frac{\gamma_1^{2-q}}{(q-2)\,a(q)}\,B^{-\frac{q+1}{2}}\,. \tag{6}$$

Using Eq.(5), we then get the equipartition magnetic field

$$B_{eq} = \left[\frac{7.44\times 10^{-21}}{1+\chi}\left(1+\frac{\kappa(q)}{1+\chi}\right)\frac{\gamma_1^{2-q}\,250^{q/2}\,\psi}{(q-2)\,a(q)}\right]^{\frac{2}{5+q}}\,. \tag{7}$$

The ratio χ is a function of energy; it depends on the injection p/e number ratio, p/e, which generally depends on the injection slope (e.g., Schlickeiser 2002), and on the gas optical thickness to pp interactions (that produce neutral and charged pions). Because of the high gas density in the starburst region the production of secondary electrons is expected to be especially important there. Following Persic & Rephaeli (2010), in the rest of this paper we shall use $\chi = 0.5$ as a fiducial value for all galaxies, starburst and quiescent alike. This value approximately matches results of detailed numerical starburst models for energies $\gtrsim 10\,\mathrm{MeV}$ (e.g., Rephaeli et al. 2010, and references therein).

Finally, using Eqs.(7,5) we readily obtain an explicit expression for $U_{\rm p}$:

$$U_{\rm p} \;=\; \frac{1}{8\,\pi}\left[1+\frac{1+\chi}{\kappa(q)}\right]^{-1}\left[\frac{7.44\times 10^{-21}}{1+\chi}\left(1+\frac{\kappa(q)}{1+\chi}\right)\frac{\gamma_1^{2-q}\,250^{q/2}\,\psi}{(q-2)\,a(q)}\right]^{\frac{4}{5+q}}. \qquad (8)$$

From Eq.(8), using the expression for the proton-to-electron energy density ratio, $\kappa(q)$, reported in (e.g.) Persic et al. (2008), and setting $\chi = 0.5$, numerical values of $U_{\rm p}$ can be obtained from the relevant observational quantities for our sample galaxies (see Table 1): they are reported in Table 2.

3. Cosmic rays and γ-ray emission.

In this section we will review some basic features of the modeling of γ-ray emission from star-forming galaxies, and the status of HE/VHE γ-ray observations of the latter. Detections of several such galaxies have enabled measurements of $U_{\rm p}$ values either in starburst cores or galaxy-wide.

In most starburst galaxies, such as the two nearby ones M 82 and NGC 253, the central starburst region (also called the source region) with a radius of $\sim 200 - 300\,\mathrm{pc}$ is identified as the main site of particle acceleration. Here, the injection particle spectrum is assumed to have the non-relativistic strong-shock index $q = 2$. A theoretical $N_{\rm p}/N_{\rm e}$ ratio, predicted from charge neutrality of the injected CRs, is likely to hold in this source region – as is also the assumption of equipartition. A measured radio index $\alpha \simeq 0.7$ in the source region implies $q = 2\,\alpha + 1 \simeq 2.4$ there. This indicates a substantial steepening of the CRe spectrum from the injection value $q_0 = 2$, due to diffusion ($D \propto \gamma^{-\delta}$) effects, that cause the steady-state particle spectral index to be $q_0 + \delta$ above some break energy.

Adopting the convection-diffusion model for energetic electron and proton propagation and accounting for all the relevant hadronic and leptonic processes, the steady-state energy distributions of these particles, in both the (active) starburst nucleus and the (passive) disk of these galaxies, can be determined with a detailed numerical treatment once the gas distribution is known (e.g., Torres 2004; Persic et al. 2008). The relevant energy loss processes are electron emissions by bremsstrahlung and Compton scattering by electrons, and γ-ray emission from π^0 decay following pp collisions. Bremsstrahlung losses dominate at lower energies, whereas π^0-decay losses dominate at higher energies. In the GeV-TeV region, emission mainly comes from pp-induced π^0 decay (e.g., Rephaeli et al. 2010).

The procedure is similar when star formation is not undergoing a burst confined to the nuclear region but is proceeding more quietly and uniformly scattered over the whole disk.

For a source with ambient gas number density $n_{\rm gas}$, proton energy density $U_{\rm p}$, and volume V, the integrated hadronic emission from pp-induced π^0 decay is

$$L^{[q]}_{\geq E} \;=\; \int_V g^{[q]}_{\geq E}\; n_{\rm gas}\; U_{\rm p}\; {\rm d}V \;\;\; {\rm s}^{-1}\,, \qquad (9)$$

with the integral emissivity $g^{[\eta]}_{\geq \epsilon}$ in units of photon $\mathrm{s}^{-1}(\text{H-atom})^{-1}(\mathrm{eV/cm^3})^{-1}$ (Drury et al. 1994). Therefore $U_{\rm p}$ can be determined, once $L_{\geq \epsilon}$ and $n_{\rm gas}(r)$ are observationally known and the particles' steady-state energy distributions have been numerically worked out by solving the convection-diffusion model for CRe and CRp propagation. By its very nature, this is a direct measurement of U_p.

The two local starburst galaxies M 82 and NGC 253 are the only non-AGN extragalactic sources that, up to now, have been detected in both the GeV (Abdo et al. 2010a) and TeV (Acciari et al. 2009; Acero et al. 2009) domains. The measured fluxes and spectra of both galaxies in the two bands agree with predictions of recent numerical models (see above) that have $U_p = \mathcal{O}(10^2)\,\mathrm{eV\,cm^{-3}}$ in the starburst nucleus. The highest-SFR galaxy in the nearby universe, Arp 220, was undetected by MAGIC (Albert et al. 2007).

HE γ-ray detections were obtained with the Large Area Telescope (LAT) on board the orbiting *Fermi* telescope for a number of low-SFR galaxies: *(i)* the Andromeda galaxy M 31 (Abdo et al. 2010b), with $U_p \simeq 0.35\,\mathrm{eV\,cm^{-3}}$; *(ii)* the Large Magellanic Cloud (LMC) whose average spectrum, either including or excluding the bright star-forming region of 30 Doradus, suggests $U_p \simeq 0.2 - 0.3\,\mathrm{eV\,cm^{-3}}$ (Abdo et al. 2010c); *(iii)* the Small Magellanic Cloud (SMC), whose integrated spectrum requires $U_p \simeq 0.15\,\mathrm{eV\,cm^{-3}}$ (Abdo et al. 2010d). Only flux upper limits exist for the Triangulum galaxy M 33 (Abdo et al. 2010b).

For the Milky Way, the modeling of the Galactic diffuse HE emission along the lines outlined above requires an average $U_p \simeq 1\,\mathrm{eV\,cm^{-3}}$ (Strong et al. 2010; Ackermann et al. 2011). This compares with $U_p \sim 1\,\mathrm{eV\,cm^{-3}}$ measured at the Earth's position (e.g., Webber 1987), and with $(6 \pm 3)\,\mathrm{eV\,cm^{-3}}$ in the $\sim$200 pc region of the Galactic center, as inferred from the measured VHE γ-ray emission (based on HESS data: Aharonian et al. 2006).

The U_p values measured through GeV-TeV observations are reported in Table 2.

4. Supernovae and cosmic rays.

The SN origin of CRs was suggested early on. As a test of this hypothesis, we obtain an estimate of U_p by combining the SN frequency with the residency timescale, τ_{res}, of CR protons that give rise to TeV emission in the star-forming region, and assuming a bona-fide value of the energy that goes into accelerating CR particles per SN event.

Two timescales define τ_{res}:
(i) the energy-loss timescale for *pp* interactions, $\tau_{pp} = (\sigma_{pp} c n_p)^{-1}$ that, for protons with kinetic energy $E \gtrsim 10$ TeV for which $\sigma_{pp} \simeq 50\,\mathrm{mb}$, can be written as

$$\tau_{pp} \;\sim\; 2 \times 10^5 \,\Big(\frac{n_p}{100\,\mathrm{cm^{-3}}}\Big)^{-1}\ \mathrm{yr}\,; \tag{10}$$

and
(ii) the CRp advection timescale, τ_{adv}, describing the removal of CRs out of the disk mid-plane region in a fast ($v_{adv} \sim 2500\,\mathrm{km\,s^{-1}}$ for M 82: Strickland & Heckman 2009) starburst-driven wind, which for a homogeneous distribution of SNe within the starburst nucleus of radius r_s writes

$$\tau_{adv} \;=\; 3 \times 10^4 \,\Big(\frac{r_s}{0.3\,\mathrm{kpc}}\Big)\,\Big(\frac{v_{out}}{2500\,\mathrm{km\,s^{-1}}}\Big)^{-1}\ \mathrm{yr}\,. \tag{11}$$

Thus,

$$\tau_{res}^{-1} \;=\; \tau_{pp}^{-1}(n_{HI}) + \tau_{out}^{-1}(r_s, v_{out})\,. \tag{12}$$

During τ_{res}, a number $\nu_{SN}\tau_{res}$ of SN explode and deposit the kinetic energy of their ejecta, $E_{ej} = 10^{51}$ erg per SN (Woosley & Weaver 1995), into the interstellar medium. Arguments based on the CR energy budget in the Galaxy and SN statistics suggest that a fraction $\eta \sim 0.05 - 0.1$

of this energy is available for accelerating particles (e.g., Higdon et al. 1998). We then express the CRp energy density as:

$$U_{\rm p} \;=\; 85\,\frac{\nu_{\rm SN}}{0.3\,{\rm yr}^{-1}}\,\frac{\tau_{\rm res}}{3\times10^{4}\,{\rm yr}}\,\frac{\eta}{0.05}\,\frac{E_{\rm ej}}{10^{51}\,{\rm erg}}\,\Big(\frac{r_{\rm s}}{0.3\,{\rm kpc}}\Big)^{-3}\;\;{\rm eV\;cm}^{-3}. \qquad (13)$$

5. Energy densities of CR protons in galaxies.

Regions in galaxies, that can be defined as *active* or *passive* with respect to CR acceleration and diffusion, can be either clearly separated or mixed up. For example, in M 82 and NGC 253 star formation goes on mainly in the nuclear region, so CRs are accelerated in the nucleus from where they diffuse out into the disk, which is a passive target for hadronic illumination: in these cases active and passive regions are easily separated, and $r_{\rm s}$ is typically the radius of the starburst nucleus. In other galaxies, e.g. the Milky Way and the Triangulum galaxy M 33, star formation proceeds throughout the disk and along the spiral arms, so active regions of CR acceleration and passive regions of CR illumination are mixed on a relatively fine scale: in these cases $r_{\rm s}$ reflects the whole galaxy disk.

For the sample of galaxies detected in γ-rays with *Fermi*/LAT (see Table 1), $U_{\rm p}$ have been measured directly by modelling the observed γ-ray emission. Using Eq.(8) and Eq.(13), in this section we now provide estimates of $U_{\rm p}$ based on the radio emission from CR electrons and the statistics of SN explosions.

Arp 220. The starburst activity takes place in a central molecular-gas disk of radius 480 pc and thickness 90 pc that accounts for SFR$\sim$120 $M_\odot$ yr^{-1}, and embeds (at a galactocentric distance of 200 pc) two extreme starburst nuclei of radii 68 pc and 110 pc that account for SFR$=$ 50 and 35 $M_\odot$ yr^{-1}, respectively. This scheme, derived from Torres (2004) and using Eq. 4 of Kennicutt (1998), is consistent with the galaxy-wide IR-inferred SFR$\sim$ 225 $M_\odot$ yr^{-1}. For the disk alone we assume $r_{\rm s} \simeq 0.25$ kpc, $f_{1\,{\rm GHz}} \simeq 0.3$ Jy, and $\alpha \simeq 0.65$ (the latter implies $q = 2.30$, hence $a \simeq 0.09$ and $\kappa \simeq 15$). From Eq.(8) we obtain $U_{\rm p} \approx 615\,{\rm eV\,cm}^{-3}$ (radio method). In Eq.(11) we use 0.09 kpc, the thickness of the starburst disk perpendicular to which the wind breaks out, to compute $\tau_{\rm adv} = 9\times10^{3}$yr; and using the latter and $\nu_{\rm SN} = 3.5\,{\rm yr}^{-1}$ in Eq.(13), we get $U_{\rm p} \simeq 515\,{\rm eV\,cm}^{-3}$ (SN method).

M 82. Following Persic et al. (2008; and references therein) we take the central starburst to be a region with a radius of 300 pc and height of 200 pc, hence $r_{\rm s} = 260$ pc, whose nonthermal radio emission has $f_{1\,{\rm GHz}} = 10$ Jy and $\alpha \simeq 0.71$. The latter implies $q \simeq 2.42$, that in turn implies $a \simeq 0.09$ and $\kappa \simeq 10$. From Eq.(8) we obtain $U_{\rm p} \simeq 90\,{\rm eV\,cm}^{-3}$ in the starburst nucleus (radio method). Because $\nu_{\rm SN} = 0.25\,{\rm yr}^{-1}$ and $\tau_{\rm res} \sim \tau_{\rm adv} \simeq 3\times10^{4}$yr (being $n_{\rm p} \sim 10^{2}{\rm cm}^{-3}$ in the starburst nucleus, and $v_{\rm adv} \sim 2500$ km s$^-$1; see Persic & Rephaeli 2010), from Eq.(13) we get $U_{\rm p} \simeq 95\,{\rm eV\,cm}^{-3}$ (SN method);

NGC 253. Following Rephaeli et al. (2010; and refs. therein) we assume $r_{\rm s} = 200$ pc, $f_{1\,{\rm GHz}} = 5.6$ Jy, and $\alpha \simeq 0.75$. The latter implies $q \simeq 2.50$, hence $a = 0.0852$ and $\kappa \simeq 8$. From Eq.(8) we obtain $U_{\rm p} \sim 63\,{\rm eV\,cm}^{-3}$ in the starburst nucleus (radio method). Being $\nu_{\rm SN} = 0.12\,{\rm yr}^{-1}$ and $\tau_{\rm res} \sim \tau_{\rm adv} \approx 3\times10^{4}$yr (being $n_{\rm p} \sim 10^{2}{\rm km\,s}^{-1}$, in the starburst nucleus, and $v_{\rm adv} \sim 2500$ km s$^-$1; see Persic & Rephaeli 2010), we get $U_{\rm p} \simeq 75\,{\rm eV\,cm}^{-3}$ (SN method);

Milky Way. Measurements of Galactic CRs indicate $U_{\rm p} \simeq 1$ eV at the Sun's position. The Galactic CRp flux measured locally is the result of the superposition of particles, accelerated in several sites scattered throughout the Galaxy, that have diffused out into the disk. By this argument, if we consider the whole Galaxy as a site of SN explosions, i.e. we use a radius of 15 kpc and a thickness of 0.5 kpc, then in Eq.(13) we use $r_{\rm s} = 4.4$ kpc, $\tau_{\rm res} \sim 2\times10^{7}$ yr (being $n_{\rm p} \sim 1\,{\rm cm}^{-3}$), and $\nu_{\rm SN} = 0.02\,{\rm yr}^{-1}$ – and obtain $U_{\rm p} \sim 1\,{\rm eV\,cm}^{-3}$ (SN method). In order to compare the SN-based estimate to available measurements of the innermost Galactic region, Persic & Rephaeli (2010) estimated $U_{\rm p} \sim 5\,{\rm eV\,cm}^{-3}$ in the central $r_{\rm s} = 0.2$ kpc (SN method).

M 31. From Beck & Gräve (1982) we derive a nonthermal radio flux $f_{1\,\mathrm{GHz}} = 4.8\,\mathrm{Jy}$ and $\alpha = 0.88$ averaged over 19.2 kpc radius. The latter corresponds to ~3.34 exponential length scales ($R_d = 5.75\,\mathrm{kpc}$), i.e. it encompasses ~85% of the mass of the corresponding exponential disk, hence most of the stellar distribution: for an assumed thickness of 0.5 kpc, we get $r_s = 5.17\,\mathrm{kpc}$. The measured spectral radio index implies $q = 2.76$, that in turn implies $a \simeq 0.08$ and $\kappa \simeq 4$. From Eq.(8) we obtain $U_p \simeq 0.9\,\mathrm{eV\,cm^{-3}}$ (radio method). Taking the whole disk as a site for SN explosions, the implied average gas density is $n_p \approx 0.5\,\mathrm{cm^{-3}}$, hence $\tau_{res} \sim \tau_{pp} \approx 3.7 \times 10^7\mathrm{yr}$. From Eq.(13) we derive $U_p \approx 0.2\,\mathrm{eV\,cm^{-3}}$ (SN method).

M 33. In this galaxy star formation proceeds scattered along the spiral arms. So we consider the whole disk as the site of SN explosions and of CR acceleration. From Tabatabaei et al. (2007) we derive a nonthermal radio flux characterized by and $f_{1\,\mathrm{GHz}} = 3.3\,\mathrm{Jy}$ and $\alpha = 0.95$ averaged over 7.6 kpc radius. The latter corresponds to 5.3 exponential lengthscales ($R_d \simeq 1.43\,\mathrm{kpc}$), i.e. it encompass 97% of the mass of an exponential disk, hence virtually the whole stellar distribution: for an assumed thickness of 0.5 kpc, we get $r_s = 2.79\,\mathrm{kpc}$. The measured spectral radio index implies $q = 2.90$, that in turn implies $a \simeq 0.08$ and $\kappa \simeq 3$. From Eq.(8) we obtain $U_p \simeq 0.8\,\mathrm{eV\,cm^{-3}}$ (radio method). Given the gas mass reported in Table 1, the average gas density is $n_p \sim 1\,\mathrm{cm^{-3}}$, hence $\tau_{res} \sim \tau_{pp} \approx 2 \times 10^7\mathrm{yr}$. Inserting the relevant quantities (see Table 1) into Eq.(13), we then derive $U_p \simeq 0.7\,\mathrm{eV\,cm^{-3}}$ (SN method).

LMC. This satellite of the Milky Way can be modeled as a truncated disk/spheroid with $r_t \simeq 3\,\mathrm{kpc}$ whose half-thickness is also ≈3 kpc (Weinberg & Nikolaev 2001), so we use $r_s = 3\,\mathrm{kpc}$ in Eqs.(13,8). The measured spectral radio index implies $q = 2.68$, that in turn implies $a \simeq 0.08$ and $\kappa \simeq 5$. Eq.(8) yields $U_p \simeq 0.1\,\mathrm{eV\,cm^{-3}}$ (radio method). There is no mass outflow in the LMC, hence $\tau_{res} \sim \tau_{pp} \approx 10^7\mathrm{yr}$. (The average gas density is $n_p \approx 2\,\mathrm{cm^{-3}}$.) Eq.(13) then yields $U_p \approx 0.2\,\mathrm{eV\,cm^{-3}}$ (SN method).

SMC. This other Milky Way satellite can be modeled as a bar with sizes $\simeq 2.5 \times 1.5\,\mathrm{kpc}$ (Wilke et al. 2003) with a l.o.s. depth of 4 kpc (following Abdo et al. 2010d), so we use $r_s = 1.53\,\mathrm{kpc}$ in Eqs.(13,8). The measured radio spectral index implies $q = 2.70$, that in turn implies $a \simeq 0.08$ and $\kappa \simeq 4$. From Eq.(8) we obtain $U_p \simeq 0.10\,\mathrm{eV\,cm^{-3}}$ (radio method). The SMC has a (galaxy-wide) SN rate of $\nu_{SN} = 10^{-3}\,\mathrm{yr^{-1}}$. There is no mass outflow from the SMC, so $\tau_{res} \sim \tau_{pp} \simeq 1.4 \times 10^7\mathrm{yr}$. (It is $n_p \approx 1.4\,\mathrm{cm^{-3}}$.) From Eq.(13) we then obtain $U_p \approx 1\,\mathrm{eV\,cm^{-3}}$ (SN method).

The U_p values estimated according to the various methods are reported in Table 2. They are only as precise as the quantities used to estimate them can be (see Persic & Rephaeli 2010 for a more complete discussion). Measuring ν_{SN} may be strongly affected by heavy optical extinction (e.g., in starburst nuclei), and radio counts of SN remnants need information on their ages in order to be turned into actual SN rates: for our sample galaxies, published observational results suggest that ν_{SN} are known to within a factor of $\lesssim 1.5$. The main uncertainty in the value of τ_{res} arises from the fast wind velocity, that is probably known to within ~50%. For the energy released by a core-collapse SN that is available for CR acceleration, we have assumed the 'Milky Way normalization' (i.e., 5% of 10^{51} erg per SN) to hold also for the galaxies in Table 1. For the starburst galaxies considered in Table 1, the starburst radii r_s are deduced from high-resolution optical and radio data, so their values are quite accurate. We conclude that the galaxy quantities in Table 1 relevant to Eq.(13) are quoted in the literature as observationally precise, to within a factor of $\lesssim 2$.

6. Discussion

The three methods discussed here are not independent. The γ-ray method and the radio method are coupled through the p/e ratio at injection, through the secondary-to-primary electron ratio (which in turn depends on the fraction of CRp that interact with the ambient matter during their residency time in the emission region), and through the imposed condition of particle-field

Table 2. Star-forming galaxies: CRp energy densities^{+}.

Object	γ-ray meth.	radio meth.	SN meth.	other meth.	r_s (kpc)	loss mode
Arp 220	–	615	515	–	0.25	adv
M 82	$200^{a,c}$	90	95	–	0.26	adv
NGC 253	$200^{b,c}$	63	77	–	0.20	adv
Milky Way	1^{d}	–	1	1^{i}	4.4	pp
	6^{e}	–	5	–	0.2	pp
M 31	0.36^{f}	0.9	0.7	–	4.77	pp
M 33	$< 3^{f}$	0.8	0.7	–	2.79	pp
LMC	0.25^{g}	0.1	0.2	–	3	pp
SMC	0.15^{h}	0.1	1.0	–	1.53	pp

$^{+}$ Values are in eV cm^{-3}.
(a) Acciari et al. 2009 (see also Persic et al. 2008 and De Cea et al. 2009). *(b)* Acero et al. 2009 (see also Paglione et al. 1996, Domingo-Santamaría & Torres 2005, and Rephaeli et al. 2010). *(c)* Abdo et al. 2010a. *(d)* Strong et al. 2010. *(e)* Aharonian et al. 2006. *(f)* Abdo et al. 2010b, with Drury et al. 1994 in the case of M 33. *(g)* Abdo et al. 2010c. *(h)* Abdo et al. 2010d. *(i)* Webber 1987.

equipartition. The SN method is not independent of the γ-ray method either, because they both depend on the residency time of CRp in the emission region - although, unlike the γ-ray and radio methods, it does not depend on the particles' radiative yields but on the statistics of core-collapse SN. Also, the thee methods are not on equal footing: the γ-ray, radio, and SN method respectively *measure*, *infer*, and *estimate* the value of $U_{\rm p}$. This, because: *(i)* π^0-decay γ-ray emission is the most robust *measure* of $U_{\rm p}$ once the distribution of target gas is known and the particles' diffusion and losses are treated accurately and self-consistently (like in published numerical models of M 82 and NGC 253); whereas *(ii)* radio emission enables *deduction* of $U_{\rm p}$ from electron synchrotron flux and spectrum once assumptions have been made on the link between energetic electrons and protons, and between such particles and the magnetic field; and *(iii)* assuming a SN origin for CRs, SN statistics for a given (region of a) galaxy leads to an *estimate* of $U_{\rm p}$ there.

A substantial agreement among estimates based on the three methods is reached for most of the galaxies in Table 1. The only exception is the SMC, whose CR confinement volume could be small, so that most CRs diffuse out and get lost to the intergalactic space (Abdo et al. 2010d). If so, the γ-ray method returns the (lower) *actual* CRp energy density whereas the radio and SN methods estimate the (higher) *produced* amount.

Once these considerations are weighed in, the discrepancy between the very high CRp energy densities ($U_{\rm p} \sim \mathcal{O}(10^2)\,{\rm eV\,cm^{-3}}$) deduced for starburst nuclei and the low values ($U_{\rm p} \sim \mathcal{O}(10^{-1})\,{\rm eV\,cm^{-3}}$) deduced for very quiet environments appears very significant. Also, our sample galaxies do conform to the notion of a SN-powered VHE γ-ray luminosity.

7. Conclusions

Using 8 galaxies for which pointed GeV-TeV data (from *Fermi*/LAT and from Cherenkov telescopes) are available, we have estimated $U_{\rm p}$ using the radio, γ-ray, and SN methods. A substantial agreement among estimates based on the three methods is reached for 9 galactic regions (3 starburst nuclei, the Milky Way's disk and central region, 4 quiescent galaxies) considered in this paper. The discrepancy between the very high values ($U_{\rm p} \sim \mathcal{O}(10^2)\,{\rm eV\,cm^{-3}}$) found in starburst nuclei like M 82 and NGC 253, and the low values ($U_{\rm p} \sim \mathcal{O}(10^{-1})\,{\rm eV\,cm^{-3}}$) found in very quiet environments like the Local Group galaxies, appears to be highly significant.

Based on the results of this study, we confirm and extend our earlier sugggestions: *(i)* star-forming galaxies are effective CR accelerators; *(ii)* CRp energy densities in star-forming galaxies range from $\mathcal{O}(10^2)\,{\rm eV\,cm^{-3}}$ in very active environments down to $\mathcal{O}(10^{-1})\,{\rm eV\,cm^{-3}}$ in very quiet ones; *(iii)* equipartition CR and magnetic field energy densities can be used as proxies of the actual quantities; and *(iv)* core-collapse SNe share a universal CR acceleration efficiency. An expanded version of this paper (Persic & Rephaeli 2011) is underway.

References

[1] Abdo A A *et al* (LAT Collab.) 2011, *ApJ*, submitted
[2] Abdo A A *et al* (LAT Collab.) 2010b, *A&A*, **523**, L2
[3] Abdo A A *et al* (LAT Collab.) 2010c, *A&A*, **512**, A7
[4] Abdo A A *et al* (LAT Collab.) 2010a, *ApJL*, **709**, L152
[5] Abdo A A *et al* (LAT Collab.) 2010d, *A&A*, **523**, A46
[6] Acciari V A *et al* (VERITAS Collab.) 2009, *Nature*, **462**, 770
[7] Acero F *et al* (HESS Collab.) 2009 *Science*, **326**, 1080
[8] Ackermann M *et al* (LAT Collab.) 2011, *ApJ*, **726**:81
[9] Aharonian F *et al* (HESS Collab.), 2006, *Nature*, **439**, 695
[10] Albert J *et al* (MAGIC Collab.) 2007, *ApJ*, **658**, 245
[11] Beck R and Gräve R 1982, *A&A*, **105**, 192
[12] Beck R and Krause M 2005, *Astr. Nacht.*, **326**, 414
[13] de Cea Del Pozo E, Torres D F and Rodriguez Marrero A Y, 2009, *ApJ*, **698**, 1054
[14] Domingo-Santamaría E and Torres D F 2005, *A&A*, **444**, 403
[15] Drury L O' C, Aharonian F A and Völk H J 1994, *A&A*, **287**, 959
[16] Haynes R F, Klein U, Wayte S R *et al* 1991, *A&A*, **252**, 475
[17] Higdon J C, Lingenfelter R E and Ramaty R 1998, *ApJ*, **509**, L33
[18] Hughes A, Staveley-Smith L, Kim S, Wolleben M and Filipovic M 2007, *MNRAS*, **382**, 543
[19] Iwasawa K, Koyama K, Awaki H *et al* 1993, *ApJ*, **409**, 155
[20] Kennicutt R C Jr 1998, *ARAA*, **36**, 189
[21] Klein U, Wielebinski R, Haynes R F and Malin D F 1989, *A&A*, **211**, 280
[22] Krolik J H and Begelman M C 1986, *ApJ*, **308**, L55
[23] Longair M S 1994, *High Energy Astrophysics* (2nd ed.; Cambridge: Cambridge Univ. Press)
[24] Paglione T A D, Marscher A P, Jackson J M and Bertsch D L 1996, *ApJ*, **460**, 295
[25] Pavlidou V and Fields B D 2001, *ApJ*, **558**, 63
[26] Persic M and Rephaeli Y 2010, *MNRAS*, **403**, 1569
[27] Persic M and Rephaeli Y 2011, in preparation
[28] Persic M, Rephaeli Y and Arieli Y 2008, *A&A*, **486**, 143
[29] Rephaeli Y 1979, *ApJ*, **227**, 364
[30] Rephaeli Y, Arieli Y and Persic M 2010, *MNRAS*, **401**, 473
[31] Sarazin C L 1999, *ApJ*, **520**, 529
[32] Schlickeiser R 2002, *Cosmic Ray Astrophysics* (Berlin: Springer), p.472
[33] Strickland D K and Heckman T M 2009, *ApJ*, **697**, 2030
[34] Strong A W, Porter T A, Digel S W *et al* 2010, *ApJ*, **722**, L58
[35] Tabatabaei F S, Beck R, Krügel E *et al* 2007, *A&A*, **475**, 133
[36] Torres D F 2004, *ApJ*, **617**, 966
[37] Tucker W 1975, *Radiation Processes in Astrophysics* (Cambridge, MA: MIT Press)
[38] van den Bergh S and Tammann G A 1991, *ARAA*, **29**, 363
[39] Völk H J, Klein U and Wielebinski R 1989, *A&A*, **213**, L12
[40] Webber W R 1987, *A&A*, **179**, 277
[41] Weinberg M D and Nikolaev S 2001, *ApJ*, **548**, 712
[42] Wilke K, Stickel M, Haas M *et al* 2003, *A&A*, **401**, 873
[43] Wilson A S and Ulvestad J S 1982, *ApJ*, **263**, 576
[44] Woosley S E and Weaver T A. 1995, *ApJS*, **101**, 181

Journal of Physics: Conference Series **355** (2012) 012039 doi:10.1088/1742-6596/355/1/012039

Correlated neutrino and gamma-ray emission from Active Galactic Nuclei - an estimation

M Doert[1], J K Becker[2], F Halzen[3], A O'Murchadha[3], W Rhode[1]

[1] Experimentelle Physik V, Technische Universität Dortmund, 44221 Dortmund, Germany,
[2] Theoretische Physik IV, Ruhr-Universität Bochum, 44780 Bochum, Germany,
[3] Department of Physics, Phenomenology Institute, University of Wisconsin-Madison, Madison, WI, USA

E-mail: marlene.doert@tu-dortmund.de

Abstract. Active galactic nuclei are among the candidates for sources which produce the extragalactic charged cosmic ray flux. However, the composition of their emission regions and the processes leading to their gamma-ray emission are still being investigated. A detection of neutrinos from these sources would prove the presence of hadrons in the emission region and thus establish AGN as sources of the charged cosmic rays. The search for neutrinos from astrophysical sources has been performed for many years now, employing e.g. the AMANDA and now the just-completed IceCube detector. However, the persistent absence of a detection suggests to venture an estimate on the neutrino flux which can be expected from AGN. The method presented here is based on bolometric considerations. We assume that the energy released through the different types of particles is related. On this basis the measured flux of gamma-rays from AGN can be utilized to estimate the possible output in neutrinos. We are aiming at an estimation of the expected neutrino flux, following basic assumptions. The resulting flux estimations show that a detection of single sources still appears to be out of reach. Still, the investigation of a stacked sample of AGN with additional constraints in time and energy seems promising.

1. Introduction

The search for neutrinos which originate from active galactic nuclei (AGN) could finally shed light on some of the most interesting matters of astroparticle physics. AGN constitute a class of extragalactic objects which produce an enormous output in photons, spanning nearly 20 orders of magnitude in energy. This leads to the supposition that the processes, which produce these highly energetic photons, have the ability to also provide an equally energetic output in charged particles. This particle output could substantially contribute to the measured flux in charged cosmic rays, whose origin is currently still under debate. But although AGN are already well-studied in the light of gamma-rays, the processes leading to the different parts of the electromagnetic spectrum are not completely understood. Amongst others, the question if leptons or hadrons are the main origin of the very high energy (VHE) gamma-ray emission is not fully solved. However, a detection of neutrinos from AGN would give more than a clue towards the composition of the emission region, as neutrino emission only takes place within a hadronically dominated acceleration scenario. Following this scenario, the flux of high energy gamma-rays and neutrinos must be closely linked. Thus, gamma-ray measurements, which are by now available for many sources and over a broad energy range, can yield information about the possible correlated neutrino flux. In this study, we use the example of two AGN, which are classified as blazars.

2. AGN / blazar emission models

The spectral energy distributions (SED) of AGN or blazars in particular are characterized by a double bump structure consisting of one bump at lower energies and one bump at higher energies. For high-frequency peaked blazars (HBLs) the first bump appears at UV to X-ray energies and the second one at gamma-ray energies. The low energy bump is well-established to be originating from synchrotron emission of accelerated electrons. The origin of the second peak, however, is not yet completely pinned down. In a leptonic scenario, this emission is caused by the Inverse Compton effect. Lower energy photons are pushed to higher energies through interaction with fast electrons. There are several different specifications of this model type. Some assume the electron population, which causes the synchrotron emission, to interact with their own synchrotron field (SSC models, see e.g. [1]). Others presume an external field of photons to interact with the electrons, such as photons coming from the accretion disk of the AGN [2]. On all accounts, these models imply a simultaneous or at least correlated change in flux of the high energy bump in case the low energy bump shows a flare. In a hadronic emission scenario, where protons and possibly also heavier nuclei are present in the emission region, the second peak in the SED is thought to be produced mainly by a different process: If protons interact with the present photon field or with other protons, Δ^+ resonances are produced which decay to neutral and charged pions

$$p\,\gamma \rightarrow \Delta^+ \rightarrow \begin{cases} p\,\pi^0, & \text{fraction } \frac{2}{3} \\ n\,\pi^+, & \text{fraction } \frac{1}{3}, \end{cases} \tag{1}$$

see also [3]. The resulting pions decay further, and therein produce highly energetic gamma-rays:

$$\pi^0 \quad \rightarrow \quad \gamma\,\gamma. \tag{2}$$

While neutral pions produce the gamma-rays which can, according to this scenario, be found in the second peak of an AGN SED, the charged pions decay further to leptons, including the possibly illuminative neutrinos:

$$\pi^+ \quad \rightarrow \quad \mu^+\,\nu_\mu \rightarrow e^+\,\nu_e\,\overline{\nu}_\mu\,\nu_\mu. \tag{3}$$

If this hadronic scenario applies to the processes leading to the strong electromagnetic emission seen by AGN, an accompanying neutrino flux should thus be present. This makes the detection of neutrinos from cosmic sources the "smoking gun" for determining the acceleration and emission processes within the source.

3. Search for neutrino sources with AMANDA and IceCube

The search for neutrinos from astrophysical sources is currently in full swing. Because of the neutrinos' low interaction probability, large detector volumes are needed to search for neutrino signals from outside our solar system. The current state-of-the-art experiments exploit ice or sea water as their detection material. By installing photomultipliers and microphones on a grid within ice or water they utilize large volumes and can therefore largely enhance the detection probability for cosmic neutrinos. At the South Pole the neutrino telescope AMANDA has been replaced by the cubic kilometer scale detector IceCube. The building of IceCube has lately been finished and the detector is in full operation since 2011.

While the detector is in operation, the field of view comprises the whole sky. However, there

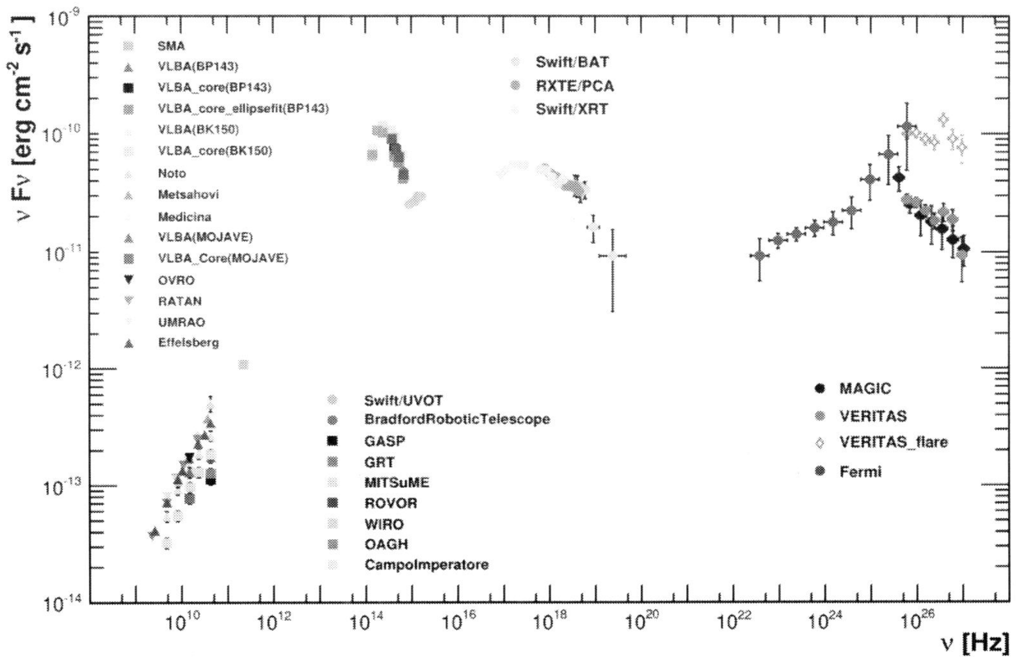

Figure 1. The spectral energy distribution for the blazar Mrk 501, ranging from radio to VHE gamma-rays (for details concerning the illustration see [4]). The typical double bump structure is visible: The first bump is superimposed by emission from the host galaxy, but is still clearly identifiable between 10^9 and 10^{20} Hz. The second bump is depicted by the gamma-ray spectral points in the range between 10^{22} and 10^{28} Hz, showing both low state and high state measurements.

are important differences regarding events seen from above or below the horizon [5]. While for upgoing events, which have travelled through the Earth before they reach the detector, the main background consists of atmospheric neutrinos, the downgoing events are dominated by a background of atmospheric muons. Also the accessible energy range in neutrinos is different for events from above the horizon, because highly energetic neutrinos are absorbed by the Earth due to the increase of the neutrino-nucleus cross section with energy.

The directional reconstruction of incoming events is carried out in bins of about $1^\circ \times 1^\circ$ in size. This way, the analysis spreads over thousands of bins. Due to statistical inter-bin fluctuations an occurrence of several accumulations of events with a statistical significance of $3-5\,\sigma$ is expected within an all-sky analysis. This complicates the search for point sources, as a pile-up on the skymap of recorded events does not necessarily point to a real signal.
So-called blind searches help to avoid this effect. The recorded events are scrambled w.r.t. their sky coordinates whilst the analysis is ongoing. This disarranging is carried out in a way which conserves the acceptance and the efficiency of the detector. The analyzer needs to define the sky coordinates, the analysis procedure and cuts for the expected signal before the data is finally "unblinded". If an accumulation of events occurs for the full data sample at the predefined source positions investigated here, the presence of an actual signal is much more significant, as the trial factor for tried positions is much smaller. If a signal from one single location cannot be expected, a stacking analysis offers the possibility to detect a signal from a whole sample of sources of the same type [6],[7]. The detection probability for each source is stacked and a

conclusion can be drawn if the whole sample of investigated sources contains a signal or not. Furthermore spectral analyses can be performed on a stacked data sample, which could reveal the presence of cosmic neutrinos. As the atmospheric neutrino flux roughly follows a spectrum with a power-law index $\alpha_{atm} = -3.7$ and neutrinos originating from cosmic acceleration processes are expected to show a spectrum with about $\alpha_{source} = -2$, a flattening of the neutrino energy spectrum towards higher energies would point to a domination of the flux by neutrinos of non-atmospheric origin. Both the single source and the stacking analysis method imply the necessity to have a good guess about which sources might yield a possible neutrino signal. To get a notion of which sources to take into account, estimations on how many neutrinos can be expected from different sources or source types are essential.

4. How many neutrinos can be expected? - An estimation

To obtain an estimation of the possible neutrino flux from gamma-ray emitting sources, which is as unrestricted as possible, the method presented here is independent from elaborate acceleration and emission models. Instead, it is based on bolometric considerations only.
The approach shown in this contribution has been discussed before in [8] and has recently been applied to gamma-ray burst studies [9]. As there is new instrumentation at hand which allows a more thorough investigation of both the gamma-ray emission and the neutrino emission, a revisit to this simple method promises to offer new insights also for AGN neutrino studies.

4.1. The method

As depicted above, the emission of highly energetic gamma-rays and neutrinos must be closely linked within the scope of a hadronic scenario, as they are both decay products of pions which are themselves produced with a fixed ratio in hadronic or photo-hadronic interactions. For the estimation given here we assume the interaction of protons with a photon field. This field could consist of synchrotron photons which are emitted by the accelerated particles themselves, in which case a correlation in time between the synchroton photon flux, the gamma-ray emission and the flux in neutrinos is expected. Another possible source of low energy photons is given by the accretion disk of the AGN. At the moment no specific scenario is chosen, but a constant photon field is assumed.
We neglect, for now, the possibility that proton proton interactions could also cause the emission of high energy neutrinos, referring e.g. to [10] where it is noted that photon-hadronic interactions dominate in the scenario investigated there. A more detailed discussion of the optical depths for both processes is aimed at in a subsequent step of this study.
Making an assumption for the initial spectral shape of the interacting protons and hence for the pions which basically follow the distribution of the protons, the spectra of gamma-rays and the first neutrino can be gained from standard two-body decay kinematics [11],[12],[13]. The second neutrino results from the decay of the muon which emerges from pion decay (see eq. 3). This is described by a three-body decay of a particle resulting from two-body-decay and is determined following [13]. For explicit calculations see also [9]. The resulting spectra are normalized w.r.t. each other, but an absolute normalization of the fluxes is required, which can be provided by gamma-ray measurements. Ground-based gamma-ray telescopes have been measuring the VHE part of the gamma-ray spectrum ($\sim$ 50 GeV - 50 TeV) for many years now. Since 2008 the satellite experiment Fermi complements these measurements by probing the energy range from 20 MeV to 300 GeV with its Large Area Telescope. These instruments thus provide measurements of more or less the entire second peak in the SED of AGN.
From their production to their measurement by Earth- or space-bound instruments, the gamma-rays undergo several interactions which leave them with an altered spectrum. An effect which has its main influence on distant sources is the interaction with the extragalactic background

light (EBL). Here the gamma particles interact with lower energy photons from former stars and galaxies and annihilate via e^+e^- pair production. However, elaborate models describing the EBL density allow us to correct the measured gamma-ray spectra for this absorption effect in case the distance of the source is known (see e.g. [14]).
An effect which is more complex to reconstruct is the cascading of gamma-rays within the source. After their production in the emission region, the high energy gamma particles have to escape from this region. In general, they are surrounded by the synchrotron photon field. Thus an interaction of high energy and low energy photons is possible and can lead to a shift of the gamma-rays to lower energies. However, as the two bumps in the SED are distinctly separated, the assumption that the cascaded photons remain within the peak is a conservative one. Hence, the energy which has originally been emitted in gamma-rays should still be contained in the high energy bump, which is visible in the data. Thus, a normalization of the pion-induced gamma-ray flux is possible using energy conservation considerations:

$$\int_{E_{min}}^{E_{max}} E\frac{dN_{\pi^0\rightarrow\gamma\gamma}}{dE}dE = \int_{E_{min}}^{E_{max}} E\frac{dN_{\rm meas}}{dE}dE. \tag{4}$$

4.2. Application to gamma-ray data
In this paper the estimation of a possible neutrino flux is carried out for two example sources: the blazars Markarian (Mrk) 501 and 1ES 1959+650, which both belong to the class of HBLs. Although it has been noted ([15]) that this type of sources might be disfavored for the emission of high energy cosmic neutrinos, the availability of recent data which cover the entire SED and especially the high energy bump recommends these sources as objects for a first application of the method presented here. The available data comprise Fermi spectral points gained with the LAT and VHE spectra from the ground-based telescopes MAGIC and VERITAS, respectively. For Mrk 501 separate data for a low state [16],[17] and a high state [4] measurement have been used here. For 1ES 1959+650 no flux variations have been observed recently, thus measurements of a constant flux level have been used [16],[18]. The low state spectra have been corrected for EBL absorption, using the model from Kneiske and Dole [14]. The Mrk 501 high state data have been provided as an already absorption-corrected spectrum. Here the model by Franceschini et al. has been applied [19], while it is shown that for Mrk 501, with a redshift of $z = 0.034$, a good agreement between the different models can be seen [4].
Polynomial functions of fourth and second order, respectively, have been fitted to the data points in double-logarithmic representation.
The integration limits for the energy normalization are chosen as $E_{min} = 10^{-5}$ TeV and $E_{max} = 10^8$ TeV. The lower limit is given by the left flank of the high energy bump. The upper limit is suggested by the expected cut-off in the spectrum of protons coming from cosmic accelerators at an energy of 10^9 TeV. By using these limits we extrapolate the fit suggested by the measurement, but as the main contribution is given by the well-measured part of the spectrum, this appears to be feasible .
The collected gamma-ray data and the corresponding fits used for the integration are shown on the left side of Fig. 2.

4.3. Results
The resulting neutrino spectra can be seen in Fig. 2 on the right. Plotted are estimations for three different initial pion spectra for the two sources (top and bottom). A comparison of the low and high state flux of Mrk 501 is shown in the central panel. IceCube 40-string configuration upper limits of the Feldman-Cousins 90% confidence intervals for an $E^{-2.0}$ flux in a livetime of 375.5 days are drawn according to [20]. The atmospheric neutrino flux is calculated according to [21].

The obtained neutrino signals span an energy range from the pion production threshold at tens to hundreds of MeV up to 10^8 TeV, where they are cut off by the suppression of highly energetic protons.
The energy range accessible to IceCube, which reaches from 100 GeV to 10^7 TeV, covers a large fraction of the expected neutrino signal. Using a standard effective area of the IceCube 80-string detector, the number of neutrino events which are expected to be seen by the IceCube detector can be derived for the different flux predictions. In table 1 the expected number of events per year are shown for an initial $E^{-2.0}$ spectrum, together with the expected number of events from the atmospheric background. It has to be noted that giving the number of neutrino events corresponding to a source in high-state flux over a full year is not true to reality, but chosen here for comparability.

Table 1. Predicted event numbers for the calculated neutrino flux from the examined blazars seen within one year with the IceCube 80-string detector.

Source	# events [year^{-1}]	$\frac{\#\text{ signal events}}{\sqrt{\#\text{ bg events}}}$
Mrk 501 low state	0.878	0.192
Mrk 501 high state	6.232	1.365
1ES 1959+650	0.778	0.170
Atmosph. background	20.84	-

5. Conclusions and Outlook

From the numbers of neutrinos which are expected to be seen for AGN ground states, it is obvious that single sources are not close to being detectable by a standard point source analysis. With a complete IceCube detector, the performance will be doubled due to the doubled detection area compared to IC-40. Also, significance will improve roughly with the square root of time. Thus, the predicted flux may be detectable with a full IceCube detector after a few years of data taking. Still, the suppression of the atmospheric background is the key issue. In an all-data analysis, single sources do not seem to be seizable. Instead, the analysis of IceCube data which are correlated in time to gamma-ray high states of the sources could be more profitable, as a reduced time window reduces the atmospheric background substantially and the flux of expected neutrinos is higher in general. For studies on the time correlation of the optical, the gamma-ray and the neutrino signal see e.g. [22]. An additional optimization of the considered energy range can improve the signal-to-background-ratio further. Ultimately a stacking analysis of several promising gamma-ray sources - possibly during their high state periods - could yield a sensitivity which is high enough for a detection of neutrinos originating from these sources.
The estimations performed here will be refined further by modeling the interactions of protons with protons and of protons with photon fields in more detail, following [23] and [24], in order to gain a more precise pion spectrum. Furthermore the cascading of VHE gamma-rays within the source will be looked into. Ultimately, the study will be extended to a larger catalog of sources. Additional to the HBLs considered here, the possibly more favorable LBL sources and the class of Flat Spectrum Radio Quasars (FSRQs) will be investigated with the method presented above. A forthcoming paper will incorporate the aspects mentioned above and aim at a prediction for a stacked neutrino detection from AGN.

Acknowledgments
This research is partially supported by the DFG through the Collaborative Research Center SFB823, project C4 and the DAAD PPP-USA program, which are gratefully acknowledged. We would also like to thank David Paneque for providing the Markarian 501 high state data points.

References
[1] Tavecchio, F. et al. 1998 *ApJ* **509** 608
[2] Dermer, C. D. and Schlickeiser, R. 1993 *ApJ* **416** 458
[3] Becker, J. K. 2008 *Phys.Rept.* **458** 173-246
[4] Abdo, A. A. et al. 2011 *ApJ* **727** 129A
[5] Abbasi, R. et al. 2009 *Phys. Rev. Letters* **103** Issue 22, id. 221102
[6] Achterberg, A. et al. 2006 *ApP* **26** Issue 4-5 282-300
[7] Becker, J. K. et al. 2007 *ApP* **28** 98-118
[8] Halzen, F. and Zas, E. 1997 *ApJ* **488** 669
[9] Becker, J. K. et al. 2010 *ApJ* **721** 1891-1899
[10] Becker, J. K. and Biermann, P. L. 2009 *ApP* **31** Issue 2 138-148
[11] Stecker, F. 1971 *Mono Book Co.*
[12] Dermer, C. 1986, *ApJ* **307** 47
[13] Gaisser, T. 1990 *Cambridge University Press*
[14] Kneiske, T. M. and Dole, H. 2010 *A&A* **515** id.A19
[15] Mücke, A. et al., 2002 *ApP* **18** Issue 6 593-613
[16] Abdo, A. A. et al., 2010 *ApJ* **716** 30A
[17] Aleksic, J. et al., 2010 *A&A* **519** A32
[18] Tagliaferri, G. et al. 2008 *ApJ* **679** 1029A
[19] Franceschini, A., Rodighiero, G. and Vaccari, M. 2008 *A& A* **487** Issue 3 837-852
[20] Abbasi, R. et al. 2011 *ApJ* **732** 18
[21] Honda, M. et al. 2007 *Phys. Rev. D* **75** 043006
[22] Eichmann, B. 2011 *JPCS* in prep.
[23] Kelner, S. R. and Aharonian, F. A. and Bugayov, V. V. 2006 *Phys. Rev. D* **74** Issue 3, id. 034018
[24] Kelner, S. R. and Aharonian, F. A. 2008 *Phys. Rev. D* **78** Issue 3, id. 034013

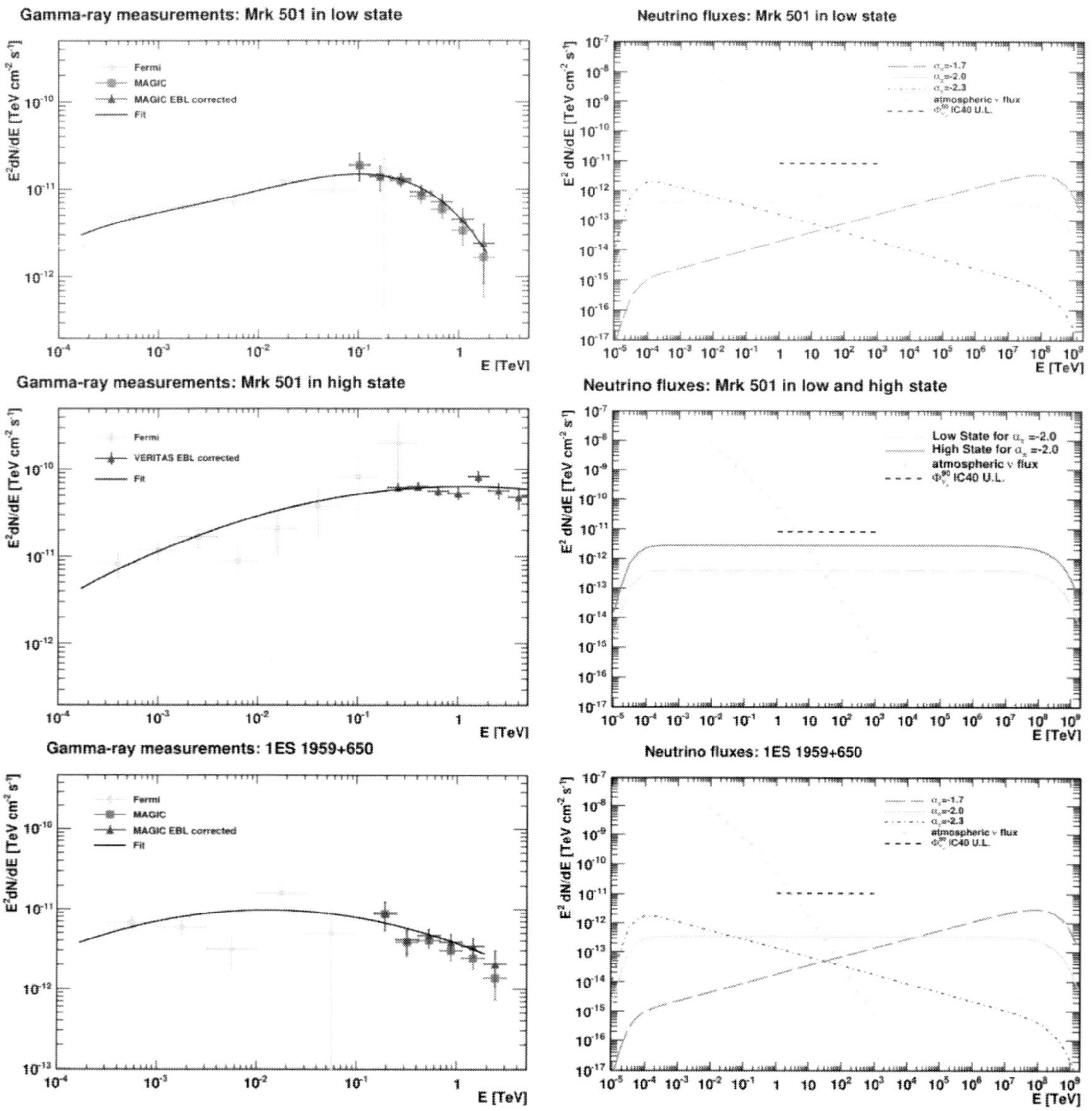

Figure 2.
Left side: Measured gamma-ray spectral points from Fermi and ground-based telescopes and the according fits to the high energy peak: Mrk 501 in low state (top panel) and high state (central panel) and 1ES 1959+650 (bottom panel). The data points for Mrk 501 have been obtained here: [16], [17] (low state); [4] (high state). For the spectral points of 1ES 1959+650 see [16] and [18].

Right side: Predicted neutrino spectra for the sources Mrk 501 and 1ES 1959+650. Top and bottom panel: Estimations for three different initial π-spectra for Mrk 501 in low state and 1ES 1959+650, respectively. Central panel: Estimations for $\alpha_\pi = -2.0$ spectrum of Mrk 501 in low and hight state. IC 40 limits are drawn according to [20]. The atmospheric neutrino flux is determined according to [21].

Search for point sources of cosmic neutrinos with the ANTARES neutrino telescope

J.P Gómez-González on behalf of the ANTARES Collaboration

IFIC - Instituto de Física Corpuscular, Edificios de Investigación de Paterna, CSIC - Universitat de València, Apdo. de Correos 22085, 46071 Valencia, Spain.

E-mail: jpablo@ific.uv.es

Abstract. The ANTARES detector is currently the largest deep-sea neutrino telescope in the Northern Hemisphere. Located on the bottom of the Mediterranean Sea, 40 km off the French coast, it consists of a three-dimensional array of 885 photomultiplier tubes which detect the Cherenkov light induced by the muons produced in the interaction of high energy cosmic neutrinos with the matter surrounding the detector. The main goal of ANTARES is the search for cosmic neutrinos and their sources. In this contribution preliminary results of the search for point sources of high energy neutrinos with the data gathered in 2007 and 2008 are presented, including some of the most stringent upper limits on the neutrino flux in the world.

1. Introduction

Neutrino Astronomy represents a new method to explore the Universe complementary to the multi-wavelenght and cosmic rays observations. Futhermore, the neutrino, as weakly interacting and almost massless particle, represent a unique probe to obtain information from the inner core of distant astrophysical objects. Candidates for cosmic neutrino sources are considered to be both galactic, as X-ray Binary systems, and extragalactic, like Active Galactic Nuclei (AGNs). In such astrophysical scenarios acceleration of hadrons up to very high energies is predicted; their interactation with the ambient matter and dense photon fields will eventually create pions, which produce neutrinos in their decay. Therefore, the detection of a source of neutrinos will distinguish unambiguously between hadronic and electronic acceleration mechanisms.

The detection principle of a neutrino telescope is based on the collection of the Cherenkov light induced by the relativistic muons produced in the interaction of high energy neutrinos with the water or rock near the detector. The main source of background comes from the atmospheric muons and neutrinos produced by the interaction of cosmic rays in the Earth atmosphere. Due to the low interaction cross section of the neutrino, large instrumented volumes are requiered. ANTARES is the first deep sea full operational neutrino telescope. Searches for cosmic neutrinos and their sources are one of the main goals of the experiment. This contribution presents a search for point source analysis using 295 days of data livetime.

2. The ANTARES detector

The ANTARES (Astronomy with Neutrinos and Abyss RESearch) neutrino telescope [1] is deployed at a depth of 2475 m in the Mediterranean Sea, about 40 km offshore the coast of Toulon, at 42°48'N, 6°10'E. The detector is an array of photomultiplier tubes arranged in

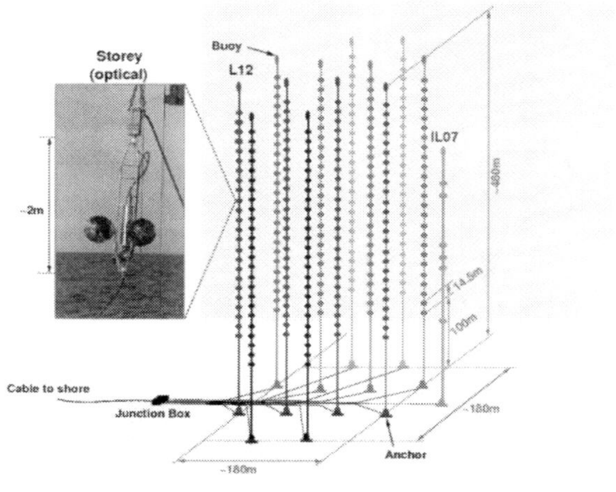

Figure 1. Schematic view of the ANTARES detector. A storey is shown in the enlarged view.

twelve detection mooring lines (a schematic view is shown in figure 1). Each line is divided on 25 storeys holding a triplet of optical modules (OM), which are pressure resistant glass spheres equipped with a 10" photomultiplier and facing downward at 45° from the line axis for an increased detection efficiency for up-going neutrinos. The lines are 450 m long and sustained vertically by means of a buoy at the top. The lowest storey is located 100 m above the sea bed. The spacing between lines is 60-70 m and the distance between adjacent storeys is 14.5 m. The Junction Box links all the lines and the Main Electro-Optical Cable transmits the power/data from/to shore station, where a computer farm applies the data filter algorithm based on different trigger criteria. ANTARES is equipped with a positioning system [2] which measures the line displacements with a precision on the relative location better than 10 cm. Time calibration [3] is performed *in situ* with a system of optical beacons distributed along the detector.

3. Data selection and track reconstruction

The analysis presented here use data gathered in the first two years of detector operation. During this time the ANTARES construction was still in progress; for most of 2007 the detector consisted of 5 lines, and of 9, 10 and 12 in 2008. A run selection criteria based on the number of active OMs (or active channels) during the run was applied considering each detector configuration, so 75%-80% of the OMs are requiered to be active when averaging over the full run. Trigger event selection requires at least 5 triplets of OMs detecting multiple photons throught the detector. The difference in the arrival time of the hits must be compatible with comming from a relativistic particle. Also events for which multiple photons are collected on two adjacent storeys are considered. Taking into account the time spent on sea operations (like the deployment and connection of new lines) and sporadic data taking problems of the detector, the total livetime of the analysis is 295 days, of which 144 days correspond to data taken with the smaller configuration of 5 lines.

The reconstruction of the muon track is achieved using the timing and position information of the hits arriving to the PMTs. A dedicated offline algorithm [4] that consists of multiple fitting steps of increasing sophistication is used. The final step consists on the maximization of the likelihood of the the observed hit times as a function of the muon direction and position. The goodness of the track reconstruction is described by the Λ parameter, which is basically the log-likelihood of the fitted track. This parameter can be used to eliminate badly reconstructed tracks by selecting an appropriate cut on the Λ value. Figure 2 shows the cumulative distribution of Λ

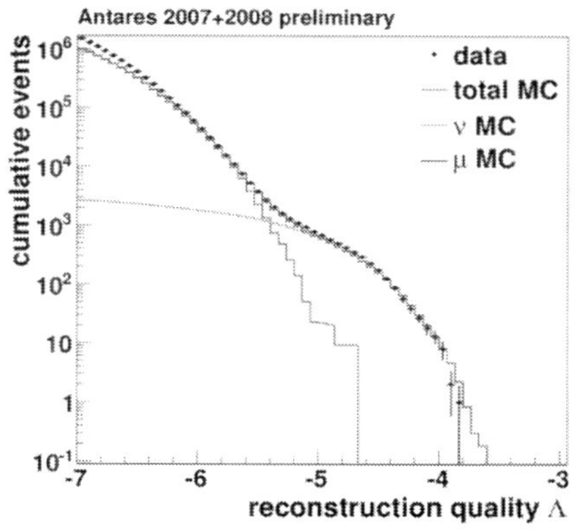

Figure 2. Cumulative distribution of the Λ parameter describing the quality of the reconstruction. The ratio between data and Monte Carlo simulation is shown in the bottom.

for data events reconstructed as upgoing, together with the contribution from the components of the expected background of atmospheric muons and neutrinos. For this analysis atmospheric muons were simulated with the CORSIKA package [5] using QGSJET [6] for the hadronic interactions and cosmic ray composition described in [7]. The atmospheric neutrinos were generated with the GENNEU [8] package according to the Bartol model [9].

Neutrino candidates events are selected requiring an upward going track and applying a cut on the quality of the reconstruction $\Lambda > -5.4$. The latter was found to be the optimal for the seacrch sensitivity obtained by optimizing the background reduction and the signal efficiency, in terms of the neutrino flux. Additionally, the uncertainty on the muon direction estimated from the fit is required to be 1°. The final sample consists of 2190 events, out of which about 60% are expected to be neutrino events and 40% downgoing atmospheric muons mis-reconstructed as upgoing.

4. Detector performance

The performance of a neutrino telescope is described by its angular resolution and effective area. Both parameters are estimated from simulation. Figure 3 shows the cumulative distribution of the angle between the reconstructed muon direction and the true neutrino direction for upgoing neutrino events assuming an E_ν^{-2} spectrum and complying the event selection criteria described in the previous section. The plot shows that roughly 75% of the signal events are reconstructed with an angular error less than 1°, being the median value of the reconstruction error, i.e. the angular resolution, $0.5 \pm 0.1°$. The uncertainty here includes all effects which can be related to a reduction of the per-OM timing resolution, and was estimated by artificially degrading the simulated timing accuracy of the detector by randomly smearing the hit times according to a Gaussian with a width of σ_t. Requiring a reasonable agreement between simulation and the observed data a lower bound for $\sigma_t = 2.0 \pm 0.5$ ns was obtained, and consequently used for simulation. Futhermore, this study revealed that an additional smearing of 3 ns was incompatible with data at the 2σ level.

The effective area is defined as the 100% efficient surface which detects the same number of events as the real one. An averaged effective area was computed considering the different detector configurations and selection criteria used for this analysis. It is shown in figure 3 as a function of the neutrino energy for a cosmic neutrino spectra of E_ν^{-2}. Based on the agreement

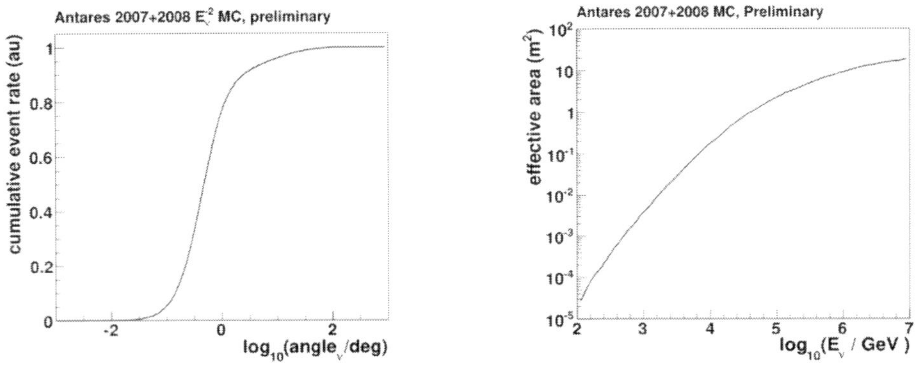

Figure 3. Cumulative distribution of the angle between the reconstructed muon direction and the true neutrino direction determined from simulations considering an E_ν^{-2} spectrum of upgoing neutrinos (left). Averaged effective area for the different detector configurations and the selection used used for this analysis (right).

between data and simulations a 15% systematic uncertainty on the detection efficiency has been considered.

5. Search method

The search algorithm used is an unbinned method based on the likelihood of the observed events as is defined in equation 1:

$$log\mathcal{L}_{s+b} = \sum_i log[\mu_{sig} \times \mathcal{F}(\beta_i(\delta_s, \alpha_s)) + \mathcal{B}i] - \mu_{tot} \quad (1)$$

where the sum is over the events, $\mathcal{F}$ is the point spread function (the probability to find the reconstructed event i at an angular distance β_i away from the true source coordinates δ_s, α_s), $\mathcal{B}$ is a parametrization of the background rate obtained from the observed declination distribution of the events, μ_{sig} is the mean number of the signal events emitted by the source and μ_{tot} stands for the total number of events in the selected sample.

The likelihood is numerically maximized yielding an estimate of the free parameters; in the case of a full sky search (see next section) these are δ_s, ra_s, and μ_{sig}, while in the candidate search only μ_{sig} is fitted as the coordinates of the sources are know. Point source searches involve testing the compatibility of the data with two hypothesis; that only background events are present in our sample (H_0), and that, over the distribution made up by the atmospheric backgroud events, there exists a cluster of signal events produced by a source of cosmic neutrinos (H_1). To this end, the so called test statistic, defined as the log likelihood ratio of the two hypotheses under consideration (equation 2), is computed.

$$Q = log\mathcal{L}_{s+b}^{max} - log\mathcal{L}_b \quad (2)$$

where $\mathcal{L}_b$ is the likelihood of the backgroud events. Lower Q values indicate that data is more likely to be compatible with background, while larger values are more probably to be produced by the presence of the searched signal. This search method has shown a better efficiency in separating signal from background than others methods (binned and unbinned) used in ANTARES.

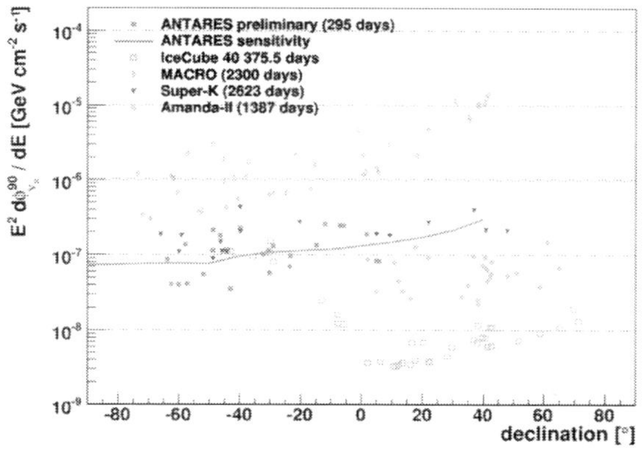

Figure 4. Limits set on the flux of an E^{-2} neutrino spectrum from the sources in the candidate list. Several previously published limits on sources in both the Southern and Northern sky are included togheter with the ANTARES sensitivity.

6. Results

Two different searches have been conducted in this analysis. The first one is a blind search where no assumptions about the source location are made. In the second approach we look for an excess of events in the direction of 24 candidate sources selected considering gamma-ray telescopes observations and the ANTARES visibility (fraction of time when the source is below the horizon.

No significant clusters of events were found in the full sky search. The equatorial coordinates of the most signal-like cluster are $\delta = -0.50°$, $ra = 43.21°$. The fit estimates 3.4 events as signal events. The value of the test statistic for this cluster is $Q = 6.8$. Such value, or a larger one, is found in the 88% of the background only experiments.

Results from the search using a candidate source list are summarized in table 1. No statistically significant excess was found near the 24 selected sources. The lowest p-value, defined as the fraction of the background only experiments that produce at least a larger Q value than the observed one, corresponds to GX 339, with a (post-trial) probability of 6.8% to occur when looking at 24 sources. Not having observed any significant point sources, limits on the E_ν^{-2} flux has been obtained at 90% CL following the Feldman-Cousins [10] prescription. Figure 4 show the ANTARES upper limits and sensitivity (the median value of the expected limit) as a function of the source declination; the systematic uncertainties described in section 3 are taken into account. Limits reported by other neutrino experiments are included for comparison. Note that in this analysis, and for the flux model considered, 80% of the selected events are in the range $3 < E_\nu < 700$ TeV.

The likelihood implementation described in the previous section was cross checked with an independent search method based on the EM-algorithm [11], using the same sample of events this search yields similar results. In addition, an auto-correlation study of the events did not reveal any structure in the data.

7. Conclusions

The ANTARES detector is the largest underwater neutrino telescope and the first full operational. Using data collected during the first two years of operation a search for point sources of high energy cosmic neutrinos has been conducted. Neither in the all sky search nor in the candidate list search has been found any significant excess of events. Many of the upper limits

Table 1. Results of the candidate source search showing the source coordinates, p-values and the limits on the flux $\phi^{90\%CL}$ obtained in this analysis (the latter is expressed in units of $GeV^{-1}cm^{-2}s^{-1}$.)

Source name	δ (°)	ra (°)	p-value	$\phi^{90\%CL}$
GX 339	-48.79	255.70	0.068	2.13
RX J0852.0-4622	-46.37	133.0	0.397	1.78
RX J1713.7-3946	-39.75	258.25	0.399	2.25
1ES 0347-121	-11.99	57.35	0.574	2.57
HESS J1837-069	-6.95	279.41	0.705	2.45
3C 279	-5.79	194.05	0.743	2.44
PSR B1259-63	-63.83	195.70	0.879	1.45
HESS J1023-575	-57.76	155.83	0.952	1.36
PKS 2005-489	-48.82	302.37	~1	1.14
RGB J0152+017	1.78	28.17	~1	1.87
Galactic Center	-29.01	266.42	~1	1.30
LS 5039	-14.83	276.56	~1	1.34
H 2356-309	-30.63	359.78	~1	1.13
PKS 0548-322	-32.27	87.67	~1	1.01
W28	-23.34	270.43	~1	0.97
HESS J1614-518	-51.82	243.58	~1	0.55
1ES 1101-232	-23.49	165.91	~1	0.70
Cir X-1	-57.17	230.17	~1	0.41
RCW 86	-62.48	220.68	~1	0.41
ESO 139-G12	-59.94	264.41	~1	0.41
PKS 2155-304	-30.22	329.72	~1	0.58
HESS J0632+057	5.81	98.24	~1	0.82
Centaurus A	-43.02	201.36	~1	0.35
SS 433	4.98	287.96	~1	0.83

on the neutrino flux obtained are the most stringent in the world. These are preliminary results (the corresponding paper is in preparation and will be published soon). Analysis of new data is underway.

[1] Ageron M *et al* 2011 Antares: the first undersea neutrino telescope arXiv:1104.1607v1
[2] Aguilar J A *al* 2011 Amadeus: the acoustic neutrino detection test system of the antares deep-sea neutrino telescope *Nucl. Instr. and Meth.* A 626-627 128-143 arXiv:1009.4179
[3] Aguilar J A *et al* 2011 Time calibration of the antares neutrino telescope *Astroparticle Physics* **34**, 539-549
[4] Heijboer A 2004 Track reconstruction and point source searches with antares (PhD thesis, Universitet van Amsterdam), available at http://antares.in2p3.fr/Publications/index.htmlthesis.
[5] Heck D *et al* 1998 Forschungszentrum Karlsruhe Report 6019
[6] Kalmykov N, Ostapchenko S, 1993 *Phys. Atom. Nucl.* **56** 346-35
[7] Bugaev E V *et al* 1998 *Phys. Rev. D* **58** 054001
[8] Bailey D 2002 Monte carlo tools and analysis methods for understanding the antares experiment and predicting its sensitivity to dark/matter (PhD thesis, Oxford University), available at http://antares.in2p3.fr/Publications/index.htmlthesis.
[9] Lipari P *et al*, 1996 *Phys. Rev. D* **53** 1314
[10] Feldman G J and Cousins R D. 1998 Unified approach to the classical statistical analysis of small signals. *Phys. Rev. D*, **57** 3873-3889
[11] Dempster A P, Laird N M, and Rubin D B, 1997 *J. Royal Statistical Soc. Series* B **38** 1314

AUTHOR INDEX

AUTHOR INDEX